Practical Well Planning
and Drilling
Manual

Practical Well Planning
and Drilling
Manual

[Steve Devereux]

Copyright © 1998 by
PennWell Corporation
1421 South Sheridan Road
Tulsa, Oklahoma 74112-6600 USA

800.752.9764
+1.918.831.9421
sales@pennwell.com
www.pennwellbooks.com
www.pennwell.com

Marketing Manager: Julie Simmons
National Account Executive: Barbara McGee

Director: Mary McGee
Managing Editor: Marla Patterson
Operations/Production Manager: Traci Huntsman

Library of Congress Cataloging-in-Publication Data Available on Request

Devereux, Steve
Practical Well Planning and Drilling Manual
ISBN 0-87814-696-2
ISBN13 978-0-87814-696-3

Printed in the United States of America

8 9 10 11   13 12

Contents

[Preface]

There are many excellent books dealing with drilling engineering, well planning, and drilling practices. Readers will note that the approach I adopt here differs from the "standard" books in three significant respects:

1. I have separated the office aspects from the rig aspects. Thus, the drilling engineer who needs to design the well and write the drilling program will find the relevant information together in the first two major sections. The wellsite drilling engineer/supervisor/toolpusher will refer more to the third major section, which deals with the practical rig site aspects of drilling the well. I hope this makes it easier for the reader to focus on his or her current area of interest. For instance, casing design information is in Section 1, notes on writing the casing part of the drilling program are in Section 2, and notes on running casing are in Section 3. For the wellsite drilling engineer, toolpusher, or drilling supervisor, much of the information given in Section 2 (Well Programming) is also relevant to the practical aspects of the work. I have tried to include extensive indexing and cross-referencing to help find all the relevant pages.

2. I have not included reference information that should be readily available in the office or on the rig. You will not find reproductions of casing design data, drillstring strength tables, cement formulations, etc. Space is limited in any paper-based media, and I would rather use that limited space for information that may not be so readily available to you.

3. I have not gone far into the deep theoretical aspects behind the work. While it is valuable to intimately understand the theoretical background, it is not strictly necessary for practical application during your everyday work. I have included references where applicable. Also, a few of the topics are covered to give some background and to show how they impact the well design and drilling program, but are not in themselves meant to be an authoritative text on the subject. For instance, completions are not usually designed by drilling engineers, but the completion requirements impact the whole well design because the completion dictates the hole sizes. Therefore, the design needs to be understood, questions need to be asked, and parts of it should be checked. Cementing is a huge topic and a nonspecialist book like this cannot cover it comprehensively; reference can be made to one of the excellent specialist books on cementing (recommended in the relevant section).

Acknowledgments

During my drilling career, I was lucky to have the guidance of many good drilling people, and the examples of many good (and a few not-so-good) people to learn from. Shell gave me my start as a trainee driller, and the time I spent with them was invaluable. I was taught how to do things properly and safely under the "old" training scheme whereby you earned your spurs on the drillfloor. My experience as a driller has been important throughout my career, even as a drilling manager. It is a great pity, and I believe detrimental for our industry, that few companies train people all the way through the ranks any more.

Stuart Smith, drilling fluids consultant, taught me a great deal when we worked together in Egypt. He contributed a lot of material for this book and continues to contribute to my *CD Drilling Manual*, enhancing the value of both.

Mark Hillman, drilling fluids consultant, contributed material on brines. Mark worked on developing potassium formate brines. He has also contributed to my *CD Drilling Manual*.

Dr. Eric van Oort of Shell and Dr. Fersheed Mody of Baroid taught me much about wellbore stability and mud design during interesting discussions and E-mail sessions. Some of the lessons I learned from these are included in this book. Fersheed has also contributed to my *CD Drilling Manual*.

Baroid Drilling Fluids, Milpark, and Tetra Technologies (UK) Ltd. allowed me to use their technical literature on muds and brines, including copying illustrations. I'd specifically like to thank Ray Grant of Baroid, Martin Ellins of Baker Hughes Inteq (Milpark), and Ian (Chalky) White of Tetra for their help.

Among my mentors I count and thank Ken Fraser. Ken was a Shell toolpusher on one of my first rigs as a trainee in Holland, and I later worked for him in Brunei as a driller. I learned a lot from him.

Another good friend and teacher is Frank Verlinden, who spent more years in Brunei than any other Shell employee. Frank talked of things forgotten in today's industry and knew from many years of experience how to handle any situation you could throw at him. Given a glass of wine and a small prompt, he could recount endless stories, which were fascinating as well as educational.

I hope my other friends and colleagues, too numerous to thank individually, will forgive me for not compiling a huge list here.

[Introduction]

The drilling industry is changing rapidly in the areas of technology, safety, environment, management, contractual relationships, training, etc. The driving forces are largely economic; there are probably few giant fields left undiscovered (especially in mature areas) and therefore the search moves to frontier areas and to exploiting smaller fields. Increased government regulations also play a large part. All of these factors increase the cost to discover and produce hydrocarbons. Add to this the pressure of low oil prices, and we are expected to continually reduce costs while improving drilling and production performance. We have to become more efficient by improving our skills and by developing new technologies and ways of working.

Computers have also caused dramatic changes for us. The computing power now available means that, if properly used, computers can help us to make better decisions. We can store, access, analyze, and summarize huge volumes of data and make complex calculations easy, even for the nonmathematically inclined. The downside is that some engineers use their PCs as a senior partner to make decisions for them rather than as a tool to help them make better-informed decisions themselves. This trend is increasing for reasons that I will come to shortly.

Early in the 1990s, operators and drilling contractors slowed down or stopped their ongoing training programs. This was largely due to low oil prices and high drilling costs. With less activity, many skilled people left the industry. The accountants decided that funds spent on training should be assigned elsewhere (perhaps on recruiting more accountants?) and so the major sources of highly trained, well-rounded drilling people dried up. To continue operating, new contracting schemes transferred responsibility from the operators to the contractors. This led to an exodus of people from the operators to the contractors and into the consulting market, depleting those skills within the operators much faster than by natural attrition and without replacing them, except by employing consultants. Some operators may end up with no one to properly supervise the core business of drilling. This will expose them to risks associated with major incidents such as blowouts without being able to manage that risk. Even on a turnkey well, the operator still has risks.

Alliancing contracts are becoming common, where a lead contractor is employed to subcontract and manage all the services needed to drill a well, but the operator still stays closely involved. Effectively, the lead contractor provides most of the resources of a drilling department, plus areas of specialist expertise.

There are some positive benefits from these strategies. If a true team spirit emerges where people work cooperatively together for achieving the same goals, costs can possibly be cut on long-term (development) projects. However, one guiding principle should be that *the operator retains the technical ability to plan and supervise the wells.* This means keeping competent drilling people in place.

There are at least three necessary factors for an alliance: commitment, communication, and competence. These take time to get in place. An alliance will not swing smoothly into action from the start therefore management also needs the commitment to see it through the initial hiccups.

Another clear trend is that many people planning and supervising wells do not have significant wellsite exposure. You can take the smartest person there is, put them through a degree program, and send them to all possible classroom courses. However, without the practical knowledge—the feel for drilling that comes from years on the rig—they are unlikely to become first class drilling people. They will tend to "use their PCs as a senior partner to make decisions for them rather than as a tool to help them make better-informed decisions themselves," and they will be unduly influenced by the people around them.

The attention paid to safety, the environment, and quality control has advanced immeasurably. Running an operation that is safe minimizes environmental impact and concentrates on all aspects of quality, which is ultimately more cost effective. Even now this is still sometimes a "hard sell"; many people pay lip service to these things but are not committed to them. I remember years ago, working offshore Brunei as a driller, being told to wait until after dark and then dump a reserve tank containing about 50 bbls of oil-based mud into the sea. The line to the pump was plugged, and we had to clean it out. We ran a hose in and used a small pump over several hours to recover this mud back into the active system so that none was lost overboard. I was chewed out for this initiative the following morning because I did not follow instructions, even though we had saved valuable mud and not polluted the South China Sea. This lip-service attitude still exists in some places.

There are two keys to drilling a cost-effective well and your well-planning efforts should be directed at these two keys. The first key is *avoiding problems,* which is chiefly related to the mud properties and to good drilling practices (but not by being overly cautious!). Do the job properly, avoid unplanned short cuts that often lead to unnecessary problems, and pay that extra bit to get the most suitable mud system. The second key is maximizing progress, which is more related to optimum bit/BHA selection, optimum drilling parameters, good forward planning, and good drilling practices.

The success of a well is determined first by the effort devoted to producing the best possible well plan, and second by the competence of the supervision while drilling, bearing in mind the two keys. This book is about those things—effective well planning and managing/supervising the drilling operation. I hope that this book will be a useful tool to drill safer, fit for purpose, cost-effective wells, and I look forward to your feedback on how well I have achieved this.

Steve Devereux, CEng, MIMinE, MIMgt
http://www.drillers.com

[List of Acronyms]

A	Area
Ac	Constant of proportionality
ADEPT	Adaptive Electromagnetic Propagation
AFE	Authority for Expenditure
AIT	Array Induction Imager Tool
AMS	Auxiliary Measurements Service
AS	Array Sonic
ASI	Array Seismic Imager
ASP	Anticipated Surface Pressure
assy	Assembly
AV	Annular Velocity
bar	Unit of barometric pressure.
bbl(s)	Barrel(s)
bent	Bentonite
BHA	Bottom hole assembly
BHF	Braden head flange
BOP	Blowout preventer
BP	Bridge plug
bpm	Barrels per minute
BSW	Bottom sediment and water
BTC	Buttress threaded and coupled
btm	Bottom
Butt	Buttress (threads)
Ca	Calcium
CBT	Cement Bond Tool
CBU	Circulate bottoms up
CCM	Circulate & condition mud
CDN	Compensated Density Neutron Tool
CDR	Compensated Dual Resistivity Tool
CEC	Cation Exchange Capacity
CERT	Correlated Electromagnetic Retrieval Tool
CET	Cement Evaluation Tool
cent	Centralizer
chk	Choke
circ	Circulate, circulation
Cl	Chloride
CNL	Compensated Neutron Log
co	Change out
COOH	Chain out of hole
cp	Centipoise
cmt	Cement

CS	Casing shoe
CST	Chronological Sample Taker
csg	Casing
Dexp	D exponent
dia	Diameter
DC	Drill collar
DHT	Dry hole tree
DIL	Dual Induction Resistivity Log
DLL	Dual Laterolog tool
DLS	Dog leg severity
DP	Drill pipe
DPT	Deep Propagation Tool
DSI	Dipole Shear Sonic Imager tool
DST	Drill stem test
dwks	Drawworks
EU	External upset
F or Fin	Finish
FC	Float collar
FCTA	First crystal to appear
FG	Fracture gradient
FL	Flow line or fluid level
FMI	Formation Micro Imager tool
FMS	Formation MicroScanner tool
fph	Feet per hour
FPIT	Free Point Indicator Tool
fpm	Feet per minute
fps	Feet per second
FS	Float shoe
FTP	Flowing tubing pressure
GIH	Go in hole
GLT	Geochemical Logging Tool
GOR	Gas/oil ratio
GPM	Gallons per minute
GS	Guide shoe
GST	Gamma Ray Spectrometry Tool
Hgr	Hanger
hd	Head
HLDT	Hostile Environment Litho Density Tool
HP	Horsepower, high pressure
HHP	Hydraulic horsepower
HTHP	High temperature and high pressure

HW	Heavyweight
IBOP	Inside blowout preventer
ID	Inside diameter
IEU	Internal-external upset
IOEM	Invert oil emulsion mud
ISIP	Initial shut-in pressure
IU	Internal upset
JB	Junk basket
jt	Joint
KB	Kelly bushing
KOP	Kickoff point
LCM	Lost circulation material
LCTD	Last crystal to dissolve
LD	Lay down
LDL	Litho Density Log
Lig	Lignosulphonate
LIH	Left in hole
LP	Low pressure
LTC	Long thread and coupling
LWD	Logging while drilling
MD	Measured depth
MDT	Modular Formation Dynamics Tester
ML	Mudline
MSCT	Mechanical Sidewall Coring Tool
mu	Makeup
MUT	Make up torque
mw	Mud weight
ND	Nipple down
NGS	Natural Gamma Ray Spectrometry Log
NML	Nuclear Magnetism Log
NPLT	Nuclear Porosity Lithology Tool
NU	Nipple up
NV	Nozzle velocity
OBDT	Oil Base Dipmeter Tool
OBM	Oil based mud
OD	Outside diameter
OS	Overshot
P&A	Plug and abandon
PAC	Poly anionic cellulose
PBR	Polished bore receptacle
PBTD	Plug back total depth

Pe	Photoelectric effect measurement
Pf	Alkalinity of filtrate
PHPA	Partially hydrolysed polyacrylamide
PI-SFL	Phasor Induction - Spherically Focused Log
pkr	Packer
PM	Alkalinity of mud
POB	People on board
POOH	Pull out of hole
ppg	Pounds per gallon
PPM	Parts per million
psi	Pounds per square inch
psia	Pounds per square inch (absolute)
psig	Pounds per square inch (gauge)
PTD	Proposed total depth
PU	Pick up
PV	Plastic viscosity
Rt	True formation resistivity.
RAB	Resistivity At the Bit
RFT	Repeat Formation Tester
RIH	Run in hole
RKB	Rotary kelly bushing
RM	Ream
RST	Reservoir Saturation Tool
RU/RD	Rig up/rig down
SFJ	Super flush joint
SHRDT	Stratigraphic High Resolution Dipmeter Tool
SITP	Shut-in tubing pressure
sk/sx	Sack/sacks
SLM	Steel line measurement
SO	Slack off weight
sow	Slip-on weld
SP	Spontaneous potential
SPM	Strokes per minute
sssv	Subsurface safety valve
sscsv	Subsurface control safety valve
St	Stand
TA	Temporarily abandoned
TCT	True crystallization temperature
TD	Total depth
TDT	Thermal Decay Time log
TFNB	Trip for new bit

TIH	Trip in hole
TOF	Top of fish
TP	Toolpusher
TTBP	Through tubing bridge plug
TVD	True vertical depth
USI	Ultrasonic Imager
VBR	Variable bore rams
VSP	Vertical seismic profile
vis	Viscosity
w/o	Without
wo	Workover
WL	Wireline, water loss
WOB	Weight on bit
WOC	Wait on cement
WOE	Wait on equipment
WOO	Wait on orders
WOW	Wait on weather
wt	Weight
YP	Yield point

[Section 1:]

Well Design

This section covers topics related to analyzing offset data and applying it with other relevant information to produce a well design and a drilling program. The major subjects that need to be covered for planning the well are described in some detail. References are made to other sources for users who may need higher levels of detail.

Preliminary Work for the Well Design

1.1.1. Planning Process Overview

Acquire and Review Data
- Well proposal
- Offset data
- Area experience
- Area reference data

→

Analyze Data
- Prepare hole section summaries
- Question/follow-up

↙

Well Design Meeting
Assemble team:
- Discuss all aspects of the design
- Agree on who does what

→

Design the Well
Final status of the well, including:
- Hardware, casing, wellhead, Xmas tree, sand control, completion design
- Directional requirements
- Document the major decisions made

↙

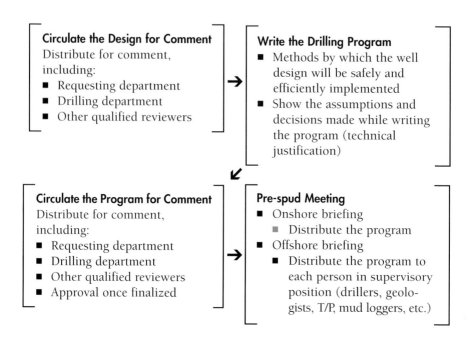

1.1.2. Data Acquisition and Analysis

The success or failure of a well, from a drilling viewpoint, is heavily dependent on the quality of well planning prior to spud. The quality of the well planning in turn is heavily dependent on the quality and completeness of the data used in planning. The successful drilling engineer is a natural detective, snooping around for every snippet of useful data to analyze.

The starting point in your data analysis trail is the well proposal. Usually the need for drilling a well starts as a request from the exploration or production department. They will put together a package of information for drilling that will define what the well should achieve and where it should be.

Well proposal checklist. The proposal should contain the following elements as relevant to the particular well:

1. Well objectives (exploration, appraisal, development, or workover)
2. Envisaged timescale (earliest/latest spud date desired)

3. Essential well design data:
   - Precompletion and completion requirements with conceptual completion design
   - Preparation work required in advance of running the completion, including permanent packers, gravel packs, completion fluid specifications, etc.
   - Perforation intervals; perforation type (if known)
   - Completion or logging sump required below the bottom perf depth (e.g., bottom perf to liner wiper plug and/or bottom of zone of interest to total depth)
   - Completion profile fully defining all elements of the completion hardware with depths; tubing, packers, subsurface safety valves (SSSVs), nipples, electric submersible pumps (ESPs), etc.
   - Completion pressure testing requirements
   - Future stimulation work envisaged, including fluids pumped, pressures used during stimulations, possible gas lift, etc.
   - Temperatures and pressures anticipated during the production life of the well
   - Likely reservoir fluid composition; any H_2S or CO_2 possible?
   - Options envisaged for future well interventions, including wireline/coiled tubing work, workovers, or recompletions (e.g., on another zone once primary zone is depleted)
   - Xmas tree and completion status on handover from drilling (e.g., plugged and depressured, killed, valve configuration, etc.)
   - Type of abandonment envisaged at the end of the well's production life
   - Any other relevant information on the completion not covered above
   - Pore pressure and fracture gradients vs. depth plot (it is useful to ask for the PPFG plot to show "best" and "worst" cases)
   - Shallow gas information (e.g., from shallow seismic surveys and offset wells)
   - Geological/seismic correlation, including all possible faults that may be encountered
   - Lithology/petrophysical correlation
   - Well directional targets (show downhole constraints to justify targets)
   - Surface location including site survey and bathymetry map, if applicable
   - Required zonal isolation of reservoirs
   - Likely temperature profile with depth

- Any known restrictions on the mud systems to be used (e.g., for logging or reservoir damage)
- Any other constraints on the well design or drilling program from the well objectives
- Any options that should be built into the well design, such as later sidetracks to different reservoirs, etc.
- A list of relevant offset wells

4. Specialist functions should specify:
   - Wireline logging program
   - Coring program
   - Geological surveying/mud logging requirements
   - Other evaluation requirements (e.g., paleontologist services, etc.)
   - Production test requirements
   - Final desired status of well; handover to production; suspended or plugged and abandoned (P&A)

5. Approval signature of the head of the sponsoring department—this is to ensure accountability. It may also be necessary for the department head to give you an account code to write the time against.

First, review the proposal and ensure that all necessary elements are present as per the above checklist. Then try to identify any surface or subsurface hazards arising out of the proposal and discuss these with the sponsoring department to see if their proposal can be modified to eliminate or reduce the hazards. Review each element of the proposal in detail. Is there any clarification required? Look in particular at the directional targets; these should be as large as possible and ideally will indicate what defines the target boundaries (faults, proximity to other wells, etc.). If "hard" target boundaries are given then you know that if the well heads outside of that target, you may have to sidetrack to get back into the target. This also gives you the largest possible target so you can later design your well to achieve the target at the lowest cost. This becomes more important if multiple targets or intermediate constraints on the wellpath are given. Often what happens is that the target is a circle of stated radius around a defined location and no indication is given as to where you can stray out of, which direction is most critical, etc.

Explorationists rarely appreciate the effect on well cost that an unnecessarily tight target can give. They know that if necessary you can drill very accurately to a target and therefore that is what they specify. In reality, if you are given the maximum area to go for, you may aim not at the center but at a place that gives you the most leeway for directional performance that does not go quite as planned.

For an offshore well (except for a platform), a seabed survey is required to check for bathymetry, seabed obstructions, seabed composition, likely leg penetration (if applicable), and shallow gas indications. Generally, this would cover a 2 km x 2 km square, centered on the proposed well location. Local currents should be checked (historical data may be available in mature areas) both at surface and at seabed level. Surface currents will affect rig positioning and marine operations; seabed currents may cause scour. Apart from the seabed survey, shallow seismic may be required to spot shallow gas anomalies and estimate leg penetration. In an area of soft seabed, the drilling contractor may require a soil boring analysis to ensure that the rig can be jacked up with minimal risk of punching through a hard crust during preloading.

Sources of offset data. Now that the location and target depths/formations are known, you can look for relevant offset wells. Except for a rank wildcat well, quite a lot could be available from company sources. This includes final well reports (which, if written properly, will be your best source of information), daily drilling reports, etc. If people who worked on the well are still with the company, make a note of it so you can contact them later if queries arise.

Other data on offset wells may be available from sources outside the company. For instance, if the mud records are missing or incomplete, ask the mud service company which was on the well if they still have information such as daily reports from that well. Bit records are often available from bit vendors. Wireline logs are usually archived for at least ten years prior to disposal by the logging contractor. IADC and geolograph reports can be more useful than daily drilling telexes because they will often hold more detailed information and are usually more accurate than the daily report telex to the drilling office. The drilling contractor may still have these somewhere.

Other outside data may include maps showing structures, surface features (for planning access to the site, locating water sources, and avoiding sensitive areas), and offset wells. In addition, government records are an important source of information. In many if not most areas worldwide, regulations demand that well information be filed with a regulatory body. It may also be possible to talk to the people responsible for wells in the area. They often have detailed personal knowledge of the wells drilled.

Hole section summaries. There is a good method of summarizing large quantities of offset data in a way that is meaningful "at a glance." These documents can be updated as further wells are drilled. They are invaluable later on for those 2 a.m. calls from the rig when problems are encountered, as well as for updating the programs of subsequent wells. The wellsite drilling supervisor can use them for reference if attached to the well plan. They are also useful to include in end of well reports. They are called *hole section summaries* because each sheet summarizes one hole section for all wells drilled at that time.

A hole section summary is created for each hole section. Information from each offset well is shown side by side in columns, which are divided in rows by formation. Comparisons can be quickly made in the same formations between several wells. On the left-hand side the formation is described to show lithologies and problem areas, with notes on compressive strengths (derived from sonic logs) and other relevant information.

For each well there are three adjacent columns. The left column shows depths of formation tops and thicknesses. The center column shows a representation of each bit run, including BHAs used, bit gradings, etc. An arrow shows each bit run and information regarding the run is written in. The right-hand column shows data on parameters, rate of penetration (ROP), mud properties, and short notes where needed.

Below each well section is a box to include comments on casing/cement jobs, overgauge hole from logs, and any problems logging or other worthwhile comments.

It is very easy to use these summaries to refer to while planning. It is recommended that you make these from the information available to you before you do anything else on the well plan. These summaries should not be swamped with too much detail or they lose their utility. You can easily use them to identify areas needing further study. (See Fig. 1—1.)

Formation Details	Well 1			Well 2		
Guitar	**Depth BRT**	**Bitrun Data**	**Parameters, Comments**	**Depth BRT**	**Bitrun Data**	**Parameters, Comments**
LSt, Sh interbeds Lst: RPM sensitive Sh: WOB sensitive	1010	ATM05, 20-13-13 Locked BHA == == == == == == == ==	KCl Polymer mud 0.49 psi/ft	1084	ATM05, 18-13-13 Locked BHA == == == ==	KCl Polymer mud 0.49 psi/ft
Compressive strength			Tight hole on conns WOB 30–55			Tight hole on conns 1171–1295
9—12,000 psi with to 18,000						Backreamed, no further problems
Shale streaks		== >	RPM 100–150		==	7–20 m/hr
Cavings @ <.5 psi/ft		Bit pulled at 1531			== -|-	Checktrip @ 1353 no drags
Losses @ >.5 psi/ft		521 @ 15.8 m/hr, PI 33.6 1/1/BT/A/E/I/JKD/PR	Pulled doing 8 m/hr		== == == == == == >	100k o/pull @ 1350 POH WOB 35–55 RPM 120–140 Ream 1 time before conns
Tight hole on trips, can pull through with care		--------- ATM22GD, 20-13-13 Locked BHA == ==	Hole good on trip Started drilling at 7m/hr		Bit pulled 1843, 759 @ 11.3 1/1/BT/M/F/I/ID/PR, PI 29.0 ---------	Pulled doing 7 m/hr

Fig. 1—1 Example of a Hole Section Summary

Plectrum			
Sst, Lst, Sltst, Sh	(892)		WOB 30–60
			RPM 100–150
Total losses possible	1902	Bit pulled 1964; 433 @ 7.3	ROP increased to 12
		2/1/NO/AE/I/NO/TQ, PI 16.7	Hole good on trip
Compressive strength 6–9000 psi with shale streaks to 15,000 psi		MF27D, 20-16-16 Locked BHA	WOB 30–55
		Bit pulled 2059; 95 @ 7.6	RPM 100–150
		0/0/NO/A/E/I/NO/ TW, PI 8.8	Existing crack in DC? No torque indication seen.
Increasing Sst with depth		MF27D, 20-15-15	WOB 10–50 (TQ)
		Bit pulled 2111; 52 @ 4.1	RPM 110–140
	(244)	3/2/CT/H/E/I/NO/TQ, PI 2.7	Crooked hole?

	ATM11HG, 18-13-13	Started drilling at 7 m/hr
	Locked BHA	7 m/hr
1870		12 m/hr
		Average parameters: WOB/RPM 30–47/110–130 GPM/PSI 600/2100 Mud 0.49 psi/ft
		Torque 5000 ft/lbs
		10 m/hr average
		10 m/hr

Fig. 1—1 Example of a Hole Section Summary, continued

Gstring						
C strth 7500–12,000 psi	2146	MF27DL, 20-15-15	Ream fr/ 2059, 24 hrs	2107	==	4.5 m/hr
Lst w/Sst streaks	(26)	Locked BHA	WOB 10–50 (TQ)		==	Tight hole conn 2115
		==	RPM 100 150		==	4.5 m/hr
		==			==	
Bridge					==	
Sst, Slsts, Sh with Lst streaks	2172	> Bit pulled 2253; 142 @ 4.5	High drilling torque	2131	== >	7.5 m/hr Stab twistoff (ran with this bit only)
Sand: 12–20 m/hr		3/3/CT/H/E/I/BT/TQ, PI 8.1	Tight hole POH to 2060		Bit pulled 2220; 377 @ 10.0 1/1/BT/H/F/I/CT/TW, PI 20.6	No o/p POH or drag RIH
Shale: 3–6 m/hr Compressive strength 7500-12,000 psi	------	SS84FD, 18/16/16			----------	Mud 0.49 psi/ft
Comments: general, problems casing and cementing, etc.						

Fig. 1—1 Example of a Hole Section Summary, continued

With the hole section summaries completed, you now have a detailed overview of all the relevant wells. Now look at each formation and list all the problems seen within that formation: tight hole, enlarged hole, kicks, stuck pipe, etc. For each problem, do a complete analysis and establish:

- What were the contributing factors that can be seen from the data?
- What other factors may have been relevant but were not noted in the records?
- How can this problem be eliminated, or at least reduced?
- What actions can be taken if the problem is seen on the next well to mitigate the effects of the problem?

If possible, avoid relying on the conclusions of other people who have reported on the problem. It is better to look at the source data yourself and make your own conclusions. Let me give a real case example.

An offshore well was being planned in the Mediterranean. According to prognosis, the pore pressure was to increase from normal pressure (hydrostatic) only 500 m below the seabed. Offset wells, even using oil-based mud, had reported very unstable wellbore conditions in shallow Pliocene shales with large quantities of cavings. A report by the previous concession holder had looked at seismic and sonic data, concluding that increased sonic transit times were due to undercompacted/overpressured shales. On the surface, this was consistent with the drilling problems that were experienced .

A closer look at the hole section summary revealed some interesting facts. Using oil-based mud in the first offset well, bottoms up after wipertrips had brought up large quantities of cavings. Mud density was increased, but the problem got worse, not better. However, they continued to increase the mud density and still the cavings level increased every time they wipertripped, which was often. The third offset well was drilled with a pilot hole through these shales and opened up with seawater with no flow from the well.

Unfortunately, samples from the shakers were not available from this interval on any of the offset wells. However, all the evidence seemed to be consistent with fractured shales, not overpressured shales. The mud used on the new well included additives to plug off these fractures as they were drilled, mud density was minimized, no wipertrips were done, and the drillers were briefed on good tripping and connection practices to minimize surge and swab pressures. This strategy was successful and the Pliocene shales were drilled and cased quickly using a KCl-PHPA-Glycol mud with only minor cavings. There were no problems tripping or running casing.

Field operational notes. Daily drilling reports often leave a lot of relevant information unrecorded. Drilling programs rarely give sufficient information to the drilling supervisor about the formations he/she is expected to drill through. Both these concerns can be overcome by writing and updating field operational notes.

All the available data relating to each formation should be summarized for future reference when planning and drilling. These should be kept up to date. The following example of field operational notes are from an actual operations manual (see Fig. 1-2). It can be seen that there is much useful information to aid in bit selection and use, mud parameters, and drilling practices.

These notes were allowed into the public domain with the kind permission of the Badr Petroleum Company, Egypt (BAPETCO).

Formation Name: Abu Roash   Type: Limestone + shale interbeds
Principal Problems: **Lost circulation, shale cavings, hydration**
                          swelling, gauge wear, and washouts

The Abu Roash formation presents a delicate problem. Too much density + ECD with KCl polymer muds lead to losses (probably in the limestone) but the mud density needed to minimize losses causes shale cavings. Tight control of mud parameters and drilling practices is needed. Ideally keep the density between 0.48-0.50 (maximum), PV as low as possible (10-15), YP in the range 17-21, and gels 3/5 to 5/8. Avoid surging on trips or after connections and minimize ECD. It is possible to live with the cavings, keep the hole clean, have low hole drags, and have few loss problems. If the hole was to be kicked off higher up, increasing hole angle would lead to problems of cuttings/cavings beds forming and consequent hole cleaning difficul-

Fig. 1-2 Example of Field Operational Notes

ties. Logging will show large washouts (off-scale in places), but the benefits in saved time with avoided losses more than compensates.

If losses occur in spite of good mud control, try reducing the circulation rate. You may find that a small reduction is all that is needed to cure the losses. After an hour or two of drilling ahead, it may be possible to slowly bring the circulation back to full rate. If total losses occur, first measure how much water is needed to fill the annulus. If the hole is static and full with water on top, slowly kick in the pumps and try to attain a circulation rate that will at least lift cuttings up the hole to the loss zone and cool the bit with very low weight on bit/revolutions per minute (WOB/RPM). Circulation of 250 gallons per minute (GPM) will give 50 feet per minute annular velocity (FPM AV) around 5 in drillpipe in 12$^{1/4}$ in hole; this should be used as the minimum. Drill ahead at reduced parameters and monitor drags and torques carefully for signs of drilled solids causing problems (potential stuck pipe). The losses are likely to cure themselves as generated cuttings act as lost circulation material (LCM) to plug the loss zone. Note that in past wells, LCM and cement have both been pumped, lost lots of time, and did not work.

The shale interbeds need a fair amount of inhibition and by experience it has been determined that if KCl is maintained at 40-42 ppb and shaledrill polymer at 1.0-1.5 ppb, there are no shale hydration problems. Keep a close eye on the mud properties and have the mud man run several tests throughout the day. The drilling engineer can be delegated the specific task of keeping an eye on this and personally supervising the tests to ensure that the tests are done properly and accurate results are given. There have been cases of mud men giving false results after a test to make it look as if the mud is in good shape when in fact it needs treatment.

In order to get the best drilling performance, the driller has to have the freedom to adjust the parameters for best ROP. The formation is quite streaky and changes constantly. The limestone is more sensitive to high RPM/lower WOB and the shales are better drilled with maximum WOB/lower RPM. If the driller is given a range of parameters to work within and is constantly experimenting for best ROP, the overall bit run will be far better.

Differential sticking is not probable in the Abu Roash formation. On trips out of the hole (through hole drilled since the last trip) the shales can give quite high overpulls. Backreaming is not necessary or helpful, just take the time to work up carefully without setting off the jar. Make sure you can come up about 3 m extra before setting the slips to break off a stand, otherwise you may not be able to go down enough to free the pipe if you pull straight into tight hole after racking the stand. Once you have wiped through it, you will probably not see significant overpull again at the same depth.

In summary, keep a very tight handle on the mud properties. Ensure that solids control equipment is kept functioning at top efficiency and check this personally several times a day. Do not rely on the crews to spot equipment problems. A centrifuge is worthwhile. Dump the sand trap and dilute as necessary. Avoid swab/surge that will increase shale instability and possibly induce losses. If losses occur it will be quicker and easier to cure with cuttings and care than pumping LCM and waiting. Run large nozzles, since maximizing hydraulic horsepower (HHP) does not seem to have a measurable beneficial effect; optimize impact force if optimization is preferred. Good bit runs have been obtained with Hughes ATM05 and Smith F1 in a 12¼ 12° in hole. Be willing to use maximum WOB. General type of bit: journal bearing, good gauge protection with diamond-coated heel row and gauge inserts, and a 4-1-7 or 4-2-7 type cutting structure. Also use locked bottom hole assemblies (BHAs) where possible. Ream before connections. Work carefully through high overpulls when tripping out, it may add a couple of hours to the trip, but you should not get into trouble.

$\begin{bmatrix} 1.2 \end{bmatrix}$

Well Design: General

The well design defines the desired final status of the well. The design, therefore, defines casing sizes, grades, weights, connections, and setting position (relative to depths or formation tops). Cement tops and particular requirements will be noted. It will define whether the well is to be completed, tested, suspended, or abandoned. Precompletion status (e.g., permanent packers, perforation intervals, downhole sand control measures such as gravel packs, completion fluid) and required completion design, wellhead, and Xmas tree will be specified. Surface location and directional requirements are also part of the well design.

Once the well design is known then the drilling program can be written to achieve the well design safely and cost effectively.

The steps to take to design the well include the following:

- Summarize and evaluate all relevant offset information. This first stage is vital if you want to write the best possible drilling program. (Refer to "Sources of offset data," "Hole section summaries," and "Field operational notes" in Section 1.1.2., Data Acquisition and Analysis)

- Identify all potential hazards (surface and subsurface) and potential drilling problem areas. For each potential hazard or problem gather as much information as possible, establish the root causes of the problem, and determine how they can be addressed. Ensure that the well design takes account of them to minimize impact and allow safe recovery. At this stage the outline casing points and mud performance requirements may start to become clear.
- Identify completion design or drill stem test (DST) requirements, including fluids, necessary sand control measures, or other downhole equipment (e.g., packers). This should be done before the casing design since it may have an impact on the casing.
- Choose casing points that allow kick tolerances to be maintained, minimize potential downhole hazards, and minimize potential drilling problems. Identify casing properties (outside diameter, weight, grade, connections, etc.) for each casing string, taking into account the directional plan. This is an iterative process because the directional plan may depend on the casing design and vice versa. If the rig is known at this stage, ensure that the conductor and casings can all be handled (rotary table inside diameter, derrick load, handling equipment).
- Specify the cementing requirements. Tops of cement slurries; any particular requirements (e.g., high compressive strength for perforating).
- Define the wellhead and Xmas tree requirements. If the well has to be suspended this might mean a mudline suspension or subsea wellhead system offshore.
- Check all items on the proposal and ensure that the well is designed to meet them; obtain dispensations or amendments if necessary from the sponsoring department.
- Issue the well design document for approval.
- Estimate times and costs to prepare an authorization for expenditure (AFE) and time/depth curve. (Note: a more accurate estimate can be made after finishing the drilling program but timescales usually dictate that an AFE is done sooner.)
- Identify long lead time items and obtain approval to place orders in good time.

- Check that the supporting infrastructure (roads, airfields, support bases, etc.) is in place and fit for purpose; flag up if infrastructure needs upgrading.
- Define the earliest possible start date (logistics, permits/approvals, rig availability, weather, AFE, infrastructure, etc.).

$$\begin{bmatrix} 1.3 \end{bmatrix}$$

Precompletion and Completion Design

"Precompletion" covers the well requirements that need to be met after drilling operations have ceased and before running the completion string. Precompletion will cover pre-installed downhole sand control measures (such as gravel packs), packers run before the tubing, etc.

"Completion" covers the tools and tubing that are run in as part of the production tubing string. This will include completion-run sand control measures (such as screens), downhole safety valves, etc.

The type of completion run will be determined by the production needs of the well. Size of tubing, types of connections, accessories to run, etc. will depend on factors such as fluids produced, gas/oil ratio (GOR), production potential, tertiary recovery techniques planned, sand control requirements, etc.

Casing sizes should be no larger than that required by formation evaluation requirements and drilling and production equipment sizes in order to drill cost-effective wells that are fit for purpose. Sizing down a conventional well by one casing size can save over 20% on the drilled well cost. A production well traditionally drilled to total depth

(TD) in 8½ in hole with a 7 in liner and 5 in production tubing could be TD'd in 6 in with a 5 in liner/5 in monobore completion with no loss of production for significantly less cost. This would not be termed "slimhole" according to the current definition (completed in 4¾ in hole or smaller) but it takes advantage of some of the slimhole developments to drill a cheaper, fit-for-purpose well while avoiding the complications of true slimhole.

1.3.1. How the Completion Relates to the Well Design

The completion will affect the entire well design, especially the casing design. The completion proposed must be considered for all stages of the well's lifecycle: running the completion, pressure testing, production, stimulation, workover, and abandonment.

Refer to the requirements of the well proposal in regard to what we need to know about the completion.

Preparation for the completion. There may be work required after the production casing or liner are cemented and before the completion is run. This work may just be a bit and scraper run or it may be necessary to install packers, perforate, and gravel pack, etc.

The following preparations may affect the production casing and/or liner string, including the cement:

- Perforated intervals require high-compressive strength cement (2000 psi is recommended) and a competent (360° coverage) sheath for zonal isolation. If it is a gas well, gas-blocking additives may be called for. Where future recompletions on other zones are anticipated, these intervals also need to have carefully tailored cement. Wells with bottom hole static temperatures above approximately 230°F require silica flour in the cement for long-term temperature stability.
- The sump required below the bottom perforation (e.g., to drop guns after perforating) will affect the final TD. Below the sump will be the shoetrack (normally two casing joints plus float equipment) and a pocket below the shoe.
- Fluid gradients, temperatures, and potential surface pressures will dictate the strength of the casing required for any treatments carried out before the completion is run.

- Permanent packers set in the well will work over a range of casing sizes and weights. The correct packer must therefore be used. If the packer is only available to fit a certain casing's inside diameter (ID), it may affect the choice of casing. Where heavier wall casing (smaller ID) is run higher up, the packer will have to clear through the smaller ID when run and there must be sufficient clearance. Special drift casing could also be used.
- If the completion is to sting into the liner polished bore receptacle (PBR) then the liner must incorporate a PBR and generally will require a polishing mill to be run before the completion. This could be combined with a bit and scraper run.
- It is essential that the liner lap seals. Liner hangers can incorporate integral packers, which are set after cementing; this may save time compared to running a tieback packer and can isolate the formation from well pressures while the cement is still fluid (e.g., when reversing out excess cement).

Other preparatory work that may affect the well design apart from the casings includes:

- Completion fluid characteristics may be dictated by the type of perforations, reservoir physical characteristics, and reservoir fluids chemistry.

Running the completion. The following circumstances may affect the well design:

- Tubing accessories outside diameters (ODs) (such as SSSV nipples, side pocket mandrels, packers, etc.) may dictate the possible range of casing IDs. In some cases a tapered casing string is required; for instance if a 7 in completion is run in $9^5/_8$ in casing, the SSSV nipple may be too large for the $9^5/_8$ in casing ID. It may be possible to run $10^3/_4$ in casing higher up, swaged down to $9^5/_8$ in below the SSSV depth. Of course this introduces further complications for running and cementing.
- If a dual completion is run, the sizes of tubings, collars, and accessories must be carefully checked to ensure that sufficient clearance exists inside the production casings. Remember that the strings

will move relative to each other during running as the telescopic joints take up the differences in joints run. Thus, tubing accessories may move opposite collars on the other string.

■ In high-angle wells, the maximum practical deviation for cable or wireline tools is about 60°. This may necessitate alternative strategies such as using coiled tubing or, if possible, pump down tools or setting nipples higher up in the well.

■ The type of completion will also dictate what kind of wellhead system to use and how it is to be configured.

Pressure testing the completion.

■ Fluid gradients, temperatures, and potential surface pressures will dictate the minimum strength of the casing required during pressure testing. Tubing and packer leaks must also be considered in terms of where the pressures may be exerted and whether in collapse or burst. Temperature correction factors (TCF) are needed in hotter wells. (Note: TCF at 200°C = 0.81 for Nippon steels!)

■ In deviated wells, consider the potential for casing wear and the effect on the pressure rating of the casing. Burst strength will be determined by the thinnest part of the casing wall.

Production.

■ Fluid gradients, temperatures, and potential surface pressures will dictate the strength of the casing required during production. Tubing and packer leaks must also be considered in terms of where the pressures may be exerted and whether in collapse or burst. Temperature correction factors are needed in hotter wells.

■ In deviated wells, consider the potential for casing wear and the effect on the pressure rating of the casing.

■ Produced fluids and temperatures could affect the grade of casings used.

■ Produced fluids could affect the completion fluid chemistry.

Stimulation including gas lift.

■ Fluid gradients, temperatures, and potential surface pressures will dictate the strength of the casing required for any treatments carried out during stimulations.

■ Injection pressures for gaslift must be considered for casing burst; also remember that if the production casing leaks, this pressure will be exerted against the previous casing.

Workovers and recompletions.

■ Provision may need to be made for a different completion in the future; for instance as the well depletes it may be desired to run gas lift valves, submersible pumps, or other tools. The casing design will have to account for these future possibilities.

■ Where other zones may be produced later on, the casing design will have to ensure that the required zones are accessible (e.g., not behind multiple casings), and that the cement sheath at that depth is high-compressive strength and provides good zonal isolation.

Abandonment. Eventually the well will be abandoned. Government regulations may require certain actions to be taken. For example, in Egypt it is required that all annuli have cement between the casing and open hole, even though no permeable zones may be present. This is not currently a requirement in the North Sea. It is important to know these details at the well design stage to avoid unnecessary work in the future.

Restoration of the site after abandonment should also be considered at the well design stage to minimize the expense later on.

1.3.2. Monobore Completions

Monobore completions (where fullbore access to the production zone is possible through the completion tubing) allow hole sizes to be decreased with no loss of production compared to traditional completions. (See Fig. 1-3)

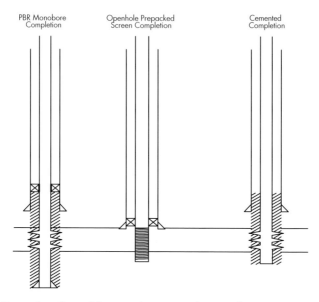

PBR Monobore
Completion

Openhole Prepacked
Screen Completion

Cemented
Completion

Fig. 1-3 Examples of Possible Monobore Completion Schemes

Advantages of slimhole, monobore wells.

1. Reduced well cost/enhanced project profitability.
2. Increased activity levels as cheaper wells make conventional, marginally economic prospects worthwhile.
3. The capabilities of existing installations can be extended.
4. Possibly reduced location size (purpose-designed slimhole drilling rig), waste, and environmental impact.
5. Ideally suited to completions through several reservoirs where the reservoirs are produced and abandoned from the bottom up or production can be commingled.
6. Increased wellbore stability in some formations (e.g., fractured shales).
7. Coiled tubing drilling—especially underbalanced—may provide sidetracking opportunities from existing wells while minimizing impairment.
8. More expensive drilling and completion fluids can be used to minimize impairment since the volumes required are much less.
9. Monobore completions allow fullbore access to the reservoir and give more flexibility in managing the reservoir. Most or all well

intervention work could be done under pressure; not killing the well will eliminate impairment due to killing for workovers.

10. May be useful for deeper wells where many casing strings are needed.
11. Reduction in required consumables and tangibles may ease logistical problems in remote areas. Offshore rigs need less resupply.
12. Exploration wells can be drilled as slim, cheap, throwaway wells instead of as expensive potential producers. The target zones can be drilled using wireline-retrievable continuous cores to give better petrophysical and geological information. Platform location can then be optimized and wells designed for production can be drilled from the optimum location.
13. Where production is not a constraint (e.g., observation, injection wells) then a slim well may fully meet the objectives at minimum cost.
14. Smaller, compact wellheads and blowout preventers (BOPs) can be used.

Disadvantages of slimhole wells.

1. Lack of contingency hole sizes.
2. A monobore completion may be less suited than a conventional completion to wells requiring the ability to select different zones.
3. Few dedicated slimhole rigs are presently available and there will be a lack of incentive for drilling contractors to invest in them, unless operators find ways to create incentives.
4. In true slimhole sizes, well control presents more challenges and requires better training and equipment to detect and deal with smaller kicks, low annular capacities, and higher annular pressure drops.
5. Commitment is required both from operator management and from contractors and service companies, where many senior people find it hard to commit to radical change. Effective and committed management of the project is a prerequisite to success.
6. Slimmer wells generally mean reduced contractor and supplier profits. New contracting strategies are required to align contractor and operator goals, which will include sharing risks and rewards of field development.
7. Planning and executing slimhole wells require high levels of drilling engineering expertise and the involvement of multidiscipline teams (including contractor and service company personnel).
8. Limited mud densities due to high equivalent circulating densities (ECDs).

9. Limited potential to reuse the original wellbore for sidetracks to other reservoirs.
10. May limit information obtained if the hole size prevents all required logging tools.

Relevant developments.

1. Monobore completions.
2. Coiled tubing drilling.
3. Developments in downhole equipment (i.e., motors, drillstring, bits).
4. Computerized systems to efficiently monitor the operation and to detect very small influxes (<1 bbl).
5. Better understanding of wellbore instability and superior muds now available compared to previously have reduced the risk of losing a well while drilling smaller holes than in the past.
6. Better understanding of downhole hydraulics.
7. All regular logging equipment exists for holes down to $4^{1}/_{8}$ in hole, 150°C, 15,000 psi.
8. Full production testing capability now exists with $3^{1}/_{8}$ in OD tools.

1.3.3. Multiple String Completions

Where a well penetrates several zones that need to be produced but cannot be commingled, it is possible to run two completion strings using special dual packers to separate the zones. Planning to use dual completions clearly introduces some special considerations.

The two strings will generally be made into a dual packer, such as the Baker RDH (retrievable dual hydraulic) hanger. There will be two strings from above this packer to the dual hangers set in the wellhead. Below the dual packer will generally be a short tailpipe on one side and a long string with a seal assembly on the other side. When the completion is run, the seal assembly on the long string will sting into a packer previously set downhole, so that production is possible along the long string from below this seal assembly depth or along the short string from between this seal assembly and the dual packer.

The long string can also selectively produce from different zones. Sliding side doors that can be opened or closed by wireline can open the long string between different seal assemblies.

Just above the dual packer, the long string side will have a telescopic joint (TJ). This is to allow the strings to be equalized before setting the slips, and because joint lengths vary slightly, if there was no TJ then much effort would have to be made to pick up joints in matching pair lengths. Sometimes the TJ may become fully opened or fully closed and a pup joint has to be placed in one of the strings to put the TJ around 50% open.

Another problem is with the placement of accessories. Side pocket mandrels can be a particular headache. Since they are not round but rather elliptical in cross section, you have to be careful not to place nipples where they could move opposite the side pocket mandrel during running (i.e., due to movement of the telescopic joint). It is possible that the side pocket mandrel, if the opposite string is aligned with its widest section, will crush nipples, collars, or sometimes even the tubing itself, depending on the clearance inside the production casing.

If you are planning a well that would benefit from a dual completion, it would be worthwhile to find somebody who has operational experience of running dual completions to aid in well design.

1.3.4. Completion Fluids

Often wells are perforated and completions are run in a packer fluid that is designed to protect the well during production. This requires careful consideration and expert input at the design stage of the well. This section covers the important points behind completion fluid decisions, but it is important to realize that this specialist area should have input from someone with expertise in completion fluids design.

Completion fluids must be as nondamaging to the formation as possible so they do not compromise productivity. Apart from being chemically and physically compatible with the reservoir and its contents, the solids content of the mud must be kept as low as possible. Any solids must also be removable by acid or other treatments. Damaging solid precipitates or emulsions can be formed downhole by chemical reaction between formation fluids and mud or brine filtrates, which also need to be prevented.

Rheology and fluid loss properties may have to be controlled. Nondamaging additives or those that can be easily removed are needed for tailoring these properties. Calcium carbonate, graded by size depending on the formation pore sizes, makes an effective fluid loss agent that is acid soluble. Yield point (YP) and gels should be sufficient to avoid solids settling out unless a solids-free brine is used.

1.3.5. Brines

Brines can offer solids-free systems with densities up to 1.07 psi/ft (20.5 ppg). Another consideration is using solids-weighting materials that are acid soluble, such as calcium carbonate and iron carbonate .

While being able to overbalance formation pressures, properly designed brines do not create formation damage, neither by plugging the reservoir with unremovable solids nor by causing reactions with formation fluids or solids. Potential interactions of brines in the reservoir include:

- Scale from the reaction of a divalent brine with dissolved carbon dioxide, producing an insoluble carbonate (divalent brines containing calcium or zinc salts, i.e., the metal ion has a valence of two)
- Precipitation of sodium chloride from the formation water when it is exposed to some brines
- Precipitation of iron compounds in the formation resulting from interaction with soluble iron in the completion fluid (most common with zinc bromide, $ZnBr_2$)
- Reaction of formation clays with the brine
- Corrosion of casings and tubulars (not such a problem with monovalent brines)

Consider corrosion and biodegradable properties for completion fluids that will remain in the well for a long time. Corrosion inhibitors are available to suit various muds and brines. The pH should also not be too high or too low to prevent damage to tubulars, cement, and elastomers. Biocides can help control bacterial activity.

Selection of a brine system. There are three main criteria to use in selecting the brine system for a particular well.

1. *Density*. Different brines have different ranges of possible densities. Downhole density can be significantly different to surface density due to the effects of pressure and temperature. This difference is greater with heavier brines. The desired density will restrict the choice of brine to use.
2. *Compatibility*. The brine system must be compatible with the reservoir solids and fluids to ensure that solid additives, precipitates, or emulsions do not form and block the reservoir; and to minimize problems with the well (e.g., corrosion).
3. *Cost*. Different brine configurations are possible to meet the two criteria previously listed, but the cost can vary significantly depending on the salt(s) used.

Additives can be used in the base brine system to control other properties such as fluid loss.

Salts used in brines. The general salts used in the oilfield for brine formulation include sodium chloride, potassium chloride, calcium chloride, sodium bromide, calcium bromide, and zinc bromide. Other less commonly used salts include magnesium chloride, ammonium chloride, sodium formate, and potassium formate.

Some of these salts can be blended together to produce the most cost-effective recipe at a certain density. This is commonly the case when mixing high-density brines using expensive bromides.

Many salts, especially calcium chloride, are manufactured at a variety of purities. When comparing costs of salts for formulating brines, base the calculations on the salt purity that will actually be supplied. The cost per unit of actual chemical is what should be compared.

Effects of temperature and pressure on brines. Often overlooked during brine planning is the effect of downhole temperature and pressure on the density of a column of brine. When calculating the required density this must be considered, especially when using high-density brines. This depends on several factors: brine type, brine density, well depth, ambient temperature, and bottom hole temperature.

Specialist companies can run a program for your specific well, plotting the temperature effect on the brine density against depth (See Fig. 1-4). This allows the average density at any depth in the hole to be calculated. This will determine the density that must be achieved at surface temperature to produce the desired density in the hole.

Bottom hole temperature	365°F
Surface temperature	60°F
True vertical depth	18,100 ft
Surface brine density	13.19 ppg
Using a calcium chloride/calcium bromide blended brine.	

Depth (ft)	Temp (°F)	Pressure (psi)	Density (ppg)
0	60	0	13.19
2,000	93	1,365	13.13
4,000	127	2,723	13.05
6,000	161	4,074	12.98
8,000	194	5,416	12.91
10,000	228	6,751	12.83
12,000	262	8,078	12.75
14,000	295	9,397	12.68
16,000	329	10,708	12.60
18,100	365	12,083	12.50

Fig. 1-4 Example of Report—Temperature Effect on Brine Density Against Depth
(contributed by Tetra Technologies UK Ltd.)

This report indicates that the average density in the hole is 12.85 ppg at TD.

Brine crystallization point and eutectic point. The *crystallization point* of brine is the temperature at which salt crystals begin to fall out of solution and thus reduce the density of the brine. The temperatures at which the brine will be transported and stored should exceed the crystallization point by at least 10°F (6°C). Crystallization can also plug lines and damage pumps.

There are three temperatures relative to crystallization occurrence: *first crystal to appear* (FCTA), *true crystallization temperature* (TCT), *and last crystal to dissolve* (LCTD).

Adjusting the density of brine using dry salts affects the crystallization point. With single-salt solutions, adding more of the same salt initially lowers the crystallization point temperature to the *eutectic point.* This is the lowest temperature of the crystallization point of a solution. For example, the lowest crystallization point obtainable for calcium chloride brine is when the density reaches 10.8 ppg. Further addition of dry calcium chloride to a 10.8 ppg brine solution raises the crystallization point, even though the density continues to increase. For two-salt brines with a crystallization point of 30°F (-1°C), the addition of dry salt raises the crystallization point temperature. (See Fig. 1-5)

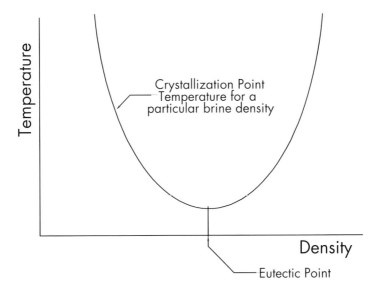

Fig. 1-5 Graph of Temperature vs. Crystallization Point

Brine additives. A solids-enhanced fluid is necessary for completion or workover operations when the use of clear brine will result in the loss of large fluid volumes to the formation. Sized calcium carbonate is often used because it is completely acid soluble.

Treating the finished brine with corrosion inhibitor, oxygen scavenger, and bactericide is recommended. Depending on the well conditions, other treatments such as a scale inhibitor or hydrogen sulfide scavenger may be required.

Since oxygen scavenger will be treated out by oxygen from the atmosphere, it should be added to the brine just before pumping on the final circulation. Ideally additions should be made using an injection pump directly into the suction line of the displacement pumps.

1.3.6. Points to Check on the Completion Design

While the drilling department does not usually design the pre-completion and completion (because this is a specialist task as can be seen from the factors mentioned in the previous section), it is important to check the design and make sure that there are no practical problems with it. Some specific points to check include:

■ Clearances inside the planned production casing or liner. Various accessories will be run on the completion tubing such as packers, side pocket mandrels, nipples, etc., that have to fit inside the casing.

■ Dual or triple completion strings can be run. If completion accessories are run at the same depth on different strings, that can cause real problems running. For instance, a side pocket mandrel is not round in cross section, but if run on a dual $3^{1}/_{2}$ in completion, it has to be aligned sideways to the other tubing string or else it can crush the other tubing inside $9^{5}/_{8}$ in 47 ppf casing. Control and injection lines may be run (i.e., to a surface controlled subsurface safety valve). Clearances in these complex completions must be examined and may require redesign and/or special running procedures.

■ In a deviated well, wireline tools get hard to run above 50° inclination. If profiles are run in the string that require slickline intervention (to operate sliding side doors or set/pull injection valves), then they need to match your directional plan to ensure that they can be reached on slickline. Otherwise, it may be possible to design these accessories to be worked by coiled tubing if repositioning is not possible.

■ Completion components should be checked under the different operating conditions for burst, collapse, and tension. The different operating conditions may include several of the following:
  ▪ Running
  ▪ Spacing out and landing
  ▪ Pressure testing
  ▪ Stimulation, including fracs and acid jobs
  ▪ Producing throughout the life of the well (initial and depleted reservoir pressures)

- ▪ Squeezing cement
- ▪ Abandonment/fishing
■ Completion tubing should be checked under the different operating conditions for possible buckling. If an anchor seal assembly or a tubing run packer is used, refer to Section 1.4.14, "Calculating for Buckling (Nb)" to calculate whether the tubing will buckle just above the packer. Include in the axial force calculations any planned setdown weight that will be left on the seal assembly.

All of the relevant operating conditions listed in the previous point should be considered. In addition, yield strength reduction due to increased temperature may be significant. Refer to Section 1.4.9. "Factors Affecting Pipe Yield Strengths." If an anchor seal assembly is not run but the completion incorporates a sealing assembly that lands in a packer, then when calculating the axial force the effect of pump out of the seal assembly has to be added. This force is the product of

(internal pressure - external pressure)x(packer sealbore area), or

$$Force = \Delta P \times .7854d^2$$

where ΔP is the pressure differential (positive if internal pressure is greater than external) and d is the packer bore diameter.
If the tubing is set down by greater than the pump out force, or if an anchor seal assembly is run, use the planned setdown force instead of the pump out force when calculating the buckling force. Buckling will put high stress on the completion and may make it hard or impossible to run wireline tools (depending on the buckling "wavelength").
Completion components should be checked for compatibility with the produced fluids, treatment chemicals, and packer fluid. The presence of H_2S, CO_2, chlorides, water, etc., will affect the material used and may dictate that alloy steels be used. Refer to Section 1.4.19, "Material Grades."

It is recommended that when considering the stresses acting on a completion tubing string, a computer program should be used to easily analyze the design and consider different scenarios. However, the engineer should always confirm the results by comparing them with some basic manual calculations. Few programs are perfect and it only takes one mistyped digit to render the computed analysis inaccurate. Worse, if you assume it is a good design when there is a fatal flaw, it may fail catastrophically in service.

The various methods for calculating physical forces on tubing and relating this to strength requirements are the same as they are in casing design. These topics are covered in the following Section 1.4 and may be used as reference for tubing.

$$\left[1.4 \right]$$

Casing Design

These two documents should be made available to anyone working on designs for casings and tubulars:

- *API Bulletin* 5C2 (Properties of Casing, Tubing, and Drillpipe)
- *NACE Standard* MR-01-75

Ensure that the current issue of each is used. The following sections do not attempt to reproduce wholesale the contents of these documents; rather references will be made and illustrations given as appropriate.

Also, the following *Bulletin* may be referred to for the formulae used by API for calculating the properties outlined in *Bulletin 5C2*. However, in most cases the formulae are not needed because the tables in *5C2* are comprehensive.

- *API Bulletin* 5C3 (Formulas and Calculations for Casing, Tubing, Drillpipe, and Linepipe Properties)

1.4.1. General Points and Definitions

There are two different jobs that casing must be designed for. The first is to allow you to safely drill the well and resist any forces or conditions that are imposed on it during drilling, without sustaining significant damage. The second is to act throughout the life of the well to meet the well objectives without requiring a workover. The design criteria for each string of casing are different during drilling and during the remainder of the life of the well. It is your job to design the most economical casing that is fit for purpose for the full design life of the well.

Computer programs make detailed casing designs routinely possible, including triaxial analyses. This can give lower casing costs. While it is recommended to use a recognized design program, it is important to check the results with some hand calculations to confirm that the results are in the right ballpark. Computer programs often have bugs and a slip in data entry can also give invalid results.

It is useful first to briefly define the terms used in this manual and to summarize the drilling and production purposes of each string of casing. Some basic mechanical properties and formulae are discussed, then each element of the design is considered and the calculations covered. Tables summarizing the suggested criteria for each casing for drilling and production are given.

Stove pipe. The stove pipe is used in onshore drilling to protect the surface soil from erosion and to allow fluid returns while drilling for the conductor. It is usually set very shallow during location preparation, a couple of meters below the cellar floor, and no diverter or blowout preventer (BOP) is nippled up on it. It does not support subsequent casing or wellhead loads and generally serves no purpose once the conductor is in place. The only design criterion is that it is big enough to allow the conductor to be run inside it.

Conductor pipe. On a land rig or bottom-supported offshore rig, the conductor may be drilled and cemented in place (hard seabed) or driven to refusal with a hammer. The setting depth has to be sufficient to withstand the extra hydrostatic pressure imposed at the bottom by bringing returns up to the flowline. The conductor then has a diverter nippled up to it for drilling the next hole section for surface casing.

On a floating rig, the conductor may be drilled and cemented in place or jetted into a soft seabed. Returns to the rig are not usually established and the next hole section (for surface casing) is therefore drilled with returns to the seabed. If the seabed is soft enough to jet the conductor in, it is not likely to support the extra hydrostatic pressure imposed by closing the circulation system. No diverter is attached; if shallow gas is encountered the rig will drop the drillstring and move off location.

If cemented, the conductor is always cemented to the surface or to the mudline. Sometimes losses may occur during cementing, and if this causes the cement to drop below the surface, a top-up job is needed where tubing is run outside the conductor from the surface and cement is pumped into the annulus once the primary cement has set. This is only possible where the annulus outside the conductor is accessible.

The conductor will have to resist the compressive loads imposed by subsequent casing strings, completion strings, wellhead, and BOP weights. Buckling may be a consideration if the conductor extends a significant height above the soil and is not supported. On a bottom-supported offshore facility it will also be subjected to waves and currents as well as severe corrosion conditions in the splash zone. If driven it must handle the driving loads. It may have to resist collapse pressures if losses are encountered or if a diverted gas kick evacuates the conductor.

Surface casing. This is normally the first pipe that can take a blowout preventer on top. The shoe must be set deep enough so that the formation fracture pressure is high enough for the well to be closed in on a kick while drilling for the next casing string. Any gas encountered before a BOP can be nippled up is termed "shallow gas."

As surface casing in some development areas is set quite deep (sometimes deeper than 3000 ft), shallow gas can be encountered fairly deep. It is not correct to refer to gas as shallow gas if it occurs after a BOP is nippled up on surface casing.

During the well life, surface casing may be subject to burst pressure from well kicks, bad cement jobs (fluids migrating up outside subsequent casings), or to collapse pressure if the fluid level inside drops due to losses or if bad cement jobs allow migration of gas outside the casing. Surface casing is normally cemented to the surface or to the mudline.

Intermediate casing. Intermediate casing is run in deeper wells where kick tolerances or troublesome formations make it unsafe or undesirable to drill from surface casing all the way to the production casing setting depth in one hole section. Therefore, its primary drilling purpose is to resist the forces imposed by kicks, losses, and mobile formations.

The top of cement may be planned below the previous casing shoe. This gives you the option to cut and pull casing and sidetrack out without losing a hole size, as long as the wellbore remains stable enough for you to pull out the casing once it is cut. However, in some areas, government regulations may force you to cement off all exposed formation when abandoning the well (even if impermeable). If this is the case then it may incur substantial extra cost on abandonment to get cement in place and this may prevent the option to sidetrack under the previous shoe.

On deeper vertical wells, buckling in intermediate casing may be a consideration. Further, during production, intermediate casings may be subjected to burst or collapse pressures as stated for surface casing. Several intermediate casings may be required in a deep well.

Production casing. Production casing may be run at the total depth of the well or it may be set above the reservoir prior to drilling to TD without setting another full casing string. In the latter case, liner(s) may be run or the well produced through an uncased hole or through a sand control screen.

The completion tubing is run inside the production casing. If the completion tubing were to leak, the production casing would be subjected to extra internal pressure. During its design lifetime this casing may need to resist high burst pressures from leaking completions, injection of gaslift gas or other fluids through valves in the completion string, frac job support pressure, etc. There may be significant effects from temperatures, wear, corrosion, and reactions to produced materials such as H_2S or mobile formations (massive salts).

If it is set across the reservoir, the production casing must withstand collapse loads due to drawdown during production. On deeper vertical wells, buckling in production casing may be a consideration.

Drilling liner. A drilling liner is run inside production casing but is set above the reservoir. This is to permit deeper drilling without the expense of running another full casing string. It forms part of the production "pressure vessel" as it could be subjected to all the loads (previously discussed in "Production casing" of this section), but not the collapse loads due to production drawdown as detailed below. It might also be set above the reservoir for a barefoot, slotted liner, or prepacked screen completion.

Production liner. This is set inside the production casing or inside the drilling liner if one has been run. It is set across the reservoir zones to provide zonal isolation and, if necessary, may incorporate sand control measures such as internal or external gravel packs. Apart from loads (previously discussed in "Production casing" of this section), the production liner will have to resist collapse loads due to the pressure drawdown while producing.

The top of a liner hanger generally incorporates a PBR. This can be used for tieback liners (to convert the liner into a full string to surface), for scab liners (a length of liner with a packer on top in case the liner lap does not seal), or for setting retrievable completions without using a packer in the casing. These tools stab into the PBR and incorporate sealing elements.

1.4.2. Hole and Casing Sizes: Considerations

"Slimhole," according to the current industry definition, is a well which is completed in $4^{3/4}$ in or smaller hole at TD. Monobore completions allow the well to be drilled in smaller diameters and also have other advantages and some drawbacks. These were discussed in the previous section on completion design.

There are substantial savings possible by drilling smaller hole sizes than would be traditionally used. While a well may not necessarily be "slimhole" according to the definition above, hole sizes can still be reduced in many cases with no loss of objectives. The current state of development at the time of writing is summarized within this section, following by the procedure for choosing the final hole size (Section 1.4.3).

As the advantages of slimhole drilling become more apparent, new tools are being developed for small holes. Standard logging tools are now available for hole sizes below 6 in and coiled tubing drilling techniques are advancing rapidly for drilling small diameter holes.

The optimum radial clearance outside casing for running and cementing is $1^{1/2}$ in. When planning hole and casing sizes, try to maintain a hole diameter at least 3 in larger than the casing diameter. Also, check that the drift size of the casing is greater than the bit diameter for the next hole section.

Planning a slimmer well. The starting point in planning is where you want to end up. What final hole or liner size is required to meet all of the well objectives? Is a contingency hole size needed (exploration) or not (development)? Can a monobore completion be used (and if not, why not)? What are the required logging tools and what is the minimum hole size in which they can be run? Are sand control measures needed and would they affect the final hole or liner size?

If company experience of drilling slimhole is limited, a dedicated team could be formed that would include drilling engineering (operator/team leader), exploration or production (operator), and key contractors (i.e., rig, drilling fluids, mud and wireline logging, logging while drilling (LWD), coring, directional, completion, and testing as relevant). It would be worth considering whether an outside consultant can be used who can import experience onto the team for the first well or two. A good consultant should be able to improve the plan, reduce the cost, and ensure that all necessary data is captured from the well to improve future slimhole planning. For instance, expert advice on wellbore stability, well control in slim high pressure/high temperature (HPHT) horizontal situations as relevant, or drilling hydraulics may help specific areas of the program. The team leader has to have the full authority to manage the project and the responsibility to create the best possible well design and drilling program.

Data capture is essential to make the learning curve as short as possible. A lot of data goes unreported or gets lost or ignored at evaluation time. As requirements are specified from each area of expertise, ensure that the data capture requirements are also addressed.

1.4.3. Hole and Casing Sizes: Selection

The first decision required is "what should be the final hole size at TD?" This has to be decided first because casing sizes, kick tolerance calculations for proposed setting depths, and directional work decisions all depend on hole sizes.

The following considerations will apply when choosing the planned final hole size, in order of importance:

1. The size of the desired production or test tubing (as defined in the well proposal).
2. Whether a monobore or a conventional completion design is anticipated.
3. Whether a contingency hole size is required (exploration well or other reason; should be justified based on the risk of having to run an unplanned casing string).
4. The requirements for gathering information from the well (e.g., logging, coring, DST tools) and any restrictions these place on the minimum hole size.
5. Possibly the rig available and whether equipment, crew capability, or other considerations apply. For instance in true slim holes, well control considerations will dictate the need for accurate kick detection and capable crews.

Having defined the hole size at TD, once the casing points are defined then the sequence of hole and casing sizes can be seen. The actual sizes chosen may depend on available casing stocks, standard company sizes, rig equipment, or other factors.

1.4.4. Pore Pressures and Fracture Gradients

Knowledge of how the pore pressures and fracture gradients are likely to change with depth is fundamental to a safe casing design. Discussed here are ways of establishing the likely trends at the casing design stage.

If the pores in all formations from the depth of interest to the surface were hydraulically connected, the pore pressure of the fluids within the formations would be determined by the hydrostatic gradients of those fluids. This would be termed "normally pressured." In the absence of more accurate data, a generally accepted average gradient of formation fluid is 0.465 psi/ft; fresh water is 0.433 psi/ft.

The overburden stress (that is, the stress caused by the weight of all the materials above a depth of interest) is supported partially by the pore pressure and the rest by grain-to-grain contact within the formations.

The overburden stress divided by the depth will give the overburden gradient. In areas of low tectonic activity, the overburden stress is generally accepted to be at around 1 psi/ft. In tectonically active areas it may be as low as 0.8 psi/ft.

Conditions may exist that give pressures that are higher than normal (termed "abnormally pressured") or may be lower than normal (termed "subnormally pressured") when compared to formation fluid hydrostatic. In predicting where abnormal pressures may occur, two conditions are necessary:

- There must be an impermeable barrier above the abnormal or subnormal zone. In normal depositional basins, a layer of clean shale (i.e., no sand within it) will commonly form this barrier.
- There must have been a mechanism for creating the abnormal or subnormal pressure regime.

Mechanisms of abnormal pore pressure generation. Abnormal pore pressures can be created by several different mechanisms. Understanding of the mechanisms is important in identifying where these may occur.

Most instances of overpressures occur in areas of fast deposition of sediments. Water held in the formation pore spaces does not have time to move out of the rock matrix as the rock becomes increasingly compressed with growing overburden. This will cause the formation fluids to bear a larger proportion of the overburden pressure as the grains of rock are prevented from increasing their contact and taking their share of the load. As porosity normally decreases with depth, any change in this trend that slows the rate of porosity decrease with depth is an indicator of possible abnormal pressure. If the formation contains salt water, then the normally decreasing trend of resistivity with depth will also slow down or stop. Abnormal pressures may start from the top of this trend change. Where overpressures are caused by this mechanism the increase is gradual with depth; ROP trends such as D exponent can be used to identify this type of overpressure as drilling continues.

Gas generated under an impermeable boundary by decaying organic matter (biogenic gas) will cause an increase in pore pressure.

Salt domes distort and compress the formations around them and high abnormal pore pressures can result.

Bad cement jobs on offset wells or faults below a sealing formation can allow gas migration into higher zones, charging those zones to abnormal pressures. A similar mechanism is where a long gas column is normally pressured at the bottom by an aquifer. Due to the low density of gas, as you move up the gas column the difference between the gas pressure and the normal pressure will increase and be highest at the top of the gas column. Since the transition is very sudden if the cap rock is not leaky, these types of abnormal pressure would not be detected by D exponent or ROP trends while drilling (see Fig. 1-6).

Some rock transformations can cause significant increases in rock volume. Montmorillonite changes to illite under pressure, releasing water. Gypsum ($CaSO_4 \cdot 2H_2O$) also releases water as it changes to anhydrite ($CaSO_4$). If this liberated water is unable to move, pore pressures could increase significantly.

Normal pressure trapped within a boundary may be moved up by tectonic activity. If the pressure cannot reduce within the boundary then it will become abnormally pressured at shallower depths. Severe kicks can be taken by drilling into a raft of fractured dolomite within a massive salt sequence. This is a good example of trapped pressure, originally normal, which migrates up inside a pressure containing system. Examples are seen in the Zechstein sequence of the North Sea; saltwater kicks may be taken at much greater pressure gradients than would otherwise be expected. The pressures can be so high that increasing the mud density to control it and then drilling ahead may not be a viable option. However, these are normally of relatively low volume and if it is possible to allow the formation to flow, the pressure can deplete fairly quickly.

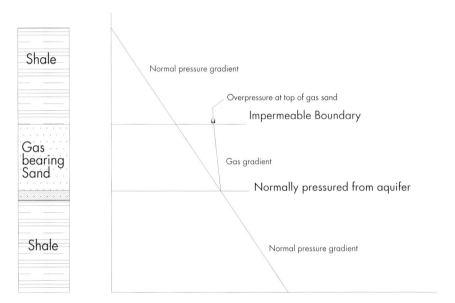

Fig. 1-6 Pressure-Depth Graph Showing Abnormal Pore Pressure Due to Light Fluid Column

Salt beds can also cause very high pressures underneath. Salt is a low shear strength material that is nonporous (i.e., no "pore pressure" to help support the overburden), impermeable (forms a seal—one of the conditions necessary for overpressure), and flows plastically under pressure. As such it can transmit hydraulically the full overburden pressure from above the salt to formation fluids below it.

Mechanisms of subnormal pore pressure generation. Subnormal pore pressures will occur as a reservoir is produced. The reservoir rock has to compress since the overburden (which stays constant) has less support from the pore fluids; therefore, as the pore pressure lessens, the rock vertical stresses increase. The formation does not have to be produced by a well; it could be that a fault could allow migration to a higher zone (overcharging that zone) or to surface, leading to less than normal pressure.

Drilling in mountainous areas may require a drilling location to be much higher than the water table in the area. Hydrostatic pressures in the wellbore may be much higher than formation pore pressures until casing is set fairly deep. Similarly, in arid areas (deserts), even without being far above sea level, the water table level may be deep.

When formations are distorted over geological time, some of the strata may be "stretched," causing subnormal pressures. Other parts of the same strata may become foreshortened, causing abnormal pressures.

High downhole temperatures will cause pore fluid to expand. If the fluid is not isolated and can move away then the fluid density and hence hydrostatic pressures will reduce.

Transition zone. A transition zone occurs where the pore pressure gradient changes with depth to a higher or lower-pressure regime. Transition zones are important. If possible, potential transition zone occurrences should be predicted before drilling into them.

If drilling goes too far into a transition zone of increasing gradient, there is risk of a kick. However, if the casing is set too early above or in the top of a transition zone, there may not be the kick tolerance needed to drill deep enough in the next hole section. This is because fracture gradient generally increases as pore pressure increases, so the deeper casing is set into a transition zone, the stronger the casing shoe strength.

Where transition zones occur due to under-compacted shales, the change in pressure gradient is likely to be gradual. This means that pore pressures can be monitored while drilling and there is less risk of drilling too far if the driller is vigilant.

If the transition takes place quickly there may not be much, or any, warning before the well kicks. It is especially important to identify potential overpressures where transitions may be rapid.

Detecting a transition zone while drilling is more difficult with a higher overbalance of mud density over pore pressure.

Indicators of abnormal pore pressure at the well design stage. Pore pressure and fracture pressure gradient prediction prior to drilling the well will use information from a variety of sources. The more field data there is available, the better the prediction should be. Data sources include:

1. Seismic surveys showing subsurface structures, gas anomalies, possible pressure communicating paths (faults), and indications of high pore pressure.

2. Offset well records:
   a) Pressures measured by sampling tools, e.g., repeat formation tester (RFT) and modular formation dynamics tester (MDT)
   b) Leakoff tests and frac job results will give infield stress levels
   c) Drilling records showing drilling trends, gas levels in mud, kicks, losses, and wellbore stability
   d) Measurements of shale bulk density (from drill cuttings)
   e) Resistivity logs that show a departure from the normal trend with depth may indicate overpressures. Thus, a resistivity LWD tool may give prior warning of a possible overpressured zone.
   f) Sonic logs can give bulk densities (allow overburden stress calculation) and maximum formation horizontal stresses. In undercompacted formations, the sonic velocity will be lower; be aware, however, that normally pressured fractured shales will also produce lower sonic velocity and so this should be correlated with resistivity log results.

Predicting overpressures from seismic surveys. Seismic surveys can give quite a lot of useful information for overpressure prediction. Sonic velocity through rock generally increases with depth as the formations become more compacted. If this compaction stops increasing with depth then the velocity will not increase with depth.

An empirical relationship was developed in 1968 by E. S. Pennebaker, which relates the interval transit time (reciprocal of velocity) to pore pressure, lithology, depth, and age. This was found to be a reliable indicator of overpressures in the Gulf Coast.

If this can be interpreted from the seismic, it may be then an indicator of potential overpressure. As noted previously, normally pressured fractured rock may produce the same result and this possibility also must be considered.

A new technique that may show promise is amplitude variation with bandwidth (AVB). The seismic wavelet is divided into bands of frequencies. The amplitude of each of these frequencies in the returning wavelet is compared with the original source signal. How the shape of these bands of frequencies is changed may give leads on where abnormal pore pressures are likely.

Structures may be seen that would indicate a column of a lighter than saltwater gradient (e.g., oil or gas) from a lower, normally pressured zone. As you move up the light column, overpressure (the pressure above normal) will increase and maximize at the top.

If rafts of other material can be seen within a massive salt, the well-bore may be moved to avoid them. Unfortunately, the sonic qualities of massive salts make it hard to interpret features within it to give an accurate assessment of the location. Where rafts are not frequent the margin of error could be accounted for. In some cases, it may not be possible to completely avoid any chance of drilling into a raft, with the attendant danger of a high-pressure kick. In this case the best that may be done is to ensure that casing is set as low as possible before reaching the critical depth so that at least a good shoe strength will be available if the well kicks.

Interpreting seismic is outside the scope of this book. The drilling engineer planning a well should spend time with a competent seismic interpreter and ensure that the proper questions are asked to obtain the fullest picture of the subsurface features.

Examining offset well records for overpressure prediction. If a planned well is in an area where offset wells have been drilled, these should provide good data for flagging areas of abnormal pressure. Offset data points should be plotted and marked with the well name on a depth-pressure graph; these will give pore pressure (PP) and fracture gradient (FG) trends for the offset wells, which should give an indication (when combined with overall field knowledge) of the likely trends in the planned well. In some cases a full pore pressure-frac gradient analysis may have been done and, if this is felt to be reliable, can be plotted directly on the graph.

As with all other data, try to evaluate raw data rather than relying on other people's interpretations. Mistakes may have been made in the earlier interpretation that could then mislead you. This does happen!

Direct measurements such as RFT or MDT pressures provide very accurate and reliable pressure data points and formation fluid gradients. These can then act as a qualitative check on other data. If a kick is taken then the formation pressures will be well known, and this also should be reliable data.

If total losses were taken and the well was filled with water to find a static level, then the pressure in the loss zone can be calculated if the depth is known. It is commonly assumed that if losses are suddenly taken while drilling that the loss zone is on bottom, but this is not necessarily the case. Before you rely on such data, try to establish whether another potential loss zone was exposed higher up in the open hole.

Just as pore pressures and fracture gradients are intimately related, frac data (from leakoff tests or frac jobs) provide data points that can be plotted on the depth-pressure graph. Unless there were dramatic changes in formation stresses or formation depths between the offset and planned well locations, these should relate fairly well to the new well. This may not apply in areas of rapid geological change with distance. In general pore pressure trends and fracture gradient trends follow similar paths on the graph.

Other drilling records that show ROP trends such as background gas levels, increasing mud chlorides (from formation water), and wellbore caving, do not give an accurate pore pressure. The indications may be that the mud in use (with ECD applied) does not overbalance the pore pressure by much and so it may possibly be said that the pore pressure is likely to be a maximum of mud hydrostatic + ECD. Remember, however, that you may not be able to tell where in the open hole the cavings or background gas are being generated. Also, remember that some background gas will be liberated in the cuttings as gas bearing formations are drilled and that a low-permeable, gas-bearing formation may not produce much gas even if the pore pressure is somewhat higher than the mud pressure. Connection gas and trip gas give reasonable indications that the formation is just overbalanced by the mud hydrostatic; the small and temporary drop in hydrostatic due to upward pipe movement has allowed a bit of gas into the well.

It is possible to measure shale bulk density from well cuttings. Where the bulk density does not increase with depth in accordance with the normal compaction trend for the area, then overpressures may be indicated. The procedure to measure the bulk density is as follows:

- Wash the sample of shale and ensure no cavings are present (these will come from another formation and would therefore affect the result)
- Add shale to the cup of a mud balance until the balance reads 0.433 psi/ft (fresh water) with the top placed on
- Top up with fresh water and take the balance reading, W psi/ft
- Calculate the shale density as follows:

$$SGshale = \frac{0.433}{0.866 - W}$$

Wireline logs can provide excellent indicators of pressure transition zones. As previously noted, the normal trends with increasing depth are for porosity and resistivity to decrease and sonic transit times to increase. If these trends change then it is likely that the pore pressure trend is also deviating from normal. These indications may be hard or impossible to interpret in streaky, nonhomogeneous formations. It is clearly very important to take all available offset data and interpret it intelligently to arrive at a picture that is likely to be close to reality.

Relationship between pore pressure and fracture gradient. When a casing is drilled out, it is normal to perform a leakoff test. This test applies pressure to the wellbore and should detect the point at which formation fracture is initiated without actually causing deep fractures. This pressure then dictates how far the well can be safely drilled ahead.

The fracture gradient is the pressure applied to the formation to initiate failure divided by the true vertical depth (TVD) of the formation being tested.

The resistance of the formation to fracturing comes from the addition of the formation fluid pore pressure and the tensile strength of the rock. The tensile strength of the rock at depth comes from the natural (unconfined) rock strength plus the supporting stresses imposed on the formation by field stresses. Therefore, by taking a formation test to the start of leakoff, information can be gained on the minimum field stress.

In an increasing gradient transition, pore pressures start to increase above normal. As pore pressure increases, so does the fracture gradient. If the lithology and the field stress gradient do not change during the transition zone, the pore pressures and fracture gradients will plot similar but diverging lines on a depth-pressure plot. So it is desirable to set a casing shoe as far into the transition zone as possible. This gives increased shoe strength for drilling the next section.

The following formula was proposed by Eaton to calculate formation fracture gradient:

$$Fg = \left(\frac{S}{D} - \frac{Pp}{D}\right)\left(\frac{v}{1-v}\right) + \frac{Pp}{D}$$

where
  Fg = Fracture gradient, psi/ft
  S = Overburden load, lbs
  D = Depth in feet (therefore S/D = overburden gradient)
  Pp = Pore pressure (therefore Pp/D = pore pressure gradient)
  v = Poissons Ratio

This formula can be simplified to:

$$\rho f = \left(\frac{\rho o - \rho p v}{1 - v} \right) + \rho p$$

where
  ρf = Fracture gradient, psi/ft
  ρo = Overburden gradient, psi/ft (generally assume 1 psi/ft)
  ρp = Pore pressure gradient, psi/ft (may be measured or estimated)
  v = Poissons ratio

1.4.5. Casing Shoe Depth Determination: General Points

Defining the casing setting depths has to take several different factors into account. In a directional well, casing points and directional planning are intimately entwined and may take several iterations to achieve a good overall design.

The first step is to decide which formations would give a competent shoe (i.e., one that will hold a reasonable wellbore pressure assuming that the cement job is good). Refer to the geological information, lithology column, pore pressure and frac gradient prognosis, hole section summaries of the offset wells, and any other available data. You are normally looking for competent shales or unfractured limestones that are impermeable and have a reasonable fracture gradient.

Of particular significance is the existence of a pressure transition zone. If casing is set just above a transition zone then pore pressure in the next section will increase shortly after drilling out and the kick tolerance will reduce as mud density is increased. It may reduce enough to prevent drilling the following section to the planned depth. On the other hand, if a transition zone is penetrated too deeply a kick may result in the worst circumstances since there will be a lot of open hole under the previous casing shoe. (See Fig. 1-7)

Ideally, casing should be set deeply enough in the transition zone to give a sufficient kick tolerance for the next hole section, while maintaining enough kick tolerance in the current hole section.

Where a drop in pore pressure is expected, casing could be set just above it to give the best shoe strength prior to drilling in to the weaker zone (see Fig. 1-7)

The next step is to note which formations may give problems relating to wellbore stability, losses, mobile formations, differential sticking, etc. Examination of these may flag where separation of problem areas is needed or where a hazard would be created by having two sets of problems in the same section (e.g., a weak "loss" zone in the same section as a hydrocarbon bearing zone). It might also show where dealing with the problems would cause serious incompatibility in the required solutions (e.g., different/conflicting mud properties required). Offset well data as presented in "Hole section summaries" in Section 1.1.2 will be helpful here.

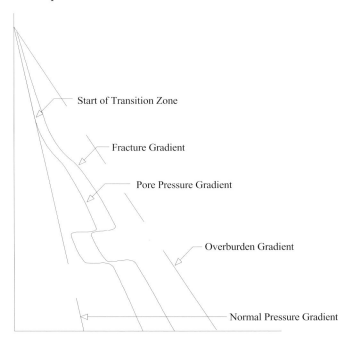

Fig. 1-7 Pressure-Depth Graph Illustrating a Transition Zone

If there are several target or reservoir zones to penetrate, it may be necessary to drill through these in one hole section for efficient completion. In this case designing the casing points to set a casing shortly above the top reservoir zone would minimize open-hole time in the reservoirs and would offer the best chance to get through in one attempt.

With the possible casing points and problem areas in mind for a directional well, draw up a rough directional plan. Formations giving wellbore stability problems should be avoided for the buildup section since hole enlargement will make directional control more difficult. Builds should be finished some distance before setting the next casing. Note those formations that should not be used for the buildup part of the well.

It is good practice to run casing to cover the kickoff section of a directional well. This consolidates the well and reduces the danger of keyseating in the dogleg section. However, the casing point should not be inside or just at the end of the kickoff section or else a keyseat could be worn in the casing shoe, which may be impossible to get free from if you get stuck in it.

In a horizontal well it is common to build to horizontal in the reservoir, run the production casing, and then drill ahead horizontally for a slotted liner or prepacked screen completion. This protects the buildup section and (assuming a good cement job) isolates production fluids from the formations above.

Now decide where suitable formations will fit in with the potential problems and the tentative directional plan. This will give the preferred casing setting points.

Starting at the bottom section, define the required hole size for each section. Once the preferred casing setting points and hole sizes are established, kick tolerances should be checked to ensure that these can be safely reached. A good target to set casing is in a transition zone giving you the shoe strength to drill ahead. Starting at the surface, calculate kick tolerances for each hole section so that, based on the assumed mud densities required and the likely fracture gradients, each section can be drilled to the preferred casing setting points with acceptable kick tolerance limits. If this is not the case, the casing shoe depths should be revised and it may be that an extra casing string is needed. (See Appendix 1 for calculating kick tolerances.)

Once the casing points allow a safe well, isolate problem areas, and with acceptable kick tolerances, revisit the directional plan. Adjust kickoff points and build rates as necessary to fit in with the casing points.

Having finalized the casing and directional program, note the TVD and measured depth (MD) of each shoe and the maximum anticipated pore pressure in the subsequent hole section.

We looked previously at summarizing offset data by using hole section summaries. Even in exploration areas these should be done with any offset wells that may be relevant even if they are quite far away—because they may give useful data. Now get together the lithology and pore pressure/fracture gradient prognoses for your well, with the hole section summaries, well proposal, and site survey (including shallow gas indicators, if any).

1.4.6. Individual Casing Points

The individual casing points will now be discussed. Criteria for physical properties of each of these casings will be summarized later.

Conductor. Start off at the top of the well by defining where your conductor pipe should be set. It protects topsoil from washing out and allows returns to a flowline. It may have to support future wellhead, casing, and completion loads. A conductor will allow a diverter to be set for drilling ahead.

The conductor depth will be determined by one of several factors:

1. On an offshore bottom-supported rig, it should be set at least deep enough to withstand the extra hydrostatic head imposed by bringing returns up to the flowline. This can be calculated using the formulae in Appendix 5.
2. If driving the conductor (whether on land or offshore), it should be driven to refusal. If the refusal depth is insufficient to allow returns to the flowline, you will have to secure the conductor and drill out of it and then recommence driving.
3. Offshore, you may drill into the seabed (perhaps with a template or guide base on a floater) and cement the conductor in place. The setting depth does not need to be greater than what is required to prevent washing out at the seabed, if the next section will be drilled for surface casing from a floater with returns to seabed. If a riser is to be nippled up or if the conductor extends to the flowline then the shoe depth should be calculated as per factor 1.
4. On a floating rig where the seabed is soft enough, the conductor may be jetted in if the surface hole can then be drilled with returns to the seabed. For jetting in, the intention is to set the pipe deep enough only to prevent washing at the seabed, and the following section will be drilled for surface casing riserless with returns to seabed. If a guideline system is used, the guidebase is made up on top of the conductor and run with it; otherwise a guide funnel is made up on top for a guidelineless wellhead system. The length of conductor is such that you can jet in to leave the guidebase about 5 ft above the mudline.

Surface casing. Surface casing performs quite a few functions. It also allows a BOP to be nippled up. It is set deeper to allow the production or intermediate casing point to be reached. Surface casing protects the wellbore from shallow gas, unconsolidated sands, sloughing shales, lost circulation zones, and key seats (if the well is kicked off in surface hole). It isolates fresh water sources.

If high-risk gas indications are present (shallow seismic/field experience), surface casing should be set not far above the danger zone and a BOP nippled up. Do not deliberately drill into high-risk gas anomalies with a diverter. A diverter should not be thought of as a means of well control; the purpose of a diverter is to allow time for the rig to be evacuated by diverting the flow away from the well for a short period of time. If you are lucky, the flow will stop by depletion or bridging; once flowing the chances of a dynamic kill are low and the risk of equipment failure during the flow is high.

In many areas, lost circulation zones (mostly unconsolidated formations) will be present during top hole drilling for surface casing. Once a competent formation is reached, casing should be set to protect the well from further losses prior to drilling ahead.

Generally surface hole is drilled as a small pilot hole, which is then opened up for the casing to be run. There are advantages and disadvantages to drilling a pilot hole first; usually the safety aspect will be paramount and this will outweigh the apparent disadvantages.

Advantages.

1. If shallow gas is encountered, the flow rate will be much lower with smaller diameter hole. This increases the (low) chance of a successful dynamic kill and it will give you a bit more escape time while diverting. It is more likely to bridge off quickly but with an extended flow the hole will erode and enlarge.
2. If the well will be kicked off in surface hole, the smaller hole will be easier to control directionally.

Disadvantages.

1. Time—two drilling runs to reach the objective.
2. Hole opening must be done carefully (especially in a deviated bore) since an inadvertent sidetrack could be drilled.
3. Hole cleaning could be a problem where a small pilot hole is drilled through a large ID conductor pipe.

Surface casing is always cemented to surface. This is usually with a drillpipe stinger so that cement can be pumped until returns are seen at surface, when the tail slurry can be displaced around the shoe.

Intermediate and production casing/liner points. Intermediate casing may not be needed on a shallow well, but one or more may be needed on a deeper well.

Often the intermediate casing is set in a transition zone where pore pressure (and usually fracture gradient) is increasing. This allows you to drill deeper for the next casing point.

Based on the pore pressures and fracture gradients prognoses, calculate how far you can now drill ahead within the acceptable limits of kick tolerance from the assumed casing point. This is the *maximum* depth you can safely drill to before setting casing, assuming that the predicted pore pressure/fracture gradient happens in practice. See Appendix 1 for calculating kick tolerances.

Now look over your hole section summary for the interval between the surface casing shoe and the intermediate shoe. You need to decide if there are any factors, which make it better to set casing higher than the maximum you can drill to.

Can you drill through all those formations in one open hole interval? What hole problems might you encounter, and would it be desirable to set casing higher to isolate one problem interval from another? Can one mud type be used throughout the section, modifying the properties as the well deepens if necessary? Where would be a good place to set your casing shoe—preferably competent shale? If you drilled to the maximum allowed by kick tolerance would this separate different target zones, which you might want to keep together in the same hole section? What are the directional requirements likely to be and might they affect your casing point? Look over the well objectives as well as lithology and problem areas to ensure that you consider everything that is relevant.

Once you have determined the optimum casing point, repeat the exercise for the next section so that you can define all the casing point vertical depths to the well TD. Then review the complete directional and casing plans so far.

Now you have decided where casings need to be set and you have finalized the directional plan. You know the vertical and measured depths of the shoes and the hole and casing sizes. Before looking at the forces that the casing has to resist, it is worth summarizing what material considerations apply. (See Fig. 1-8)

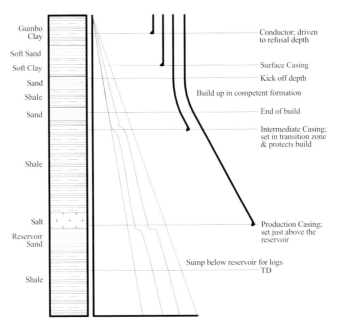

Gumbo Clay — Conductor; driven to refusal depth
Soft Sand
Soft Clay — Surface Casing
Sand — Kick off depth
Shale — Build up in competent formation
Sand — End of build
— Intermediate Casing; set in transition zone & protects build
Shale
Salt — Production Casing; set just above the reservoir
Reservoir Sand
— Sump below reservoir for logs
— TD
Shale

Fig. 1-8 Example of Casing Point Selection

1.4.7. Mechanical Properties of Steel

Steel is an elastic material, up to a limit. If a tensile load is applied to steel (*stress*), the steel will stretch (*strain*). If you double the load, you will double the amount that the steel stretches.

Stress is defined as *load ÷ cross-sectional area*. Units are usually pounds per square inch. Stress is usually given the symbol S.

Strain is defined as the *amount of stretch ÷ the original length*. Stress does not have any units, being a ratio. Strain is usually given the symbol e. Strain can be due to applied stress or it can be due to thermal expansion.

Hooke's Law states that, up to the elastic limit, stress is proportional to strain If this is the case, then *stress ÷ strain* is a constant. This constant is called *Young's Modulus of Elasticity,* symbol E, and for steel is approximately 30,000,000 (or 30×10^6). E for aluminum is approximately 10,500,000 (or 10.5×10^6).

Within the elastic limit, a load will stretch steel by a calculated distance. Removal of the load will restore the steel to its original dimensions. Once the elastic limit is exceeded, the structure of the steel is changed and it will not return to its original dimensions once the load is removed. Its behavior is now termed plastic. If more load is applied the steel will deform further and eventually fail (see Fig. 1–9).

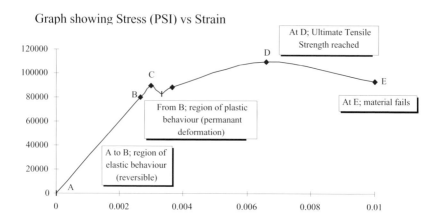

Fig. 1-9 Behavior of Steel Under Load; Stress vs. Strain

If a section of steel is stressed, it will get longer in the direction of stress. It will also get thinner perpendicular to the direction of stress. The ratio of strain in the direction of stress to the strain perpendicular to the stress is called *Poisson's Ratio*. Its symbol is usually v and the value for steel is approximately 0.3.

All of the properties mentioned are important in understanding casing design. They are also relevant when discussing rock mechanical properties for bit selection, fracture gradient calculation, and wellbore stability. For completeness it is also worth mentioning fatigue here, although fatigue is not usually a problem in casings.

Steel fails in tension if enough stress is put upon it. The elastic limit is exceeded, behavior becomes plastic, and the steel elongates and breaks. However, steel can also fail at stresses well within the elastic limit due to *fatigue*. Fatigue failures can sometimes be seen when drillstring components break. The mechanism is that exerting a cyclic stress on steel (load - unload - load - unload, etc.) induces eventual changes in the crystalline structure of the steel. It "work-hardens" and will eventually break. The higher the range of stresses exerted in the cycles (especially if the cycles alternate between tension and compression), the fewer cycles will cause the steel to break. However, there is a stress limit below which an infinite number of cycles will not cause failure.

For various materials, a graph can be constructed of cyclic stress (that is, maximum stress - minimum stress) against the number of cycles to fatigue failure. This is known as an S-N curve. It varies among steels depending on the alloy, heat treatment, etc. Normally the Y axis (stress) is a linear scale and the X axis (number of cycles) is a logarithmic scale.

An example of an S-N curve is shown below. In this example, 60,000 psi would cause almost immediate failure (10 cycles), 30,000 psi would fail at 800 cycles and 20,000 psi would allow an infinite number of cycles before failure (see Fig. 1-10).

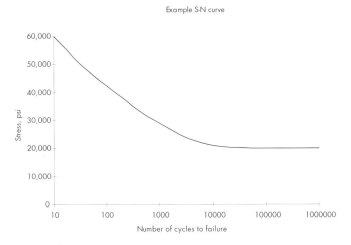

Fig. 1-10 Example of an S-N Curve

S-N curves should be available from the product manufacturer.

If there is a notch in the steel (such as a slip mark) then the stress concentrates at the end of the notch and can be much higher than would be calculated simply by looking at the load and the remaining cross-sectional area. This notch is known as a "stress raiser" and is very significant in fatigue failures since the actual stress is far higher than the average stress. The concentration of stress at the tip is increased by increased notch length and decreased notch tip radius. If a crack is carefully filed out and the edges smoothed, the stress will not concentrate and fatigue life will be extended, though the load required to fail the steel will still be reduced due to the smaller cross-sectional area supporting the load. In a corrosive environment, there is no lower fatigue S-N limit.

While examining a drillstring fatigue failure, an initial crack or defect can often be recognized that may be rusty on the faces. The steel tends to hold for some cycles while the stress causes localized work hardening, then the crack will suddenly extend a little and stop. This process repeats until finally the remaining steel cannot take the load anymore. Thus, you can see the rusty original crack and a set of failure marks, resembling rings in a cut-off tree. The final failure is a tensile failure, which will show very rough edges characteristic of this over the area where the final failure occurred. It is vital to clean with a bristle brush and soapy water, dry off, and oil the failure faces lightly to preserve these indications. Cleaning with a wire brush and/or allowing continuing corrosion may make identification of the failure mode impossible later on.

Finally, thermal strain should be mentioned as it becomes relevant to buckling in casing design. The Coefficient of Thermal Expansion (a) gives the thermal strain in a uniform body subjected to uniform heating. The commonly accepted Coefficient of Thermal Expansion for steel is given by

Strain $e = 6.9 \times 10^{-6}/°F$ $(1.24 \times 10^{-5}/°C)$

So for every °C uniform increase in temperature, steel will expand by 0.0000124 of its original length.

1.4.8. Safety Factors

Safety factors are arbitrary figures that have evolved with experience. A safety factor is applied to casing yield strength by dividing the yield strength by the safety factor. (Note that some authorities quote safety factors as a percentage (e.g. 90%) or as a number less than 1 (e.g. 0.90). In this case, multiply by the safety factor so as to reduce the available strength.)

Recommendations in various drilling literature and operator policies on safety factors can be quite confusing and widely varying. Reasons for various safety factors and clear recommendations are discussed in the following section.

Many uncertainties exist about the actual forces that a casing may be subjected to during its design life. Also, the casing strength is likely to deteriorate with time due to wear, erosion, corrosion, and reaction with produced fluids. Therefore, when designing casings, the expected forces are calculated and compared to the casing strength as stated in *API Bulletin 5C2* or other authoritative document that is downgraded by a safety factor. The actual safety factors used may be stated by company or government policy and they should be substituted for the safety factors assumed below if applicable. Following is a discussion on the most common safety factors.

Burst. Casings may be subjected to burst pressure throughout their design life. With time, the burst performance of casings may degrade due to wear, corrosion and other factors. API recommends 90% of the minimum internal yield. Experience has shown that this safety factor is sufficient. Biaxial effects increase the burst strength of casing under tension but this is not usually accounted for; casing tension can change with time due to thermal expansion. API does not account for biaxial effects on burst though it can of course be calculated.

Collapse. As with burst, casings may be subjected to collapse pressure throughout their design life and the same factors may degrade performance with time. A safety factor of 1.1 applied to the minimum collapse pressure is recommended. Experience has shown that these safety factors are sufficient. In deeper casings, biaxial effects reduce collapse strength of casing under tension (biaxial effects are covered below) and in certain circumstances, they should be evaluated or the safety factor should be increased. (Note that Table 4 of *API Bulletin 5C2* has corrections for biaxial effects in collapse.)

Tension. A quick review of safety factors in tension recommended by various authors and operating companies shows more variation than for any other safety factor. This varies from 1.3 to 1.8, some accounting for buoyancy and others not. Neal Adams, for instance, recommends a factor of 1.6 applied using the buoyant load or 100,000 lbs of overpull, whichever is more conservative. Preston Moore recommends 1.8 using buoyant load. API recommends 90%, which interestingly does not reduce tension below the elastic limit. It makes sense to examine what the safety factor covers under different circumstances and to define safety factors for those particular circumstances. Using too high a safety factor may result in overdesign of the casing, which may increase costs unnecessarily. It can be seen from the recommendations below that the more effort put into the design, the lower the safety factor needed and, perhaps, the lower the casing's grade, weight, and cost will be.

If a simple casing design on a deviated well is done using only uniaxial loads related to casing weight and pressure testing, then the uncalculated factors that will increase the overall casing load will include bending through doglegs, hole drags (overpull) when picking up, and shock loads. In this case apply a safety factor of 1.75 to the lesser of API minimum yield or connection tensile strength before comparing to the tensile force due to weight + force due to pressure testing (cross-sectional area x test pressure).

If calculations are also made to allow for bending forces due to directional profile or if the well is vertical, apply a safety factor of 1.5 to the lesser of API minimum yield or connection tensile strength before comparing to the maximum calculated tension.

If a full triaxial analysis is done (e.g., by using a casing design program incorporating triaxial calculations) and a temperature correction factor is also applied (see "Temperature correction factors for steel" within this section), then apply a safety factor of 1.25 to the lesser of API minimum yield or connection tensile strength before comparing with the maximum calculated stress.

For all the cases mentioned, specifying an *overpull allowance* based on the casing weight in air and compared to the lesser of API minimum yield or connection tensile strength with a 1.1 safety factor should also be calculated. This assumes stuck pipe (no buoyancy) and new casing. Often 100,000 lbs is used as a reasonable minimum overpull to be able to apply on casing and still be 10% below the minimum yield. If this overpull allowance is more severe than the relevant safety factors applied above, the casing weight and/or grade and/or pipe mix (in a mixed string) should be changed accordingly to allow this overpull with the casing on bottom. This will also ensure that shock loads will be covered, unless the driller handles the casing very roughly.

The driller should always know how much to safely pull on the pipe and it is a good idea to mark this on the casing running tally— especially with a mixed string.

Compression. API figures for minimum yield in tension calculate the load required to produce a stretch of 0.5% in normal strength steel (grade E). In compression, *Young's Modulus of Elasticity* is still the same and so the same load when in compression would produce a contraction of 0.5%. For pure compressive failure, therefore, it is reasonable to assume the same minimum yield strength as in tension. In most cases the only time that compression has to be calculated is for a conductor that supports subsequent casing, completion, BOP, and wellhead loads. Use the figure for minimum tensile yield without modification (e.g., safety factor is 1).

1.4.9. Factors Affecting Pipe Yield Strengths

Corrosion. Corrosion may occur during production due to electrochemical reaction with corrosive agents. It may also occur as uniform reduction in wall thickness, localized patterns of metal loss, or pitting. Of the three, pitting causes the greatest problem.

Corrosion rates increase with higher temperatures (rates approximately double for every 31°C increase in temperature), higher fluid velocities and/or abrasive solids (eroding away films forming on the metal surface), and higher concentrations of corrosive agents such as oxygen, carbon dioxide, and hydrogen sulfide.

Collapse and burst pressures are determined by the thinnest part of the wall, while tensile strength is determined by the remaining cross-sectional area.

Corrosion in a particular fluid can be measured using a corrosion coupon. This is usually in the form of a ring of metal that is weighed accurately, exposed to the fluid under realistic conditions of temperature, etc. and re-weighed after a period of time. The metal loss is recorded in weight loss (lbs) per square foot exposed per year.

If corrosion is considered a potential problem, then various treatments of the completion fluid can be made to reduce or eliminate it.

Wear. If significant wear is expected during drilling (monitoring metal in the returns; using tooljoints with rough hardfacing; high/shallow doglegs, etc.) then a Kinley caliper, ultrasonic imaging tool (USIT) log, or other tools can be run to measure wall thickness reduction. Collapse and burst pressures will be determined by the thinnest part of the wall while tensile strength will be determined by the remaining cross-sectional area.

Wear is influenced by side forces (in turn influenced by dogleg severity, inclination, and tension), number of rotating hours, roughness and hardness of the rotating surface, and number of round trips carried out.

Wear reduction measures may include protectors (standard or those that rotate on the pipe), using downhole motors, ensuring hardfacing is smooth and has been run in open hole first, and restricting dogleg severities and inclinations.

Wear mitigation measures may include using thicker wall sections and/or higher grade steels over the areas of potential wear. Experience indicates that using thicker wall sections is preferable to using higher grades of steel.

Temperature correction factors for steel. Correction factors have an engineering basis, unlike safety factors that are arbitrary. Correction factors are applied as well as the relevant safety factors.

The yield strength of steel usually decreases with increasing temperature. A temperature correction factor can be applied to the minimum yield strength before applying the safety factor as mentioned. This correction factor should be obtained from the casing manufacturer. Apply the temperature correction factor as noted here.

Tension. Tensile load will decrease with depth so that as the casing gets hotter, it is also subjected to less tension. If the top joint of an unmixed casing string is strong enough in tension, it should be fine lower down. If a mixed string is used (different sections of the overall string that have different weights and/or grades) apply the factor when evaluating the tension applied to the top component of each section.

Compression. Compression is unlikely to be relevant. Helical buckling is more likely to occur than failure in compression in the hot part of the string.

Burst. Burst could be very relevant at depth, especially in a high-pressure, high-temperature well. This is not likely to be a problem while drilling but may be a problem later in the life of the well. If a frac treatment or other procedure can be used where significant surface pressure may be applied, this burst pressure will be imposed down the exposed casing string (i.e., above any packers set).

Multiply the burst strength by the temperature correction factor and apply the safety factor before comparing the amended burst strength to the calculated burst pressure.

Collapse. Collapse could be a problem while drilling if severe losses are taken, high drawdowns are used during production, and the reservoir becomes depleted. Also if massive salts are covered, we generally assume that the salt transmits the full overburden pressure against the casing in collapse (1 psi/ft).

Multiply the collapse strength by the temperature correction factor and apply the safety factor before comparing the amended collapse strength to the calculated collapse pressure.

The following table is supplied by Nippon Steel. Figures for other manufacturers should be requested from the manufacturer of the pipe under consideration.

Temperature °C	Temperature °F	Yield Strength Correction Factor
20	68	1.0
50	122	0.95
100	212	0.88
150	302	0.84
200	392	0.81

1.4.10. Methods of Applying Buoyancy Effects

Buoyancy creates an upward force on objects immersed in a fluid. There are various ways to calculate buoyancy, depending on how you want to use the information.

The upward force that we call buoyancy is caused by the hydrostatic pressure imposed by the mud on an object increasing with depth. For an immersed object, the pressure below the object is greater than the pressure above. This pressure differential is what gives the net upward force.

1. *Apply a buoyancy factor*
   If a hollow object is immersed in fluid and if the hollow part can be filled with the surrounding fluid (or if the object is solid), then the weight in air of that object can be multiplied by a buoyancy factor. The buoyancy factor depends on the specific gravity (SG) of the object and the SG of the fluid. The formula for buoyancy factor for a solid immersed in a liquid based on the relative SGs is:

$$BF = 1 - \frac{SGliquid}{SGsolid}$$

For steel (SG 7.87), the buoyancy factor when immersed in mud can be simplified to:

$$BF = 1 - (0.2936\rho) = 1 - (0.0153\,ppg)$$

where r is the mud gradient in psi/ft or ppg is the mud weight in pounds per gallon.

If aluminum were in use (e.g., aluminum drillpipe) then the buoyancy factor can be calculated from the SG for Al, which is 2.75.

Buoyancy factor can be used for calculating the immersed weight of a drillstring (which fills as it is run in) or of casing that is filled with mud of the same density as the fluid outside.

Using buoyancy factor to calculate axial stress in a string of tubing only yields an accurate result at the top of the string.

2.   *Subtract the weight of displaced fluid from the total weight of the object and contents (see Fig. 1-11).*

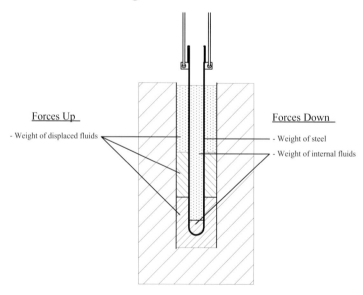

Forces Up
- Weight of displaced fluids

Forces Down
- Weight of steel
- Weight of internal fluids

Fig. 1-11 Calculating Buoyancy with the Weights of Displaced Fluids

If casing is not open to the surrounding fluid and it contains fluids of different densities to the surrounding fluid (e.g., after a cement job), then it is easier to use Archimedes' principle by calculating the weight of casing + contents and subtracting the weight of the displaced fluids.

Inside the casing there is cement tail slurry in the shoetrack and mud above the float collar. Outside the casing there is (from bottom to top) tail slurry, lead slurry, spacer, and mud. If the tail slurry is 500 ft measured depth interval and weighs 15.8 ppg outside 9⅝ in casing then the weight of displaced slurry is (500 ft x 15.8 x 3.78 [gals/ft displacement]) or 29,862 lbs. Make the same calculation for all of the fluids outside the casing, total these figures up and this is the total buoyant force acting upwards on the bottom of the casing.

In a vertical well, the total force up acts on the bottom end of the casing. The weight of the fluids inside the casing act downwards on the cement plug at the bottom of the casing. The difference between these two forces puts the bottom of the casing in compression.

The neutral point for tension can be calculated by working out the length of casing in air that equals this net force up and subtracting that length from the total length. This will give you *approximately* the depth of the neutral point, above which will be in tension and below which will be in compression. This will not be the case in a deviated well since the buoyancy force acts perpendicular to the surface of the immersed object, causing a net buoyant effect on the sides of an inclined pipe as well as at the end. This is explained in point #4. If different weights of casing are used, this will also change the neutral point, as explained in point #3.

Buoyancy force calculated as described above is useful to determine the buoyant weight of a closed-end object that has different fluid densities inside and outside; e.g., casing after bumping the cement plug. In a casing string where one weight of casing is used for the whole string, axial stress can be calculated at any depth by subtracting the weight in air of the pipe below the depth of interest from the buoyancy force.

3. *Calculate hydrostatic pressures and multiply by cross-sectional areas* (see Fig. 1-12).

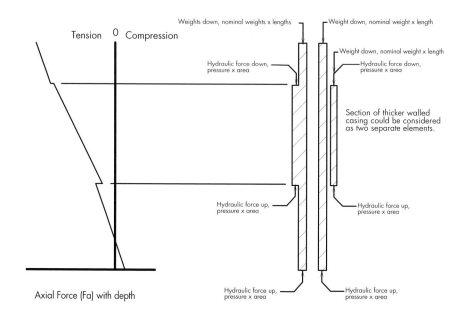

Fig. 1-12 Calculating Buoyant Forces Due to Pressure x Area

If a tubing string contains different cross-sectional areas, and if it is important to calculate accurately the actual axial force at a particular depth, then the individual buoyancy forces at the depths where the cross-sectional areas change have to be calculated. In line with convention, compressive axial force is shown positive and tension is shown negative.

Each buoyancy force element acts up or down on the exposed surface. The force magnitude equals hydrostatic pressure x area.

Axial force F_a at any depth equals the sum of forces up minus the sum of forces down below that depth. Forces up are buoyant, forces down are weight, and also hydrostatic forces (negative buoyancy) apply downwards if the exposed cross-sectional area increases with depth.

Consider a vertical mixed casing string of 9⅝ in casing, 40 lbs/ft from surface to 4000 ft, 53.5 lbs/ft from 4000 ft to 5000 ft, and 40 lbs from 5000 ft to 10,000 ft. Heavier casing is run across a plastic salt zone. Cross-sectional areas are 15.55 in² and 11.45 in² respectively. Mud gradient is 0.6 psi/ft. Calculate axial force at 9000 ft and 3000 ft.

Buoyancy force acting at the shoe is 6000 psi x 11.45 in² = 68,700 lbs. Buoyancy force will also apply upwards at the change of cross-

sectional area at 5000 ft.; this will be 3000 psi x (15.55 - 11.45) in²
= 12,300 lbs. Hydrostatic force will also apply downwards at 4000
ft. This will be 2400 psi x (15.55 - 11.45) in² = 9840 lbs.
To calculate axial force F_a at 9000 ft, subtract forces down from
forces up:

F_a = [68,700] - [1000 x 40] = 28,700 lbs (positive, so in compression).

To calculate axial force at 3000 ft, subtract forces down from forces up:

F_a = [68,700 + 12,300] - [(5000 x 40) + (1000 x 53.5) + (1000 x 40) +
(9840)] = -222,340 lbs (negative, so in tension).

4. *Calculate hydraulic force acting sideways only.*

If the side force exerted by a casing lying in an inclined wellbore is
to be considered, then the hydraulic forces acting perpendicular to
the axis need to be calculated. This may be needed for instance to
calculate drags or torques when running in a deviated well. Refer to
Figure 1-13.

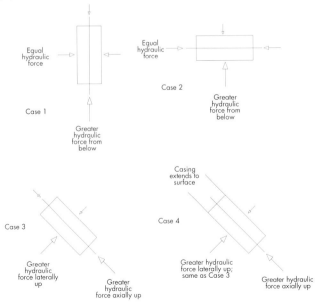

Fig. 1-13 Applying Hydraulic Forces in an Inclined Wellbore

In Case 1, the net upward force is equal to the difference between the upward and downward pressures acting on the cross-sectional area. This net upward force can in this case be calculated by using the (weight of steel x [1 - buoyancy factor]). The net upward force is the same in Case 2, except here the force results from varying differential pressures acting on the diameter of the casing instead of on the end.

In Case 3, the net upward force is equal to Cases 1 and 2, the difference is that the net force comes from the vector addition of the net lateral force and the net axial force. Again the total buoyant weight can be calculated by weight of steel x buoyancy factor. Since calculating the net axial force is easy (cross-sectional area x difference in hydrostatic head), we can deduce the net lateral force by vector subtraction. Calculating it directly would be difficult, involving differential calculus.

In Case 4, we have taken the section of casing from Case 3 and joined it on to casing extending to surface. This has no effect on the lateral force. Therefore, by treating the section of casing independently (as if it were freely suspended in mud), the net axial force subtracted using vectors from the total buoyant weight gives us the net lateral force.

For example, let us assume that you want to consider a section of casing, $9^5/_8$ in x 40 lbs/ft, 100 ft long, immersed in a mud of 0.5 psi/ft gradient (buoyancy factor 0.85; ppg 9.62). Inclination is 40°. True vertical depth of the lower end is 5000 ft. Calculate the net axial and lateral forces.

1. For the section of casing in question, use the buoyancy factor to establish the buoyant weight as if it were freely suspended in mud.
    This would be 40 lbs/ft x 100 ft x 0.85 = 3400 lbs.
    Net upward force = weight in air - buoyant weight = 600 lbs.
2. Calculate the axial forces acting at the bottom and at the top of this section if freely suspended in mud.
    TVD at bottom (at centerline) is 5000 ft; cross-sectional area is $.7854(D^2 - d^2)$ = 11.454 in²; hydrostatic pressure is 2500 psi so the total force action upwards axially is 28,635 lbs. This figure can be used later to calculate net axial force throughout the casing string.

    TVD at top is 5000 ft - (100 Cos 40) = 4923.4 ft. Hydrostatic pressure is 2462 psi and axial load acting downwards is 28,196 lbs.

Net axial force = 439 lbs.

3. Use vectors to calculate the net lateral force (Pythagoras, $\sqrt{(600^2 - 439^2)}$). (See Fig. 1-14)

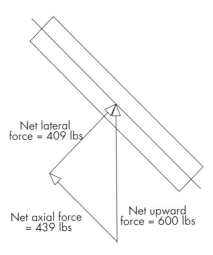

Net lateral force = 409 lbs

Net axial force = 439 lbs

Net upward force = 600 lbs

Fig. 1-14 Resolving Net Axial and Lateral Forces in an Inclined Wellbore

1.4.11. Casing Design Criteria: Definitions and Methods of Calculation

So far you have determined what the directional plan will be, where the casing shoes will be set, and what hole and casing sizes are needed. The next step is to calculate the physical forces that the casings will be subjected to so that these can be compared to the casing strengths, as modified by safety and correction factors.

Refer to *API Bulletin* 5C2 for minimum performance properties of casing, including burst, collapse, collapse under axial load (biaxial), and tensile strengths.

Note that lower tensile grades of pipe are required for an H_2S environment at temperatures below 175°F. Refer to 1.14.19, "Material Grades."

1.4.12. Calculating Burst and Collapse Loads, Including Biaxial Effects

Burst and collapse loads can be applied to the casing from various causes. Each situation should be calculated and combined to give the net burst or collapse loading on the casing in that situation. Burst and collapse loadings need to be calculated together and the net burst or collapse obtained for the worst case while drilling, producing, stimulation, or other operations to which it may be subjected.

Burst loads may occur from combinations of:

- Hydrostatic head. For a fluid of single density, the burst load equals the vertical depth multiplied by the fluid gradient. If casing is full of mud of 0.5 psi/ft gradient, the burst pressure from hydrostatic at 8000 ft TVD will be 8000 x 0.5 = 4000 psi. Where multiple fluids are involved (e.g., in cementing), the hydrostatic pressure of each is calculated separately and added up.
- Applied surface pressure. This may occur from pressure testing, from a kick while drilling, from a leaking production string, from support pressure while fracturing, from injection pressure for gaslift, or chemical injection in the production casing. This surface pressure is added to the hydrostatic head to give the total internal pressure at any particular depth. For instance, with 500 psi surface pressure applied to the previous example, the internal pressure at 8000 ft TVD is 4500 psi.

Collapse loads may occur from:

- Hydrostatic head of mud or cement outside the casing. Calculated for burst as previously discussed in this section.
- Flow from lower zones migrating upwards outside the casing, exerting extra pressure in addition to hydrostatic head of annulus fluids. For this reason annular pressures are monitored with a gauge on the wellhead side outlet. Calculated for surface pressure, as previously discussed in this section.
- Drawdown pressure while producing in the production casing or liner.
- Mobile formations such as massive salts. It is normally assumed that mobile salts can impose overburden pressure (1 psi/ft) on the casing. This may require extra heavy wall pipe across a massive salt.

Biaxial effects account for the change in burst and collapse resistance due to tension or compression of the casing. A pipe under tension will have increased burst and reduced collapse resistance; conversely a pipe in compression will have reduced burst and increased collapse resistance.

Biaxial effect could be used to reduce the required weight and/or grade of the casing string as net burst forces are usually highest at surface (assuming there are higher supporting densities outside the casing than are inside). As tension is also highest at the surface, an increase in burst strength is seen where it is needed. However, care should be taken in using this to improve a marginal design because production-casing tension at surface could reduce due to thermal expansion under production, reducing the tension and therefore the benefit gained.

The revised collapse resistance of casings accounting for biaxial effects is shown in Table 4 of *API Bulletin* 5C2. There is normally no need to use the formulae to hand calculate biaxial effect. A good casing design program would account for biaxial using the formulae.

1.4.13 Calculating Axial Loads

Once the burst and collapse requirements are known, the weight and grade of the casing needs to be estimated before tension and compression can be calculated. Use casing design tables (e.g., *API Bulletin* 5C2, *Halliburton* book, casing manufacturer's data, etc.) to decide which of the available casings will handle the worst case burst and collapse pressures. Use the lowest available weight/grade of casing that is strong enough, apply the desired safety factor, and (for a mixed casing string) apply the temperature correction factor to the minimum burst pressures before comparing it to the calculated maximum pressures. In a mature area, you will probably have a stock of your commonly used casings, otherwise you need to know what casings are available in time for the earliest possible spud date.

Tension due to weight in a deviated wellbore. In a vertical well, clearly the tensile force at the top of the casing equals the entire buoyant weight of the casing. If the entire string was placed in a horizontal bore, the tensile force at the "top" end would be zero. In a deviated well greater than 0° and less than 90° inclination, calculate the actual tensile force, accounting for the wellbore support and buoyancy.

Tensile force due to the weight of the casing in a deviated well, in air and ignoring friction, resolves to (TVD x weight/foot). Forces due to the mud acting axially will equal (TVD x casing cross-sectional area x mud gradient). (See Fig. 1-15)

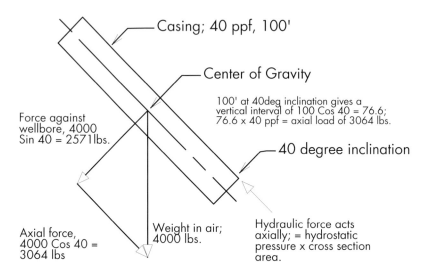

Fig. 1-15 Vector Diagram Illustrating the Effect of Inclination on Axial Load

The following example calculates the effect of deviation on a steel pipe 100 ft long weighing 40 lbs/ft (i.e., 4000 lbs total weight in air), immersed in a mud of 0.5 psi/ft gradient (buoyancy factor 0.85; ppg 9.62).

In a vertical well, no support for the casing weight is given by the wellbore. Tensile force at the top due to the weight in air is 4000 lbs and in mud is 3400 lbs.

In a horizontal well, total support for the casing weight is given by the wellbore. Tensile force at the "top" is 0 lbs in air or in mud.

In a 40° inclined wellbore, the weight of the casing (acting vertically down) can be resolved into two vectors; one perpendicular to the wellbore (force exerted by the formation on the pipe) and one along the wellbore (= tension at the top). Taking the weight in air, 4000 lbs acting vertically down is resolved into a vector component of 4000 Sin 40 = 2571 lbs against the wellbore and 4000 Cos 40 = 3064 lbs along the pipe axis.

The vertical interval of our 100 ft pipe when inclined at 40° = 100 Cos 40 = 76.60 ft. If we multiply the vertical interval by the weight,

76.60 x 40 = 3064 lbs. Note that by multiplying the vertical interval by the nominal weight in air, we arrive at the same figure as resolving the vectors for axial load.

Therefore, for weight in air, to calculate the tension due to the weight of the pipe in an inclined bore, it is only necessary to multiply the *weight per unit length by the vertical interval.*

The axial force due to the effect of fluid acts at the bottom of the pipe. This puts the bottom section in compression by: (hydrostatic pressure x cross-sectional area).

Tension due to overpull if stuck. For an unmixed string, calculate the tension of the top joint due to the weight in air of the casing. For a mixed string, calculate the tension of the top of each component of the combined string (using weight in air). In a deviated well the tension at the top due to the weight of the casing equals the nominal wt/ft multiplied by the TVD for each casing below. As is discussed above, casing in a deviated well is partially supported by the wellbore and the resultant axial force can be calculated simply by using TVD x nominal weight. The difference is significant in high-angle wells.

The minimum yield strength divided by a 1.10 safety factor should exceed this tension by an overpull allowance (usually at least 100,000 lbs) to account for overpull while running. Weight in air is used rather than buoyant weight because if the casing gets stuck the effect of buoyancy is lost and the overpull allowance will need to be added to the weight in air.

For a mixed string of different weights/grades, the same calculation should be made for the top joint of each section. For the top section joints at depth, divide minimum yield by the safety factor (1.10) and then by the temperature correction factor; ensure that this exceeds the calculated tension by at least the overpull allowance.

Tension due to buoyant string weight plus test pressure. Calculate the tension at the top joint of each different section of casing string, when in mud and cement (refer to Section 1.4.10, "Methods of Applying Buoyancy Effects"). Add to this tension the force exerted when testing to the anticipated test pressure after bumping the plug. This will equal the cross-sectional area inside the casing in square inches multiplied by the test pressure in psi, or

$$\text{Load} = .7854d^2 \times \text{psi}$$

If you are not calculating further tensile forces then compare the total tension to lesser of [API minimum yield or connection tensile strength] ÷ safety factor (use SF of 1.50 for a vertical well, 1.75 for a directional well).

For instance, intermediate casing string of 9⅝ in casing, assuming first choice is 47 lbs/ft N80 from TD at 15,000 ft measured depth; 80 ft shoe track. Test pressure will be 3000 psi. Cement is 500 ft of 0.82 psi/ft (15.8 ppg) neat "G" cement from the shoe and 3500 ft of 0.70 psi/ft (13.2 ppg) cement above that. Mud in use is 0.65 psi/ft (12.5 ppg). Connection is buttress. Well is vertical.

Weight in air is 705,000 lbs, plus 100,000 lbs overpull allowance equals 805,000 lbs. Minimum yield strength is 1,086,000/1.1 (safety factor for overpull) = 987,000 lbs so the tension due to its own weight plus overpull is within limits. Note also that for a regular buttress connection, pipe body strength is less than connection strength. If the connection were a special clearance buttress, then the connection would be weaker than the body strength and that would be the limiting factor—in this case, 983,000 lbs, which is still adequate for this example.

Weight of casing contents; casing capacity is 3.073 gals/ft. Weight of mud therefore is 14,920 x 3.073 x 12.5 = 573,100 lbs. Weight of cement in the shoetrack = 3900 lbs. Weight of steel = 705,000 lbs. Total downward force = 1,282,000 lbs.

Casing displacement is 3.778 gals/ft. Weight of displaced fluids = (500 x 3.778 x 15.8) + (3500 x 3.778 x 13.2) + (11000 x 3.778 x 12.5) = 724,000 lbs total upward buoyant force. So the net tension due to buoyant weight at the top is 558,000 lbs.

The force exerted by a 3000 psi test on an ID of 8.681 is 0.7865 x 8.681^2 x 3000 = 178,000 lbs.

So after cementing the casing and bumping the plug, the tension at the top joint will be 736,000 lbs. Applying a safety factor of 1.50 (safety factor for a vertical well) to the minimum body yield strength of 1,086,000 lbs gives an allowable tension of 724,000 lbs—not enough in this case.

Without further calculations to reduce the safety factor or dispensation to allow a lower safety factor, a mixed casing string would have to be run. P110 grade 47 ppf run at the top would give an allowable tension of 995,333 lbs at a safety factor of 1.50. If a triaxial analysis were to be done, it is possible that a mixed 53.5 or 47 ppf N80 string could work. It can be seen that it would be worth the extra effort to calculate it out if the casing string could be kept to the lower grade as a result, 53.5 ppf N80 usually being cheaper than 47 ppf P110.

Tension due to shock loading. Various authors have proposed formulae to calculate shock-loading forces on casings while running. By experience, if the above safety factors and a reasonable overpull allowance are used, they are adequate to cover shock loadings if the drill crew follow normal procedures. In any case, calculating actual shock loads probably cannot be done accurately due to the many assumptions that have to be made or mitigating effects that have to be ignored. The latter will include the damping effect of mud, frictional forces against the wellbore (especially if deviated), and actual running speed at the time that the casing is arrested.

Axial loads due to bending forces in a deviated wellbore. In a deviated well, you could now account for extra tension due to bending. Assume that our 15,000 ft well is kicked off at 1000 ft vertical depth, angle is built at 2.5°/100 ft to 25° inclination. The well is now drilled tangent to 15,000 ft TVD. (See Fig. 1-16)

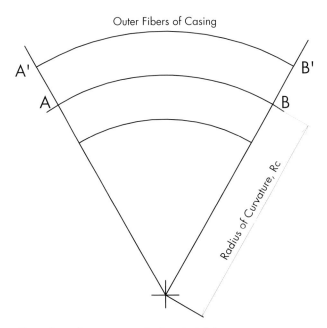

Fig 1-16 Effect of Bending Pipe on Casing Axial Stress

The radius of the dogleg is calculated by 5730 ÷ dogleg severity (DLS) in degrees per hundred foot; so the radius of curvature of the casing center Rc is 2292 ft, assuming that the casing is perfectly centralized in the hole. The radius of curvature at the outside of the bend is Rc (2292 ft) plus radius of casing (B to B¢ ft) = 2292.401 ft, which is where the greatest strain will be. The circumference of these two curves is 2pr and is 14,401.06 ft and 14,403.58 ft respectively. Strain therefore is stretch ÷ original length or

e = 2.52 ÷ 14,401.06 = 0.000175

As 2p is a constant, the formula can be simplified to:

$$\varepsilon = \frac{r_c \times DLS}{5730}$$

where Rc is the radius of the casing in feet and DLS is dogleg severity in °/100 ft.

Stress S = Young's Modulus *E* x strain e, or

S = 30,000,000 x 0.000175 = 5248 psi, in the outer fibers due to bending only.

Referring to the calculation above, the total tensile force *at the top of the kickoff point* when bumping the plug to 3000 psi is calculated using TVDs not measured depths due to the supporting effect of the wellbore (see Fig. 1–14):

Tension due to casing weight below KOP without buoyancy = 14,000 ft (TVD) x 47 = 658,000 lbs.

Use TVD and actual ID/OD to calculate internal weights and external buoyancy forces:

Shoe at 15,000 ft TVD, plugs at (15,000 - [80 Cos 25]) = 14,928 ft TVD. Internal pressure is therefore (14,928 x 0.65) + (72 x 0.82) = 9762 psi, acting down on an area of 59.19 in² = 577,789 lbs.

Shoe at 15,000 ft, vertical interval of tail slurry is 500 Cos 25 = 453 ft, vertical interval of lead slurry is 3500 Cos 25 = 3172 ft, therefore, top of cement (TOC) is 11,375 ft TVD. Weight of displaced fluids = (11,375 x 0.65) + (3172 x 0.686) + (453 x 0.82) = 9941 psi, acting on an area of 72.76 in² = 723,321 lbs.

Net axial hydraulic force at the shoe = 145,532 lbs.
So total tension due to buoyant casing weight at kickoff point (KOP) = 658,000 - 145,532 = 512,468 lbs.
Add tension due to pressure test (as previously calculated), tension at KOP = 690,468 lbs.

This is over a cross-sectional area of $0.7854(D^2 - d^2) = 13.57$ in² so the stress at this point (load ÷ area) is 50,881 psi. Tensile stress due to bending in the outer fibers of the casing is 5248 psi. Therefore, the total stress at the outside of the curve is the sum of the two, or 56,130 psi. The minimum yield stress of 47 ppf N80 casing is 80,000 psi so the safety factor at this point is 1.43 (too low). With a safety factor of 1.5, the minimum yield stress should be 84,195 psi so 47 ppf C90 run over the deviated section would suffice, or dispensation could be requested to accept the lower safety factor. In practice, the tension in the steel along the outside of the curve would be transmitted to the adjacent casing, taking some distance to dissipate the load. Running 100 ft of 47 ppf C90 above and below the dogleg section will suffice.

Axial forces in casing due to friction in a deviated wellbore (tangent section). In order to calculate the effect of friction, we need to calculate the actual force exerted between the wellbore wall and the casing.

■ Refer to Section 1.4.10, Methods of Applying Buoyancy Effects, which describes how to calculate the effect of hydrostatic (buoyancy) forces acting on an inclined pipe so that the force against the wellbore wall can be found.

■ Refer to the previous section, which shows how to calculate the effect of increasing hole angle on sidewall forces in air.
■ Subtract the buoyancy forces from the sidewall forces to give the actual force pressing the pipe against the wellbore wall.

In general,

Force due to friction between surfaces = coefficient of friction (m) x area x force pressing the surfaces together.

However, when calculating frictional forces between casing and wall, length is used rather than area and m is generally assumed to be 0.35 in open hole and 0.20 inside another casing.

Therefore, to calculate the frictional force acting against pipe reciprocation, calculate the actual force pressing the pipe against the wellbore wall, multiply by m, and then multiply by the length (in feet) under consideration.

Compression loads. Casings are mostly in tension, except for conductors. However, pure compression (as opposed to buckling that is covered next) is rarely a problem leading to failure of pipe.

Compression will occur during running and cementing pipe (due to buoyancy forces acting at the bottom). It can also occur if the casing is set down on bottom or on an obstruction.

Bending through a dogleg will place the inner fibers of the casing in compression; the effect is identical in magnitude and opposite in effect to that experienced by the outer fibers. The calculation is the same as was done for tension due to bending, covered previously. In most cases this compressive effect will serve to reduce tension in the inner fibers rather than lead to compression, except in the lower part of the casing string when running it through the dogleg section (i.e., little weight below the curved section of casing).

If a section of pipe is fixed in two places and free in between, increasing temperature will cause the casing to expand and this will then lead to compressive forces. If the pipe is in tension to start with, the tension will reduce and could eventually go into compression, depending on the initial tension and the change in temperature. (Refer to Section 1.4.7, Mechanical Properties of Steel)

1.4.14. Calculating for Buckling (N_b)

A long, unconstrained string of thin walled pipe (casing, tubing) will buckle if it is put in compression. The inherent capacity of casing itself to resist buckling is effectively zero due to the lengths involved. Conductors may be put in compression and not buckle until a significant load is imposed, but in the case of conductors, the diameter and wall thickness are significant when compared to the unsupported length.

The better the pipe is supported (e.g., centralized in gauge hole; lying against the hole in deviated wells) the higher the load that can be applied before buckling occurs. Where pipe is well supported, high loads can be carried without buckling.

For practical purposes, buckling in casings need only be considered in vertical wells where the temperature and/or internal pressure will significantly increase after cementing. In a deviated well the casing lies against the wall and is supported by it.

Buckling in freestanding conductors. Conductors are mostly either very short (e.g., land well, subsea wellhead) or are well supported (run in platform guides). Occasionally a conductor is designed to be freestanding for a significant height above the seabed while carrying the wellhead loads at the top. In these circumstances a full engineering analysis of all the service and environmental loads is required. The circumstances where a long freestanding conductor is required are not common and the calculations are complex. Therefore, these calculations are outside the scope of this manual.

Buckling in casings. There are two sets of forces to consider when calculating whether an unsupported casing will buckle. These forces can be calculated for any tubular; tubing or drillpipe can also be considered in this way.

1. The force tending to make a casing buckle—the axial (compressive) force at the depth of interest. In these calculations, compressive force is given a positive value, tensile force is negative. This is consistent with the signing convention in the literature and is termed the buckling force (F_b).
2. The force that resists buckling forces—called the stabilizing force (F_s). This is the average of the radial and tangential stresses in the casing at the depth of interest.

$$Fs = A_oP_o - A_iP_i$$

where

A_o = Area of a circle of tubing OD size = $0.7854D^2$
P_o = Pressure outside tubing
A_i = Area of a circle of tubing ID size = $0.7854d^2$
P_i = Pressure inside tubing

Buckling will occur in unsupported tubing when the compressive force exceeds the stabilizing force. The point where $F_b = F_s$ is the neutral point for buckling (N_b). *It is not the same as the neutral point for axial force* (i.e., where tensile stress = 0).

An open tube suspended in a fluid will not buckle, as long as the fluid density does not exceed the density of the tube material. *This is not going to happen for casing unless you are using mercury instead of mud!* N_b is at the bottom of the tube. At the bottom end, F_b = tube cross-sectional area x hydrostatic pressure (buoyancy force) and F_s = (area of outside diameter - area of inside diameter) x hydrostatic pressure. $A_o - A_i$ = cross-sectional area, F_b clearly equals F_s, which is the neutral point for buckling, N_b.

A suspended, unsupported tube will buckle if the bottom end is closed and internal pressure is increased. This will not occur during a cement job when the cement is inside the casing because the bottom end is not closed, fluid is exiting the bottom, the weight of the slurry will increase tension, and the casing will almost always be supported by centralizers.

In practical terms then, we do not need to consider buckling of casing during cementing. Once cemented, however, the picture changes. F_b will increase with increased temperature (as the bottom end is fixed and the casing expands). F_s will decrease with increased internal pressure—say as heavier mud is used while drilling the next hole section. N_b will move up and if it moves above the point where casing is well supported, buckling will occur.

For a deeper cemented casing string in a vertical well, carry out the following calculations.

1. Identify the top of the supported casing. This may be at the top of cement (if no centralizers above TOC) or at the top centralizer, if centralizer spacing above TOC is relatively small—say one per two joints in gauge hole.
2. Calculate F_b and F_s once the cement has set at the top supported depth, with the actual external and internal pressures. Unless very heavy mud is inside then F_s should exceed F_b.
3. Calculate or estimate the maximum internal pressure in service (due to heavier mud only; or completion fluid hydrostatic plus surface pressure if applicable) and use this to recalculate F_s.
4. Calculate the increase in F_b due to thermal expansion (refer "Revised F_b due to thermal expansion" following in this section). For tubings, also calculate the increase in F_b due to pump out force at the packer if a nonanchored seal assembly is used.
5. If the revised F_b exceeds the revised F_s then the casing will buckle. In this case calculate how much extra tension applied at the surface would cause F_b to reduce below F_s; this force can then be applied as an overpull after the cement has set by setting the casing in slip type hangers.

Revised F_b due to thermal expansion. The coefficient of thermal expansion (a) for steel gives the thermal strain in a uniform body subjected to uniform heating (refer to Section 1.4.7, "Mechanical Properties of Steel").

Strain e due to uniform thermal expansion of steel = 6.9 x 10^{-6}/°F (1.24 x 10^{-5}/°C)

$$\text{Young's Modulus of Elasticity E} = 30 \times 10^6$$

With these two pieces of information we can calculate the effect on the neutral point if we assume a uniform increase in temperature throughout a length of pipe.

Assuming a length of 13³/₈ in 72# N80 casing with the top of cement at 10,000 ft, subjected to an overall temperature increase of 20°C. The thermal strain (stretch ÷ original length) = 20 x 0.0000124 = 0.00025, and if this is multiplied by the original length of 10,000 ft then the increase in length = 20 x 0.0000124 x 10,000 = 2.48 ft.

Knowing Young's Modulus, we can calculate the equivalent stress by multiplying strain (0.00025) by E (30 x 10⁶), which equals 7440 psi.

The cross-sectional area of this casing is 26.44 in^2, therefore, a stress of 7440 psi represents a load of 7440 x 26.44 = 196,700 lbs due to the effects of thermal expansion.

Applying extra tension of 196,700 lbs (which would require stretching the pipe by 2.48 ft as calculated above) before setting the slips would cause the "heated" neutral point to be back to the original position before the "unheated" casing was stretched.

Possible courses of action. If the neutral point is above the planned top of cement, then several actions can be taken to prevent buckling taking place after cement has set.

- Change the cement program. If using an extended lead slurry, then reduce the depth of the planned TOC and/or further extend the slurry to reduce the buoyant force. Casing supported by set cement can be below N_b but will not buckle.
- Calculate the effects of temperature increases after setting cement; land the casing under extra tension once cement has set such that at the top of the supported casing interval, $F_b < F_s$.
- Place centralizers higher up to cover the length from the neutral point down to the TOC, calculating where the new neutral point will be after accounting for temperature effects any extra tension applied when landing the casing.

1.4.15. Calculating Torsional Loads

The limiting factor for torque is the casing connection. Makeup torque will always be less than the casing strength in torsion. This only becomes a factor to consider when the casing (or more commonly liner) will be rotated during cement displacement to improve mud removal.

On the last trip out of the hole, stop with the bit at the liner hanger depth. Record rotating torque. Limiting torque for the liner is then 50% of liner makeup torque + string rotating torque.

Torque can be estimated, given the assumed coefficient of friction, the radius of the casing, the force acting to press the pipe into the wellbore wall and the length. Generally the assumed coefficient of friction in open hole is 0.35 and in casing it is 0.20. These can be used unless the actual coefficients have been derived from data obtained while drilling on the same or an offset well.

Actual m will depend on hole rugosity, lithology, wallcake thickness and lubricity, mud lubricity, and relative sizes of casing and hole.

To calculate the torque:

■ Refer to Section 1.4.10, "Methods of Applying Buoyancy Effects," which describes how to calculate the effect of hydrostatic (buoyancy) forces acting laterally on an inclined pipe.
■ Refer to Section 1.4.13, "Calculating Axial Loads," which describes how to calculate the effect of increasing hole angle on sidewall forces in air.
■ Subtract the buoyancy forces from the sidewall forces to give the actual force pressing the pipe against the wellbore wall.
In general,

Force due to friction between surfaces = coefficient of friction
(m) x area x force pressing the surfaces together.

However, in calculating frictional forces between casing and wall, length is used rather than area and m is generally assumed to be 0.35 in open hole and 0.20 inside another casing.

Therefore, to calculate the frictional force acting against pipe rotation, calculate the actual force pressing the pipe against the wellbore wall F_S, multiply by m and then by the length L (in feet) under consideration; divide by the pipe radius r (in feet).

$$\text{Torque, ft lbs} = \frac{L \times \mu \times Fs}{r}$$

1.4.16. Triaxial Stress Analysis

Triaxial analysis of your casing design may be done in normal cases and should be done in critical cases. What triaxial analysis aims to do is to compute the combined effect of all the forces acting on the casing and compare the resultant stress to the minimum yield strength of the material. In order to carry out a triaxial analysis you have to have completed your design using the previously stated criteria. The casing design itself defines many of the loads; for instance the length and thickness of the casing will be the main determinant of tensile stress and this must be known before the analysis can be carried out.

Triaxial analysis is a complex procedure and, if necessary, should be done using a computer program. For this reason it is not covered in detail. In the vast majority of cases, the axial and biaxial procedures mentioned above will be sufficient. However, if you do have such a program available, the lower safety factors necessary may allow lighter grades/weights and give worthwhile cost savings.

1.4.17. Design for Casing Off Massive Salt Formations

Massive salts have two properties that may cause serious problems during the life of the well. Salts may flow plastically when field stresses exert a force on the salt. Also, significant enlargement of the wellbore will occur through salt if unsaturated, water-based mud is used.

Salt mobility is affected by several factors, such as water content, impurities, grain size, temperature, overburden pressure, and mud density. Salt can be so mobile that the bit can get stuck in it during drilling if the salt has closed around the bit gauge. Increasing mud density is one way of slowing this down, but salt will never be "pushed back out" by increasing mud density.

From a casing design viewpoint, salts can exert large forces on the casing. First, if we regard plastically deforming salt as a hydraulic fluid, it will impose overburden pressure uniformly on the circumference of the casing. Thus, collapse design for massive salt assumes a 1 psi/ft collapse pressure. However, the fact that the salt may flow in a particular direction can also cause problems of ovalled or even sheared off casing sometime after the well has been completed. Also, if the salt moves unevenly, then point loading will occur; the cement must be competent to prevent this as no casing can resist the uneven loading that will result.

If the casing has been designed for the 1 psi/ft collapse gradient *and* if the casing has a good cement sheath throughout the salt interval then failure probability is minimal. It is important to note that for preventing distortion and shear, a complete and competent cement job is as important as the casing strength. Refer to "Cementing against massive salts" in Section 2.7.4.

1.4.18. Casing Properties and Other Considerations

Having calculated the minimum strength requirements and preferred weights and grades of the casing, you now have to check against other considerations. These include:

1. *Inside diameter for running completion tools.* In production casing the ID is important to ensure that all required completion components can be run. For instance if a dual $3^1/_2$ in completion was to be run, $9^5/_8$ in 57# casing may not give the required clearance to run the completion accessories.
2. *Seamless pipe vs. seamed, electric resistance welded (ERW) pipe.* The seamless method is most common for pipe production. Historically seamed pipe was not used for casings below surface casing due to considerations of quality of the pipe. Modern ERW pipe can now be produced in quality equal to seamless pipe and because it is cheaper, ERW pipe can save a lot on the cost of a well. Major operators such as Shell have decided that seamed pipe can be used as casing for deeper strings where quality control is assured.
   Whether or not you can use seamed pipe will probably be dictated by company or government policy. It is certainly worth the effort to consider seamed pipe.
3. *Availability.*
4. *Cost.* Of the casings that are both suitable and available in time, the lowest cost string can be chosen.

1.4.19. Material Grades

API defines the characteristics of various steels and assigns letters to identify those grades; refer to *API Specification* 5CT for complete definitions. Some manufacturers also define non-API grades; refer to the manufacturer's literature for details.

The different grades may vary by chemical composition and/or heat treatment. The grade of steel needs to be suitable for the application. If a range of suitable grades is available, then the cheapest can be used.

Minimum yield stress as defined by API is the stress required to elongate the specimen by 0.5% in normal strength steel, 0.6% in Grade G and 0.7% in Grade S.

Exposure of high-tensile grades of steel to partial pressures of hydrogen sulfide (H_2S) greater than 0.05 psi at less than a threshold temperature (which varies by steel grade) can lead to catastrophic failure. Refer to *NACE Standard* MR-01-75 (current version) for specific information on suitable steels for use in an H_2S environment. *API Specification 6A* also covers selection of wellhead equipment for H_2S exposure.

$$Pp = \frac{ppm \times psia}{1,000,000}$$

where P_p is partial pressure, ppm is the concentration of H_2S (or other gas) in parts per million and psia is the absolute pressure of gas at the depth of interest. For instance at 1000 psi, 50 ppm gives a partial pressure of 0.05 psi.

Resistance to CO_2 corrosion (alone or combined with H_2S) is improved by using chromium alloy steels such as L80 13% Cr. Cr-Mo alloys improve performance in low pH conditions.

Where the application is for severe service (combinations of wet gas, combined CO_2/H_2S, higher pressures, high flow velocities, presence of chlorides in brines, etc.) then a thorough investigation will be needed once the application parameters have been defined. Expensive alloys with high levels of chromium, molybdenum, and nickel may be needed.

These alloys are very costly, require special handling to and on the rig, and tend to gall more than carbon steels.

Following is a brief summary of some currently available common API grades and their characteristics. Further reference should be made to the literature (see references in Section 1.4.24, "References for Casing Design") and to manufacturer's data to obtain specific and up-to-date information.

API H40	Carbon steel. Strength characteristics given by normalizing (heat to 1650°F and air cooling). Suitable for H_2S service at all temperatures for tubings up to 80,000 lbs minimum yield strength or for all tubings above 175°F.
API J55	Carbon steel. Strength characteristics given by normalizing (heat to 1650°F and air cooling). Suitable for H_2S service at all temperatures.
API K55	Carbon steel. Strength characteristics given by normalizing (heat to 1650°F and air cooling). Suitable for H_2S service at all temperatures.
J and K	Have the same minimum yield strength (55,000 psi) but J has an ultimate tensile strength (UTS) of 75,000 psi and K has a UTS of 95,000 psi. The UTS is what dictates the connection strength and so API gives higher tension values for K55 pipe. Note that for most other steel grades, the ratio of minimum yield to UTS is 1.36 but for K55 it is 1.727.
API L80	Carbon steel. Suitable for H_2S service at all temperatures.
API L80 13Cr	Alloy steel with 13% chromium. Suitable for CO_2 service. Susceptible to handling damage, galling, and work hardening.
API N80	Carbon steel. Quenched and tempered to produce a fully martensitic crystal structure; gives higher strength, reduced carbon, and minimizes austenite structure to reduce susceptibility to sulfide stress corrosion cracking. Suitable for H_2S service at temperatures over 150°F.
L and N	Have the same minimum yield strength (80,000 psi) but L has an ultimate tensile strength of 95,000 psi and N has a UTS of 110,000 psi. The UTS is what dictates the connection strength and so API gives higher tension values for N80 pipe.
API C75/90/95	Carbon steel. Quenched and tempered to produce a fully martensitic crystal structure; gives higher strength, reduced carbon, and minimizes austenite structure to reduce susceptibility to sulfide stress cor-

rosion cracking. C75 can be used for H_2S service at all
temperatures, C95 at temperatures over 150°F.

API P105/110   High strength steel. Suitable for H_2S service only above
175°F.

API V150   High strength steel. Minimum yield stress 150,000 psi.
Not suitable for H_2S service.

1.4.20. Casing Connections

All connections serve to join lengths of casing together. However,
other characteristics may be important to your well.

1. *Sealing mechanism.* Is a premium connection required for the production string? Buttress casing connections do not have a metal
   to metal or elastomer seal; the seal is formed from the presence of
   API HP modified thread compound in the thread roots. However,
   the buttress thread form can be tin plated and then give an effective gastight seal rather cheaper than paying for premium
   threads. There are Teflon-based thread compounds available (e.g.,
   Liquid O Ring) which claim to form a gastight seal in a non-
   premium thread.
      Premium threads (VAM, NSCC, Hydril PH6, the various Atlas
   Bradford, Fox, etc.) are mostly designed with a metal to metal
   and/or elastomer seal that is energized in the last part of the tubing
   makeup. Most operators require a premium thread for production
   casing, liner, and completion tubing on gas wells.

2. *Compatibility.* Is it compatible with float equipment, hangers, etc.?

3. *Strength.* Does the connector impose limitations on tension or
   burst? Refer to *API Bulletin 5C2* or check the relevant casing design
   table. Some connections are weaker than the pipe body in which
   case the minimum connection yield strengths for burst and/or tension should be used in casing design calculations.

4. *Availability.* Can the thread be cut in locally or in the country? You

may need pupjoints to be made up or to recover joints where the connection has been damaged.

5. *Clearance.* If an external coupling is used, there may sometimes be limitations due to wellhead design, especially with a subsea wellhead or mudline suspension system. Couplings can be turned down and/or the shoulders beveled to ensure passage through such a restriction. This may affect the connector strength in tension or burst and must be checked out before a final decision. In cases of restricted clearance, also bear in mind the possibility of fracturing weak formations if the casing is run too fast when couplings pass through the restriction.

6. *Practicality.* In severe conditions it may be hard to run a premium threaded connection or a fine thread. In areas with regular sandstorms or high winds, you may want a well-protected seal area or a robust connection that is hard to mis-stab. For instance VAM threads have the seal area exposed at the very end of the connection, where it is most likely to be damaged. NS-CC threads from Nippon Steel have the seal area well protected by a torque shoulder where it is not so likely to be damaged.

7. *Cost.* Of the connections that meet your criteria, the lowest cost can be chosen.

1.4.21. Casing and Liner Accessories

Centralization. Effective mud removal is essential to a good cement job. In an eccentric casing annulus, mud will preferentially flow through the large side instead of in the narrow part of the annulus. Good centralization is needed at the shoe and anywhere that a good bond is important (such as for zonal isolation).

The restoring force of a spring centralizer is quoted by the manufacturer. Given this, the correct centralizer placement can be calculated for a particular casing string, mud weight, and hole deviation. Service companies can recommend centralizer placement, but these recommendations should be examined closely to ensure that they meet your requirements.

In between centralizers in a deviated well, the casing will sag down towards the low side. The amount of this sag can be calculated as follows:

$$Dmp = \frac{3.0558 \times 10^{-6} \, wL^4 \sin \phi}{D^4 - d^4}$$

where

Dmp = Deflection at midpoint, in inches
w = Weight of casing, in lbs/ft
L = Length between centralizers, in feet
f = Inclination, in degrees
D = Casing OD, in inches
d = Casing ID, in inches

Of course, the amount of sag will have to be added to the distance from the wellbore center to the casing center at the centralizers if the total amount of eccentricity is needed where the centralizer OD is less than hole diameter, such as in a washed out hole.

In a mudline suspension system the manufacturer sometimes suggests that centralizers not be used, to avoid wear on the inner profile. An uncentralized casing will give you little chance of successful cementing. Minimize the number to those run over critical sections (especially the shoetrack), but not running any at all will compromise the cement job.

Placing a centralizer over a stop collar will reduce wear on the wellhead, casing, and centralizer, and will also reduce drags. Some small diameter centralizers do not have enough clearance for this and so must be run between two collars. Rigid centralizers are usually placed between two collars.

Centralizer placement. Rigid centralizers must not be run in open hole. Following is a general guide to use when the centralization program is not calculated. It may be modified as necessary. Where a subsea wellhead or suspension system is in use and the manufacturer recommends no centralization, this will normally be to reduce wear on the wellhead inner profile and sealing surfaces. However, centralization is so important to getting a satisfactory cement job that it may be thought necessary to run some centralizers, even if it is the minimum number at critical depths and around the shoe only.

Casings 18⁵/₈ in or larger:

- Two springs on the bottom two joints
- One spring per joint in the cemented section
- One rigid inside the conductor shoe and one at 10-20 m below the top of the previous casing or conductor
- For subsea wellheads run one rigid in the joint below seabed

Casings smaller than 18⁵/₈ in and larger than 9⁵/₈ in:

- Two springs on each of the shoetrack joints
- One spring per two joints over the cemented section, straight hole. Use one per joint if the calculated centralization is close to or below the API recommended minimum of 67%.
- One per joint in the buildup section only if cemented
- One rigid inside the previous casing shoe
- One rigid at 10-20 m below the hanger

The number of spring centralizers in the open hole may be reduced if wear is likely in a subsea wellhead inner profile.

Casings 9⁵/₈ in or smaller (springs between stop collars if small annular clearance):

- Two springs on each of the first two joints
- One spring per two joints over the cemented section, straight hole. Use one per joint if the calculated centralization is close to or below the API recommended minimum of 67%.
- One rigid inside the previous casing shoe
- One rigid at 10-20m below the hanger

The number of spring centralizers in the open hole may be reduced if wear is likely in a subsea wellhead inner profile.

Liners 7 in or smaller (springs between stop collars due to the small annular clearance):

- Two springs on each of the first two joints
- One spring per two joints in the open hole
- One rigid inside the previous casing shoe

Float equipment and accessories. For a normal one-stage cement job, the usual setup is a floatshoe with a float collar at two joints up the string and bottom and top plugs. Other equipment will be needed for multistage jobs, subsea operations, etc.

Correctly sized top and bottom plugs are both vital. If no bottom plug is run, the top plug will wipe a film of mud ahead off the casing, which may be enough to contaminate the cement around the shoe. For deep, critical jobs two bottom plugs may be used to separate mud, spacers, and slurry. This may take longer to drill out but an hour of extra drilling is worth paying for a better cement job.

If a mixed casing string of more than one inside diameter is used, special plugs may be needed. Generally the plugs will fit one casing OD for a range of casing weights, but it is possible that use of extremes of weights in the same size casing or different casing sizes cannot be covered with the standard plugs, and you may have to arrange special order plugs. Do not accept standard plugs modified by sticking on rubber strips around the top fins; the strips can become detached and plug the floats.

Types of liner hangers. A liner hanger generally uses a set of slip elements to grip the wall of the previous casing. Mechanical and hydraulic setting systems are used, though hydraulics are used more often.

In addition to the basic liner hanger, some systems contain integral packers that may be set after cementing. These allow reverse circulating out of excess cement without imposing circulating pressures on the formations if they are set immediately after finishing cement displacement.

These integral packers are quite reliable and can be routinely used.

Liners may also be set on bottom and cemented without hanging off on the previous casing, though this is uncommon. A cementing shoe with side passages is needed instead of the more conventional type.

Hangers (surface wellhead). The choice of hangers will be limited to those compatible with the wellhead system. On a nonfloating rig, a solid (mandrel) hanger may be used to allow the BOP to be safely nippled down without having to wait on cement, if other circumstances allow. Setting these hangers is simpler and safer than lifting the stack and setting a slip and seal assembly, trimming the casing, and running a seal bushing for the next spool. More care is needed to space out, and a landing joint of appropriate length to reach from the top of the hanger to the drillfloor is necessary.

If a solid hanger is to be run, a slip and seal backup will still be needed to land the casing if, for some reason, the casing is not run all the way to planned depth. Therefore, a slip and seal assembly and a seal bushing should be kept handy for emergency landing. A solid hanger cannot be used to land the casing under extra tension, which is sometimes done to prevent buckling.

Other accessories. If casing is to be reciprocated while cementing, scratchers may help remove wall cake and induce turbulence.

External casing packers (ECPs) are designed to seal off the annulus to stop gas or fluid migration through the setting cement. The danger here is that below the ECP, hydrostatic pressure will drop very rapidly as the cement sets unless special additives are used, so that the cement below the ECP could become porous due to gas flow from the formation into the setting cement. Consult the cement contractor and consider all the factors carefully before choosing to use an ECP.

Special cement heads are available to allow you to rotate the casing during displacement. However, experience shows that reciprocation during displacement is more effective in displacing mud. Reciprocation must be used with caution in sticky hole conditions, especially if using a fixed hanger that may be in the BOP if the casing gets stuck while reciprocating.

1.4.22. Wellheads: General Descriptions

Spool type surface wellhead. On a land or platform rig where the wells are drilled from the platform, surface wellheads will generally be used. The conductor pipe extends to surface (land rig cellar or platform wellhead deck). Surface casing is cemented inside this, and a weld or screw-on spool is attached after cutting the conductor to a convenient height.

On a land rig, surface casing will generally be cemented to surface. On a platform, small tubing is usually run in the casing-conductor annulus and cement flushed out to seabed.

Subsequent casing strings are landed in the previous spool, and so the surface casing spool will ultimately bear loads from all the casings and the completion and Xmas tree.

Compact surface wellhead. The compact wellhead offers several advantages over the spool type wellhead, including:

1. Reduced time nippling up/down
2. Safer—no need to flange up an extra spool after each casing job

The compact wellhead functions much like a subsea wellhead or mudline suspension.

Mudline suspension system. A mudline suspension system may be used when a well is drilled offshore for later completion after installing a platform. It can be run into a template set on the seabed that allows several wells to be drilled before the platform is installed. If drilling and platform construction can take place together, the time to get the field on stream is reduced.

The conductor is cemented in a drilled hole or driven to depth. Below the mudline within the conductor is a landing ring that has an internal profile allowing hangers to be set. As subsequent casings are run they locate on top of the previous hanger. The hangers are run on a running tool attached to a string of casing back to the rig.

To move the rig off location the casing strings above the hangers are backed out and recovered, with a suspension cap put in place after the last casing has been pulled. The conductor is cut above the landing ring. This leaves the well suspended at the mudline. When the platform is installed, the casing strings are rerun after recovering the suspension cap and the well can be completed.

Subsea wellhead. In deep water it may be decided to have the well completed with the wellhead on the seabed. The wells may produce through a manifold to a floating production station or into a pipeline.

Guideline subsea wellhead. Either a template is set on the seabed or a temporary guide base run. Either way, posts are positioned on the structure to allow guidelines from the rig to be latched. The drilling assembly is attached to the guideline with soft rope, which breaks after the bit has entered the slot. When the conductor is run, it is guided in the same way.

An underwater camera system (either ROV or run on wireline) allows entry to be visually monitored. The conductor (run on drillpipe) locates in the guidebase; it may be latched in place depending on seabed conditions. It is also possible to drill the conductor hole without a guide, where the permanent guidebase is run attached to the conductor to guide subsequent activities.

Guidelineless subsea wellhead. A template or reentry funnel may be set on bottom or the well may be simply spudded straight into the seabed. Care must be taken spudding; starting with very low drilling parameters until the bit is a few feet below the mudline. Excess WOB/RPM can easily cause the bit to "walk" across the seabed, especially if it is rocky. It may be necessary to wait for slack water in tidal areas. High flow rates will wash out a large crater on a soft seabed. An underwater camera system is necessary to monitor the operation.

A reentry funnel will be run fixed to the top of the conductor to guide drilling assemblies and the surface casing string, if a template or reentry funnel were not already run.

The following considerations are important in choosing a wellhead:

- Type of system as outlined above
- Maximum anticipated wellhead pressure and temperature
- Conductor and casing sizes and connections
- Wellbore fluids (e.g., H_2S, CO_2, or other conditions requiring special service materials)
- For a subsea wellhead, type of connector on top and type of connector on the subsea BOP stack
- Any other special requirements
- Cost/availability

1.4.23. Casing Design Criteria

The following Tables 1-1 through 1-6 provide drilling and production criteria for casing design.

1.4.24. References for Casing Design

Adams, N. *Drilling Engineering*, Tulsa: PennWell Publishing Company.

API Bulletin 5C2, "Properties of Casing, Tubing, and Drillpipe."

API Bulletin 5C3, "Formulas and Calculations for Casing, Tubing, Drillpipe, and Linepipe Properties."

API Specification 5CT, "Casing and Tubing (U.S. Customary Units)," see 5CTM for Metric Units.

API Specification 6A, "Specification for Wellhead Equipment."
Goins, W.C. "Better Understanding Prevents Tubular Buckling Problems"; two-part article in *World Oil*, Jan/Feb 1980.

Kane, R.D., and J.B. Greer. "Sulfide Stress Cracking of High Strength Steels in Laboratory and Oilfield Environments," *JPT* Nov 1977 and *SPE* 6144.

Moore, Preston L. *Drilling Practices Manual*, 2d ed, Tulsa: PennWell Publishing Company.

NACE Standard MR-01-75.

Sumitomo Metal Industries Ltd., "Sumitomo Products for the Oil and Gas Industries" brochure, Oct 1993.

Table 1-1 Casing Design Criteria—Conductor

	Drilling Criteria	Production Criteria
Burst	Resist a pressure at the shoe of 1 psi/ft with a gas gradient to surface (0.1 psi/ft inside). Maximum burst at surface; depth (ft) x 0.9 psi/ft.	
Collapse	Resist a normal external pressure (0.465 psi/ft) with the conductor evacuated (0.1 psi/ft inside). Maximum collapse at shoe; depth (ft) x 0.365 psi.	
Tension	Weight of conductor in air plus overpull allowance of 200,000 lbs. Safety factor 1.1.	
Compression	Withstand loads of subsequent casings, completion, wellhead, and BOP. Use a safety factor of 1.0 with the tensile minimum yield strength.	
Buckling	Long, freestanding conductor; commission full engineering buckling analysis.	
Other	1. Resist driving forces if driven. Check that the planned connections are suitable for driving. 2. Set deep enough to allow returns to the flowline, except on floating rig where the next section drilled with returns to seabed.	Offshore; corrosion protection in the splash zone.

Table 1-2 Casing Design Criteria—Surface Casing	
Drilling Criteria	
Burst	1. Resist the anticipated formation breakdown pressure at the shoe with a gas gradient to surface (0.1 psi/ft inside). Maximum burst at surface; depth (ft) x [formation gradient - 0.1 psi/ft]. 2. Casing test pressure at top of casing. 3. Casing test pressure plus mud hydrostatic to shoe less normal external pressure (0.465 psi/ft) at shoe. Safety factor 1.1.
Collapse	1. Calculate collapse pressure at shoe when cement displaced. Use planned fluid heights and densities inside and out. 2. Assume losses taken at next hole TD. Normal external pressure (0.465 psi/ft) at the loss zone; planned mud in use drops to height supported by thief zone pressure. Maximum collapse may be at the top of the mud column or at the shoe. Calculate external pressure exerted by mud in hole when the casing was run. Safety factor 1.1.
Tension	1. Weight of casing in air plus overpull allowance of 100,000 lbs. Safety factor 1.1 2. Buoyant weight plus test pressure. Safety factor 1.75 if deviated or 1.5 if vertical. 3. Buoyant weight plus test pressure plus bending stress if deviated. Safety factor 1.5. 4. Triaxial analysis incorporating temperature correction for yield strengths. Safety factor 1.25.
Compression	N/A.
Buckling	N/A; cemented to seabed or surface.

Table 1-3 Casing Design Criteria—Intermediate Casing	
Drilling Criteria	
Burst	1. Resist the anticipated formation breakdown pressure at the shoe with a gas gradient to surface (0.1 psi/ft inside). Maximum burst at surface; depth (ft) x [formation gradient - 0.1 psi/ft]. 2. Casing test pressure at top of casing. 3. Casing test pressure plus mud hydrostatic to shoe less external pressure at shoe exerted by the mud used to run casing. Safety factor 1.1.
Collapse	1. Calculate collapse pressure at shoe when cement is displaced. Use planned fluid heights and densities inside and out. 2. Assume losses taken at next hole TD. Expected external pressure at the loss zone; planned mud in use drops to height supported by thief zone pressure. Maximum collapse may be at the top of the mud column or at the shoe. Calculate external pressure exerted by mud in hole when the casing was run. 3. If there are any plastic squeezing formations, resist 1 psi/ft outside at the relevant depth(s). Ensure competent cement sheath. Safety factor 1.1.
Tension	1. Weight of casing in air plus overpull allowance of 100,000 lbs. Safety factor 1.1. 2. Axial loads due to buoyant weight plus test pressure. Safety factor 1.75 if deviated or 1.5 if vertical. 3. Axial loads due to buoyant weight plus test pressure plus bending stress if deviated. Safety factor 1.5. 4. Triaxial analysis incorporating temperature correction for yield strengths. Safety factor 1.25.
Buckling	If deep string is in vertical hole and where temperature and/or internal pressure will significantly increase while drilling next hole section, calculate revised F_b and F_s and take necessary action to prevent buckling.
Other	If cementing against massive salts, cement sheath competence is as important as the casing strength properties.

Table 1-4 Casing Design Criteria—Production Casing

	Drilling Criteria	Production Criteria
Burst	1. Casing test pressure at top of casing. 2. Casing test pressure plus mud hydrostatic to shoe less external pressure at shoe exerted by the mud used to run casing. Safety factor 1.1.	1. Assume leak at top of production tubing; calculate burst at the production packer with completion fluid in the hole minus pressure due to the mud used when casing run. 2. Calculate loadings imposed by fracturing, injection, artificial lift, or other production processes. Safety factor 1.1.
Collapse	1. Calculate collapse pressure at shoe when cement is displaced. Use planned fluid heights and densities inside and out. 2. If there are any plastic squeezing formations, resist 1 psi/ft outside at the relevant depth(s). Safety factor 1.1.	1. Assume losses when perforated. Expected external pressure at the loss zone; planned mud in use drops to height supported by thief zone. Maximum collapse may be at the top of the mud column or at the shoe. External pressure exerted by mud in hole when the casing was run. 2. Expected drawdown pressure during production internally and reservoir pore pressure externally, up to production packer depth. Safety factor 1.1.
Tension	1. Weight of casing in air plus over-pull allowance of 100,000 lbs. Safety factor 1.1. 2. Axial load at the top due to buoyant weight after displacing cement plus overpull, if extra tension applied before setting hanger. 3. Axial loads due to buoyant weight plus test pressure. Safety factor 1.75 if deviated or 1.5 if vertical.	

Table 1-4 Casing Design Criteria—Production Casing *continued*		
	Drilling Criteria	Production Criteria
	4. Axial loads due to buoyant weight plus test pressure plus bending stress if deviated. Safety factor 1.5. 5. Triaxial analysis incorporating temperature correction for yield strengths. Safety factor 1.25.	
Buckling		If deep string is in vertical hole and temperature and/or internal pressure will significantly increase during production, calculate revised F^b and F^s and take necessary action to prevent buckling.
Other	If cementing against massive salts, cement sheath competence is as important as the casing strength properties	

Table 1-5 Casing Design Criteria—Drilling Liner

Drilling Criteria

Burst	1. Resist the anticipated formation breakdown pressure at the shoe with a gas gradient to top of liner (0.1 psi/ft inside). 2. Liner test pressure and mud hydrostatic inside, mud and cement as planned outside. Safety factor 1.1.
Collapse	1. Calculate collapse pressure at shoe when cement is displaced. Use planned fluid heights and densities inside and out. 2. Assume losses taken at next hole TD. Normal external pressure (0.465 psi/ft) at the loss zone; planned mud in use drops to height supported by thief zone pressure. Maximum collapse likely to be at the top of the liner. Calculate external pressure exerted by mud in hole when the casing was run. 3. If there are any plastic squeezing formations, resist 1 psi/ft outside less normal pressure inside (0.465 psi/ft) at the relevant depth(s). Safety factor 1.1.
Tension	1. Weight of liner in air plus overpull allowance of 100,000 lbs. Safety factor 1.1. 2. Axial loads due to buoyant weight plus test pressure. Safety factor 1.75 if deviated along liner length or 1.5 if vertical. 3. Axial loads due to buoyant weight plus test pressure plus and bending stress if deviated along liner length. Safety factor 1.5. 4. Triaxial analysis incorporating temperature correction for yield strengths. Safety factor 1.25.

Table 1-6 Casing Design Criteria—Production Liner

	Drilling Criteria	Production Criteria
Burst	1. Resist the anticipated formation breakdown pressure at the shoe with a gas gradient to top of liner (0.1 psi/ft inside). 2. Liner test pressure and mud hydrostatic inside, mud and cement as planned outside. Safety factor 1.1.	1. Assume leak at top of production tubing; calculate the burst pressure at the production packer with completion fluid in the hole minus pressure due to the mud used when liner run. 2. Calculate loadings imposed by fracturing, injection, artificial lift, or other production processes. Safety factor 1.1.
Collapse	1. Calculate collapse pressure at shoe when cement is displaced. Use planned fluid heights and densities inside and out. 2. Assume losses taken at next hole TD. Expected external pressure at the loss zone; planned mud in use drops to height supported by thief zone pressure. Maximum collapse likely to be at the top of the liner. Calculate external pressure exerted by mud in hole when the casing was run. 3. If there are any plastic squeezing formations, resist 1 psi/ft outside less normal pressure inside (0.465 psi/ft) at the relevant depth(s). Safety factor 1.1.	1. Assume losses taken when perforated. Expected external pressure at the loss zone; planned mud in use drops to height supported by thief zone pressure. Maximum collapse may be at the top of the mud column or at the shoe. Calculate external pressure exerted by mud in hole when the casing was run. 2. Expected drawdown pressure during production internally and reservoir pore pressure externally, up to production packer depth. Safety factor 1.1.
Tension	1. Weight of liner in air plus over-pull allowance of 100,000 lbs. Safety factor 1.1. 2. Axial loads due to buoyant weight plus test pressure. Safety factor 1.75 if deviated along liner length or 1.5 if vertical.	

Table 1-6 Casing Design Criteria—Production Liner *continued*	
Drilling Criteria	**Production Criteria**
3. Axial loads due to buoyant weight plus test pressure plus bending stress if deviated along liner length. Safety factor 1.5. 4. Triaxial analysis incorporating temperature correction for yield strengths. Safety factor 1.25.	

Directional Design

Drilling a well directionally is more expensive than drilling to the same reservoir depth vertically. The smaller the target(s) and the more complex the resulting wellpath, the more it will cost. Discuss the targets with the reservoir engineers and establish the largest possible target—even on a vertical well.

It may be that your target is significantly smaller than that what would be acceptable. Exploration departments sometimes ask for a target of, say, 100 m radius simply because of "tradition." However, it is possible that there is room to err in a particular direction or that the acceptable size is larger than given. Reservoir parameters rarely give simple circles or rectangles, therefore, always ask what conditions define the target boundary. You want to know the "hard" target, outside of which is unacceptable and would if necessary call for redrilling part of the well to achieve. Mark on the deviation plan all the boundary conditions that constrain or affect the well, such as other wells, faults, etc.

Many people assume modern surveying tools and calculation methods are very accurate. While this is generally true, there are limits on performance that may come about for various reasons such as magnetic anomalies. Minimum curvature calculations assume a perfect arc between survey stations, although that will rarely be the case. These factors limit how accurately you can know the actual wellpath and gives a "cone of uncertainty" that increases in size as it gets deeper and within which should lie the true wellpath. This also limits how close you can plan to go to any particular boundary condition. How quickly this cone expands depends on your surveying instruments and, if it is critical, should be discussed with the surveying equipment vendor.

1.5.1. Planning the Wellpath

To start planning the deviation (or asking the directional service company to present a proposal), the following information must be available:

- Surface coordinates and the system used in defining them. (Lat/longs on a certain projection; local grid coordinate system, etc.)
- Target TVD and the outside boundary that defines the target area
- Any other limitations on the wellpath, such as adjacent wells
- Proposed hole sizes and casing setting depths (vertical)
- Offset information from other wells showing directional performance of BHAs through the same formations and formation dips/directions, if known
- Geological sequence
- Rig information: maximum derrick loads, mud pump details, drillpipe, DCs and directional equipment available
- Any other relevant information including your own requirements (such as type of well profile, wellbore inclination through the reservoir, etc.)

Keep it simple. The simplest will be the easiest and cheapest to achieve. Further, involve the directional company from the earliest stages of planning. Most directional companies have their own dedicated engineering departments to assist in well planning. However, beware of expensive, high-tech suggestions if you think that cheaper techniques are viable.

Wellbore profile. In the past, well profiles often built up, held tangent for a while, then dropped off into the target (S profile). A simple build and hold to target (J profile) is often better than an S profile because:

1. An S profile is more complex with more directional work (rigtime, rental tools expense).
2. More hole needs to be drilled.
3. The drop-off part will restrict your WOB, therefore, ROP will be less. Also, rotary drop-off BHAs will lose some directional control (due to their flexibility) and motor assemblies with bent subs sometimes tend to "flip over" when orientated to drop. This is worse when run lower down due to the length of flexible drillstring above the BHA.
4. Maximum inclination is higher and there are more doglegs, with increased hole drags. Good cement jobs are harder to achieve. Wellbore stability becomes more of a problem at higher inclinations. Wear on casings and drillstrings is increased. Fatigue may become a problem.

The S well may be called for if:

1. Intermediate targets are specified, which force you to adopt this path.
2. The target is offset and the wellbore needs to be close to vertical through the reservoir.

The wellbore directional profile has a major influence on the torques and drags encountered during drilling, as well as the drags that occur while running logs and casing. In high-angle wells or those with significant changes of azimuth as well as inclination, the frictional forces can cause serious problems. Mud lubricating additives and non-rotating protectors can be used to somewhat reduce torques and drag. A computer program can also help to tune the wellpath to reduce frictional forces; the directional company should have access to this kind of software so as to optimize the wellpath.

In-situ field stresses. The combination of in-situ stresses and hole orientation govern the tendency to destabilize the rock (collapse mode) surrounding the borehole. In a tectonically relaxed area, a vertical well has equal horizontal stresses acting along the cross section compared to a deviated well that will have unequal stresses (vertical and horizontal) acting along the cross section. The stress concentration at the borehole wall is higher in inclined wellbores, making it more prone to collapse, and since deviated holes are likely to be open for longer, there is more time for the wellbore to become unstable.

In an area where the horizontal stresses vary with direction (tectonically stressed), a deviated well tends to be more stable when drilled in the direction of the highest horizontal stress, and least stable when drilled perpendicular to it. The higher the wellbore inclination, the more pronounced the effect. If the surface location is not fixed then the surface location could be relocated to allow an azimuth that gives the best stability.

The optimum mud density in a tectonically relaxed environment tends to increase with hole angle. In general, field experience indicates an approximate increase of the mud pressure gradient by 2 ppg (0.11 psi/ft) between vertical and horizontal. In a tectonically stressed environment this relation can be different.

1.5.2. Dogleg Severity Limits—Combined Buildup and Turn Rate

Dogleg severities have to be considered at this stage. DLS has to be restricted (especially higher up in the hole) to avoid drillpipe damage, fatigue, and casing wear. This impacts on the directional well design because you will specify a kickoff point depth to reach a known target position. How quickly you build angle will determine the final inclination (i.e., a faster build rate gives a lower inclination) and measured depths of the casings.

This becomes more of a problem the deeper the well is. Say your planned kickoff depth was 1500 ft. You initially want a build rate of 3° per 100 ft of drilled hole to a 30° inclination. Deeper down, the build section is cased off. Drilling at 10,000 ft TVD (that would be at 11,212 MD) in 12 1/4 in hole with 0.65 psi/ft mud, the drillstring tension at the kickoff depth could be in the order of 180,000 lbs, using 5 in grade S drillpipe and 8 in collars. (Refer to "Tension due to weight in a deviated wellbore" in Section 1.4.13, Calculating Axial Loads.) This tension will pull the drillpipe into the inside of the curve and force the drillpipe against the casing. Of course, you will be tripping and rotating while operations continue, which will cause wear on the casing. The higher the dogleg severity, the more the sideforce generated and the greater the wear.

Tool joint damage. This sideforce also imposes a lateral loading on the tool joints that can cause damage; Lubinski suggested a limit of 2000 lbs lateral force to avoid damage to the tool joints. The dogleg severity for a given lateral force and drillstring tension can be calculated by:

$$c = \frac{108{,}000\ F}{\rho LT}$$

where c is dogleg severity in °/100 ft, F is the lateral force, L is half the length of a joint of drillpipe in inches and T is the drillstring tension at the depth of interest. Using a maximum lateral force of 2000 lbs as suggested by Lubinski and assuming 31 ft joints of drillpipe, the DLS causing this lateral force would be (108,000 x 2000) ÷ (3.142 x 186 x 180,000) = 2.05°/100 ft. From this, it can be seen that our initial assumption about the desirable dogleg severity is ambitious and is likely to cause tool joint damage. We can also calculate the lateral force for the initially planned dogleg severity by turning the above equation round, so that:

$$F = \frac{\rho \times L \times T \times c}{108{,}000}$$

and for a DLS of 3°, the lateral force F would be 2922 lbs.

In examining these limiting factors, a practical point must also be made. We run directional surveys while drilling, but these surveys inevitably give an average dogleg severity over the interval between survey points. The most common method of calculating the wellpath between surveys, "minimum curvature," assumes a perfect arc between survey points. In practice the actual dogleg severity will be greater in some places than others, imposing a point loading at those places. If the limit for dogleg severity were 2.05°/100 ft, you could plan on an average 1.5° dogleg severity to allow for this variation.

There is also a practical solution to allow higher dogleg severities than the limit calculated above. If drillpipe protectors were to be positioned at the midpoint of each joint of drillpipe, and if the OD of those protectors were similar to the tool joint OD, you would effectively halve the length of the drillpipe joint. The load would be taken by the protectors that would reduce the load on the tool joints. As the factor related to the length of the drillpipe joints L is on the bottom half of the formula, halving the length would double the allowable dogleg severity. Therefore, by using drillpipe protectors, one per joint on drillpipe being rotated through the build section, the allowable DLS will double to just over 4°. Two protectors per joint, equally spaced at one-third and two-thirds inches along the pipe, will further reduce the load and allow a larger DLS.

Drillstring fatigue. The area of the drillpipe subjected to the severest cyclic bending stresses when rotated in a dogleg is where the drillpipe body joins the tool joint. Here the stiffness of the drillpipe changes very quickly between the rigid tool joint itself and the flexible pipe body.

Calculation of fatigue is fairly complicated. Calculations for fatigue limitations of dogleg severity gives greater dogleg severities than the maximum found by calculating for preventing tool joint damage, except at very low drillstring tensions (below about 75,000 lbs or lower). Therefore, as long as doglegs are limited by the 2000 lbs lateral force for tool joint damage, pure drillpipe fatigue is not likely to be a problem.

Reference can be made to the graphs in Section B4 of the *IADC Drilling Manual* and also in API RP7G These graphs show the maximum dogleg severity for commonly used drillpipes. Preston Moore's *Drilling Practices Manual* also has some graphs illustrating fatigue limitations of dogleg severity. The most commonly referenced paper on the subject is "Maximum Permissible Dog-legs in Rotary Boreholes," by A. Lubinski.

Fatigue failures can occur at other areas on the drillpipe. If the pipe is not sufficiently torqued up so that the shoulders are compressed together, fatigue failure of the pin will occur very quickly. Also, if the pipe body has internal corrosion, external slip marks, or other damage, the effect of these stress raisers will lower the fatigue resistance of the pipe substantially, and a failure of the body may occur there rather than at the upset to the tool joint.

Casing wear. In addition to tool joint damage and fatigue considerations, wear on the casing as a result of lateral forces also has to be considered. Wear from tripping is much less than that from rotating and in medium drillstring tensions in doglegs below 6°/100 ft, the pipe body does not touch the casing. Therefore, wear arises mainly from rotating tool joints that are pushed against the casing.

Wear is affected by many different factors:

- Contact pressure between the tool joint and casing (depends on lateral force and contact geometry)
- Type of fluid in the hole
- Number of rotations
- Presence or absence of hard facing, whether it is smooth or not and whether it stands proud of the tool joint or not
- Presence or absence of tong marks or other sharp edges that would cause abrasion
- Materials in contact

and may occur as a result of three different mechanisms:

- Two-body abrasive wear—sharp edges on the tool joint act like a file on the casing. Produces small cuttings, shiny on one side, like file or lathe cuttings, and very high wear rates.
- Two-body adhesive wear—a galling mechanism where the two bodies become friction-welded together momentarily. Produces flakes of metal and moderate wear rates.
- Three-body abrasive wear—solids in the mud become embedded in rubber protectors and act as a fine abrasive. Produces fine metal powder and very low wear rates.

In new casing on the first bit run, the contact area between tool joint and casing is very small. Wear rate is very high. Since the inside of the casing is worn, contact area will rapidly increase and for the same lateral force, the lateral pressure (in psi) will decrease. The initial very high wear rate will quickly become moderate; abrasive wear will decrease as the tong marks on the pipe are worn down and contact pressure drops. After a trip there will tend to be a temporary increase in wear because new tong marks will be present.

The hard banding on tool joints is very important. It must be smooth, hard, and flush with the tool joint. In the "old days" hard banding could be very rough and stood proud of the tool joint—an efficient rotary file. If it is not flush then all of the lateral force is taken on that small area so that contact pressure is extremely high. Never run rough hardbanded tool joints inside casing while drilling.

Dull tong dies will tend to make marks worse on the tool joints as more closing pressure is required to make these dies grip. Apart from the safety aspect of slipping tongs, using dull dies is false economy. Slip and tong dies should be inspected after every round trip and replaced as soon as they become worn.

Casing wear should be monitored by placing two ditch magnets in the return mud flowline or possum belly tank. At the same time each day (usually midnight) the magnets are cleaned off and the metal recovered. Make sure that the mud particles and crud adhering to the metal is removed and then weigh the sample. The daily drilling report should note the daily and cumulative amounts of metal in lbs or kgs. Any sudden increase in the return metal trend should be investigated. Examination of the metal from the ditch magnet should indicate which kind of wear is taking place.

Lubinski proposed a limit of 2000 lbs of contact force, below which damage to tool joints would not be substantial (as discussed above). Wear rates should be moderate below this limit using solids-weighted mud, with smooth hardfacing that is level with the rest of the tool joint OD, and using sharp-tong dies, wear rates should be moderate. Use protectors to reduce lateral forces as described above to below this limit.

If a solids-free mud or brine system is used then wear rates will be much higher. Extra precautions in this case may include using nonrotating protectors (i.e., free to rotate on the drillstring), down-hole motors (to minimize rotating the drillstring), minimizing the dogleg severities and running heavier wall casing over the build and below the wellhead.

1.5.3. BHA Performance Considerations

It is better to complete the deviation work in the upper hole sections unless the target displacement is small. Aim to kickoff below surface or upper intermediate casing and finish the build before setting the next casing string. You can then use rotary locked assemblies for the rest of the well. This approach (for a typical, one-target deviated well) helps maximize ROP and minimize total footage drilled. The final inclination should be over about 17° from vertical, otherwise direction will be hard to control with rotary assemblies.

Deviating bottom hole assemblies only work predictably in an in-gauge hole. If you are likely to have hole enlargement problems in the kickoff part of the hole and if mud design and good drilling practices cannot solve these instability problems, the kickoff may be hard to control, depending on how fast the wellbore enlarges and how fast you drill. It would be better to get below these troublesome zones and case them off before starting directional work. For example, if you preferred to kickoff fairly shallow but had unconsolidated sand bodies at that depth range, there might be problems not only in controlling direction but in keeping the hole clean, problems tripping, or inadvertently sidetracking the well when reaming in with a rotary BHA through the sand. Directional work is better done in good wellbore conditions.

Avoid setting casing either immediately above or within the kickoff section. This may lead to keyseating of the casing shoe, which is tough to get out of if you get stuck in it.

Rotary tangent (locked or packed) drilling assemblies often have a slight tendency to turn to the right (check your offset wells). This can be compensated for by finishing your build section slightly to the left of the planned azimuth to the target. Also, with a tangent BHA, using high weights to maximize ROP normally gives a slight build tendency, as does drilling into formations with dips below 40°. Leaving the wellbore 2-3° below the planned inclination to the target center at the end of the kickoff before locking up the BHA would compensate for this. With good directional offset data you may consider aiming to finish the build section with the wellbore pointing a bit above the lower edge of the target, knowing that this will maximize penetration and allow application of high WOB and capacity for a slow build and the maximum tolerance. With a circular target area, since the angle is slowly built in the tangent section, the left-right tolerance that is at a maximum when aligned with the target center is also increased.

One other thing to remember when planning to finish the build a degree or two low is that building to the line (or above it) and later having to drop angle will compromise ROP with the low weights required for pendulum assemblies. If, however, building is needed, a short build assembly needs weight to work so ROP will not be compromised. It is better to be too low than too high.

If corrections are required while drilling, they should be made sooner rather than later. Corrections deeper down take longer and are harder to control.

If possible, avoid having to drop angle lower down in the hole. If circumstances dictate this, then try to avoid dropping to less than 10° from vertical. It will become increasingly hard to drop more angle the closer to vertical you get.

1.5.4. Horizontal Well Design Considerations

There is a naming convention for the build rate of a horizontal well. Long radius is less than 8°/100 ft build rate, medium radius is between 8° and 20° and short radius is over 20° build rate. Long and medium rate builds use conventional deviation equipment.

The build rate to horizontal you choose is very important. In general the longer the horizontal section you want to drill, the lower the build rate (longer the build radius).

The majority of horizontal wells, probably as many as 95%, are drilled using medium radius builds.

Take the following into consideration when planning the wellpath:

1. If possible, select the surface location for the simplest wellpath to the horizontal section. If you can build in the same plane as the alignment of the horizontal section, this is preferable to designing combined turns and builds to the horizontal. Hole drags and wellbore instability will be less. However, with fixed surface locations (i.e., platform) this may not be possible.
2. Avoid drilling updip horizontally if possible; drilling updip at

greater than 90° inclination gives you problems in attaining and maintaining angle. There are also adverse well control implications since gas bubbles may be impossible to circulate out of the hole. Finishing just below horizontal would be much better.

3. Computer modeling can be used to predict likely drags and torques. Sometimes a small change to the wellpath can make a significant difference in drags. The final design should be analyzed to ensure that it is possible to drill and trip with the rig and drillstring in use. Minimizing hole drags will increase the chances of drilling and casing the well successfully.

4. It is common (and recommended) to build to horizontal in the reservoir, set casing and then drill ahead for the horizontal. This protects the buildup section and allows isolation of the reservoir from the formations above. If the completion does not call for cemented liner to be run (i.e., slotted or prepacked liner is planned) then placing the production casing in the reservoir allows isolation from higher zones. Further, having steel into the reservoir should make it easier to run prepacked screens while minimizing potential damage to the screens.

5. When aiming to hit a narrow depth range that is small in proportion to the accuracy of depth estimates, consider drilling a pilot hole through the reservoir first, logging it, and then plugging back to kickoff to horizontal. This will be far cheaper than building up close to horizontal, going slightly out in relation to target depth, then having to either drill many feet to get back on track (if high), or trying to build too fast causing a dogleg and possibly going over 90° (if low).

6. Directional measurements come from some distance behind the bit. If it is necessary to drill along a certain section of the reservoir and "follow the geology" rather than by hole angle alone, some means of identifying location in the reservoir is needed at or closely behind the bit. This technique is called "geosteering," where LWD tools are used to measure formation parameters that allow real-time decisions for steering the bit. Resistivity and gamma ray measurements will show the distance from shale (cap rock) so the drill bit can be steered along the top part of the reservoir with reasonable confidence. Since new LWD techniques are being developed all the time, it is important to determine which available technique suits the characteristics of the target reservoir.

7. Plan to use less build rate than is possible with the downhole equipment. This gives some leeway if an increase in build rate is needed.

1.5.5. Multilateral Wellbores

At the time of this writing, the technology of drilling a number of "daughter" wellbores from a single "mother" well is developing rapidly. These techniques are applicable if the cost of carrying out the procedure is less than the net present value of the extra hydrocarbon production rate and/or enhanced recovery volume. In some cases, existing wells can be reentered and laterals drilled out from the main bore.

Multilateral wells may be applicable for:

■ Exploiting stacks of layered but separated reservoir sands
■ Allowing more footage in the reservoir than would be the case with a single horizontal wellbore
■ Using existing wells to recomplete in new zones that could not be exploited from the existing wellbore direct
■ Downhole water separation and reinjection in a different lateral

If this might be applicable to your planned well, early discussions with the reservoir engineers and directional service company should be initiated. At present the technology is still quite young.

1.5.6. Slant Rig Drilling

One other consideration is perhaps worth mentioning here. Special rigs can be used that allow the well to be spudded at a surface inclination up to 30°. A good example of this was the British Gas development of Morecambe Bay. The reservoir was shallow but required fairly high stepouts. Drilling conventionally deviated wells would have required two platforms, which would have made the economics marginal. Instead a special rig package was put on the platform where the derrick could be inclined. Three conductors were driven at an angle in each corner of the platform through which the wells were spudded. Current high angle and horizontal drilling technology gives us much greater flexibility in exploiting these fields economically; however, there is still a place in our armory for innovative solutions such as slant rigs. (There is the story of the derrick man who got off the chopper in Morecambe

Bay, took one look at the leaning derrick, and climbed straight back on the chopper again!)

1.5.7. Targets and Wellpath

To summarize, when choosing a planned wellpath, consider the following:

1. Target locations and boundaries
2. Planned casing points
3. Natural formation tendencies (development area with good off-set information)
4. BHA considerations and how they relate to directional and drilling performance
5. Offset wells or geological features that are to be avoided and that come close to your well
6. Lithologies in the buildup part of the hole
7. Orientation of the wellbore with respect to in-situ field stresses
8. Dogleg severity limitations

[Section 2:]

Well Programming

The first stage in drilling a well is to design what the well needs to look like when it is completed, which was covered in Section 1, "Well Design." This section discusses how to write a drilling program, which defines the methods necessary to ensure that the well design and objectives will be met as safely and cost effectively as possible. Section 3, "Practical Wellsite Operations and Reporting," covers practical wellsite-related operations; however, much of this section is also directly relevant to the rig. Cross-references between the two sections are used where appropriate.

$$\begin{bmatrix} 2.1 \end{bmatrix}$$

Preliminary Work for the Drilling Program

As was stated in the introduction, there are two keys to drilling the most cost-effective wells: minimizing problems and maximizing progress. A well-researched, properly written drilling program will set the stage. Simply put, the better the program, the cheaper the well will be.

2.1.1. Drilling Program Checklist

The drilling program should provide the wellsite personnel with everything they need to know to drill the well. Therefore, avoid cluttering up the drilling program with extraneous information, such as detailed casing design calculations.

A technical justification should also be written and attached to the drilling program. Its purpose is to document major decisions made while well planning. Make references where appropriate to the technical justification within the drilling program.

The program should contain the following elements as relevant (use this as a checklist).

General information:

- Country, concession, rig, program issue date, and who it is written by
- Drilling reporting requirements
- General points on safety and inspection requirements
- Offset wells used for input
- Statement on shallow gas (e.g., whether considered likely to be present or not)

Well identification:

- Field, block, permit, location coordinates, platform, conductor/slot number, well number, and well type (exploration, appraisal, development, infill, injector, etc.)

Well objectives:

- Priority order of objectives (short text); include well objectives as well as production or exploration
- Planned drilling time breakdown (time/depth curve)
- Estimated cost breakdown (cost/depth curve and costs by category)

Potential hazards or problems during drilling operations (not specific to one hole section):

- Note any hazards that remain; how to monitor; recovery procedures if the potential hazard occurs (could be suspected presence of H_2S or drilling HPHT for instance)
- Concurrent operations (e.g., drilling on a platform while production or construction work is ongoing) would also be a potential hazard and must be identified with references to safe working procedures

Well positioning:

- Surface coordinates (with reference system) and tolerances
- Datum level (lowest astronomical tide, mean sea level, etc.)
- Rig floor elevation relative to datum level
- Water depth

General notes:

- References to government, company, and oilfield standards and procedures
- Reporting procedures
- Quality control requirements (e.g., sampling and data recording)
- Diagram of the completed well, showing casing depths, liner tops, etc.
- Checklists of equipment and who supplies it (rig, company, or vendor)

Drilling notes (for each hole section):

- Wellbore stability—potential problems and how to minimize/recover from
- Particular hole problems and how to avoid them
- Required drilling practices
- Recommended operational sequence outline (avoid writing out normal rig operations in detail)
- Kick tolerance at anticipated pore pressures and fracture gradients when the next casing point is reached
- Proposed bits and BHAs together with recommended drilling parameters and performance expectations (compare to offset bit runs where applicable)
- Special requirements

Mud engineering and supervision:

- Mud gradients, types, required properties, pH, test requirements, and any special requirements (such as shale extrusion tests to measure inhibition effectiveness)
- Monitoring cavings levels and sampling, describing, and preserving cavings
- Mud sampling requirements (e.g., times, sample sizes, how preserved, etc.)
- Wellbore stability requirements
- Other mud formation requirements
- Reporting requirements (daily and post-well)
- Solids control requirements

Deviation:

- Vertical or deviated well
- Kickoff depth
- Build/drop/turn rates with inclinations, azimuths, and depths to define complete wellpath
- Target depth, coordinates, and boundaries
- Horizontal displacement and azimuth of target
- Target constraints showing the actual target area that would be acceptable, outside of what would be unacceptable
- Total well depth MD and TVD
- For a horizontal well, more information will be required such as final hole section azimuth, whether geosteering is to be used, etc.
- Surveying requirement—types of tool at which stages, distance between surveys, computing method to be used, and magnetic variation for the area
- For a well of more than 30° maximum planned inclination or with high anticipated dogleg severities
  - Drag and torque calculations, used to optimize the wellpath if necessary
  - Casing wear predictions, used to specify what precautions may be necessary (such as protector types and quantities, heavy wall casing at critical points, and special monitoring)
- Any relevant data on offset wellpaths that may affect the planned well
- Hole and bit sizes with section depths (including pilot hole sizes where applicable)
- Proposed bits and BHAs together with recommended drilling parameters and performance expectations

Conductor, casings, and liners—general notes:

- Notes on any potential high casing wear problems, how to monitor, preventative actions (such as protector requirements)
- Reporting requirements (by job and post-well)

Conductor—driven:

- Size, weight, grade, connections, minimum setting depth, and final blow count (state power setting if adjustable)

- Type of hammer

Conductor—drilled and cemented:

- Size, weight, grade, types, connections, setting depths, centralizer requirements, single or multistage, and additional jewelry
- Notes on any potential problems running casing and how these can be mitigated

Casings:

- Size, weight, grade, types, connections, setting depths, centralizer requirements, single or multistage, and additional jewelry
- Notes on any potential problems running casing and how these can be mitigated

Liners:

- Type of liner hanger, whether tieback packer is required, whether to be rotated and/or reciprocated during displacement
- Size, weight, grade, types, connections, setting depths, centralizer requirements, single or multistage, and additional jewelry
- Notes on any potential problems running liner and how these can be mitigated

Cementations:

- Cement tops, types of cementations, slurry types, gradients, and special requirements
- Plugs to be used for casings and liners
- Mix water types and additives
- Mix methods for each slurry
- Anticipated bottom hole static temperature, bottom hole circulating temperature, slurry densities, and yields
- Compatibility between mud, spacers, and cement
- Estimated cement volumes (could state as percentage over gauge or percentage over caliper)
- Specific advice on obtaining maximum mud displacement includ-

ing required mud properties prior to cementing, spacers, flushes, scavenger slurries, any reciprocation or rotation during displacement, and displacement regime
- Reporting requirements (by job and post-well)
- 24-hour compressive strength
- Minimum pumpable time

Wellhead, BOP equipment, and testing:

- Wellhead specifications
- Diverter/BOP configuration for each hole section
- BOP requirements and specifications including specific test requirements for each hole section
- Drills required
- Kick tolerance calculation assumptions made (e.g., state how much higher overpressure would be vs. mud gradient)
- Acceptable levels of influx after kick tolerance calculated
- Any special precautions (e.g., controlled ROP at certain points, flowchecks, increased kick drills, etc.)
- Shut-in procedures required
- Leakoff or limit tests to be used and procedure
- Minimum value of equivalent mud gradient and action to take if not attained

Geological prognosis:

- Expected lithology sequence with names and descriptions of formations, also information on anticipated hole problems (e.g., fractured, sloughing, washouts, etc.)
- Anticipated pore pressure and fracture gradients with depth—note also the level of confidence in the figures given
- Geological characteristics of expected formations: permeability, fluid type, hydrocarbon depths, gas zones, etc.

Wireline logging and petrophysics:

- Required logs (from sponsoring department)
- Required logs (drilling department requirements for drilling evaluation)

■ Sidewall sampling, formation sampling (RFT/MDT), and coring requirements
■ Cement logs
■ Reporting requirements (daily and post-well)

Mud logging:

■ Sampling and preservation methods required for cuttings, cavings, mud, produced fluids, metal, etc.
■ Recording requirements and formats
■ Type of unit (e.g., off or online)
■ Monitoring services required, types of alarms/alerts, routine calculations (e.g., monitor current kick tolerance, D exponent, etc.)
■ Reporting requirements (daily and post-well)

Well completion/testing:

Normally, detailed completion/testing programs will be sent out closer to the time. General notes should be made to allow some preparation to take place.

■ Precompletion requirements anticipated (e.g., bit/scraper runs, gravel packs, fracs, screens, packer setting, and completion fluid specifications)
■ Tubing sizes and surface wellhead configuration
■ General list of types of downhole completion/testing tools to be run with the completion tubing (e.g., side pocket mandrels, safety valves, packers, downhole sand screens)
■ Reporting requirements (daily and post-well)

Well suspension/abandonment:

■ Anticipated well configuration on rig departure (diagram useful)
■ Zonal isolations required
■ Whether casing will be cut and pulled
■ Cement plugs and permanent bridge plugs depths, etc.
■ Whether suspension caps will be required
■ Refer to government and company regulations or policies concerning abandonment
■ Reporting requirements (daily and post-well)

Miscellaneous:

■ Equipment checklists

Approval signatures:

■ Signatories need to state clearly what responsibility they
are accepting by signing. If the program is later amended then
those areas of responsibility will show who needs to accept the
amendment.
■ If company policy dictates that certain managers sign without
responsibility, then it should be noted as "Signed to acknowledge
program seen."

2.1.2. Technical Justification

The technical justification document explains the reasoning
behind decisions made while creating the drilling program. It also provides references to other sources of information where relevant.

In some areas, government regulations require that all technical
decisions are documented. This is good practice in any case.

There are five important reasons for writing a technical justification document:

1. Having to justify all major decisions in the drilling program forces
the program author to use a proper engineering approach to problem solving and drilling optimization.
2. During program approval, the signatories can satisfy themselves
that the program has a good engineering basis.
3. During drilling, decisions made can be re-examined in the light of
new information and this can be done with full access to the original reasoning. Therefore, better decisions are possible during
drilling, even if the original program author is not available.
4. Being able to justify all significant points in the drilling program
on engineering and cost grounds should lead to fuller cost recovery in some areas. (Cost recovery is a mechanism for the operator recovering the cost of drilling the well by offsetting against

taxes or by taking part of the government share of produced hydrocarbons.)

5. After the well, decisions made can be reviewed in the light of the intentions behind the decisions. This allows a full evaluation of how successful those decisions were and will help in improving future drilling performance.

The technical justification refers to the drilling program and not the well design upon which it is based. Therefore, items such as casing design will be referenced as part of the well design justification. (The well design refers to the final well status that is to be reached for the objectives of the well. The drilling program shows how this will be achieved.)

Example from a technical justification (from a mythical exploration well, Monty 1). This shows the kind of information that can be usefully included.

Drilling an 8¹/₂ in Hole Section

Hole condition did not seem to cause any problems on any of the offset Brecon wells in the Lower Cretaceous. Pore pressures were fairly high with RFTs over 0.72 psi/ft. Brecon 2 had a kick at 3176 ft below the top of the Lower Cretaceous (0.797 psi/ft @ 10,833 ft); Monty 1 is also prognosticated to drill 3176 ft into the Lower Cretaceous. Bearing in mind that Brecon field is about 13 miles away from the Monty field, the potential for high pressures certainly exists on Monty though *Brecon can only be used as a rough guide*.

The mud will maintain the same density as the end of the previous section initially and will be increased during drilling to 0.80 psi/ft at TD for the same reasoning as on the previous section. The sequences of shale and sandstone caused no serious hole problems (overpulls or enlargements) with gyp mud or KLM mud on the Brecon 2 and 3 wells. Uneven hole enlargement was noted in some shales on Brecon 2; with a 4-arm caliper, the major axis was out to18.5 in with the minor axis at about 10 in in 8¹/₂ in hole in places.

Losses occurred in Brecon 2 after taking a 0.797 psi/ft kick at 3176 ft into the Lower Cretaceous. The loss zone and bottom part of the drillstring were cemented off.

The mud should not be treated (expensively) to maintain high inhibition unless dictated by hole condition. YP can decrease slightly to 15-18.

The bit selection is important due to the long trip times and potentially low ROP. Unfortunately, the offset bit records are not complete though some conclusions can be drawn. Gauge protection does not seem to be an issue with *most* of the offset bits being in gauge, however, a few of the bits were significantly undergauge (up to 1/2 in). For this reason it is worth paying more for enhanced gauge protection. Insert bits with journal bearings should allow good durability with high WOBs.

There is also a possible polycrystalline diamond compact (PDC) application here. This will be looked at prior to starting this section and the final decisions on which bit to run will be made at the time.

The bottom hole assembly again will be fully stabilized. This is to allow high WOB to be run without buckling and should reduce hole drags and problems by drilling a straighter hole.

2.1.3. Formatting the Drilling Program

The program and technical justification should be written as a technical document. Therefore, they should be written for easy understanding and reference. Following are some guidelines:

- Use clear, concise writing style; avoid long words and long sentences
- Do not include irrelevant material that does not help the rig
- Paginate and include table of contents
- Use diagrams such as well status
- Print two sides (use less paper; lighter; easier to read)
- Refer to standard documents that would be on the rig, where applicable (e.g., company procedures)

2.1.4. Time Estimates

Any estimate will be improved with more detail. Spreadsheets are invaluable because they allow a great amount of detail, which can then be totaled, diced in many different ways, and easily updated.

The cost of a well is, of course, intimately related to the time it takes. In making a time estimate, break down the overall work into small chunks and give reasonable time estimates, based on local experience, for each chunk. As the work progresses, the spreadsheet can be updated by substituting the estimated times for the actual times, which will update the forecast.

The time estimates will vary depending on the purpose of the estimate. When planning a well, an estimate should give forecasts for logistics. In this case, the shortest possible time should be estimated for each operation—it is much easier to allow things to slip than to pull supply of equipment forward. For cost estimates, use a reasonable "average" time, not the shortest possible. This will give a different overall estimate.

I have created a spreadsheet (planner.xls) that takes a start date, start time, and times for individual operations, and produces a forecast for getting things to the rig. This spreadsheet can be downloaded from my web site at *http://www.drillers.com*, or you can build your own. Figure 2-1 is an example of this spreadsheet.

2.1.5. Cost Estimates

Well cost estimates are made at different stages in the planning process. These will vary by the level of detail possible and therefore by the level of accuracy. Following are cost estimates that will usually be needed as the project progresses.

Budget estimate. The accuracy of a budget estimate is ±25%. The level of information usually given will include approximate location and water depth, approximate target location and TVD, desired total depth TVD, and desired completion status. Other information may be included. For this type of estimate the best way is to look at the costs of offset wells and make simple adjustments for depths, rig day rates, etc. Since little engineering work can be done at this stage, you are making an educated guess. The spreadsheet discussed below can be used by taking a previous estimate and revising it for best guess times, rates, and depths.

Fig. 2-1 Spreadsheet "planner.xls" Example

Operational Lookahead Planner—Maximum Well

Start Date	Start Time	Minimum Duration	Description	Notes
6-Mar	06:00 AM	32.00	Drive conductor	*Load out 13-3/8 in casing*
7-Mar	14:00	18.00	N/U diverter, clean to shoe 9-5/8 in casing ready to load	
8-Mar	8:00	14.00	Pilot hole	
8-Mar	22:00	18.00	Open hole	
9-Mar	16:00	18.00	Run and cement 13.375 in casing	
10-Mar	10:00	18.00	N/D diverter, N/U and test BOP	
11-Mar	4:00	10.00	RIH, drill shoe, circ, FIT	*Logging engineers to rig*
11-Mar	14:00	24.00	Drill ahead to 975 m TVDSS, POH	
12-Mar	14:00	20.00	Logs inc. w/trip	
13-Mar	10:00	20.00	Run and cement 9.625 in casing, WOC	
14-Mar	6:00	6.00	N/U and test BOP	
14-Mar	12:00	8.00	MU new BHA, RIH, gyro inside DP	
14-Mar	20:00	5.00	Drill out, condition mud, FIT	
15-Mar	1:00	48.00	Drill ahead to TD	
17-Mar	1:00	120.00	Run logs APP, including 2 trips	
22-Mar	1:00	28.00	Run and cement 7 in liner	
23-Mar	5:00	6.00	Run 8-1/2 in bit to TOL	
23-Mar	11:00	14.00	Run 6 in bit / 3-1/2 in DP; drill flapper valve; RIH to landing collar.	
24-Mar	1:00	10.00	Run scrapers and PBR mill	
24-Mar	11:00	12.00	Run USIT in 7 in and CBL in 9-5/8 in	
24-Mar	23:00	5.00	VSP	
25-Mar	4:00	17.00	RIH PH6 tubing; clean/circ, rack back	
25-Mar	21:00	16.00	Clean TOL, set tieback packer, pressure test well, POH	
26-Mar	13:00	3.00	L/D excess pipe from derrick	
26-Mar	16:00	24.00	Rig up for well testing	
27-Mar	16:00		Start first well test	

Initial AFE estimate (prior to completing the drilling program). Most companies require an estimate to be within 10% for AFE purposes. A confident estimate to this accuracy level can only be given after full engineering work is complete, the well design fully defined, and the drilling program finished. Often the AFE estimate is required before that can be done. In this case (as long as all the information as per the Well Proposal Checklist in Section 1.1.2, "Data Acquisition and Analysis," is in hand) actual costs can be used for things that will be known (such as rig day rate, logs) and estimates made of times, casing setting depths, and directional work to add in

By using a set of linked spreadsheets, AFE cost estimates can be done fairly quickly. If similar estimates are available from previous wells, open the file, save to another name, and modify costs as appropriate. This also helps ensure that all the necessities are accounted for in the estimate.

Revised AFE estimate. Once the drilling program is completed and approved, the AFE should be revisited to see whether it is still realistic. If a detailed spreadsheet has been used, this can be done quickly and easily to see whether the AFE should be supplemented prior to spud.

The costs for individual items are often known from existing contracts. Where they are not known either contact the relevant supplier for a cost estimate or look at offset well costs and adjust if rates in general have moved since.

An Excel workbook example can be downloaded from the Files area of *http://www.drillers.com*, modified, and used for customized estimates (see Fig. 2-2). Following is an explanation of how this spreadsheet is set up.

Spreadsheet format. The spreadsheet is set up with several layers of detail. These are described individually below.

Cost summary. At the front is a summary sheet, which gives totals from the detailed sheets and shows the casing depths and drilling time estimates that the spreadsheet uses in calculations.

Formulae in the summary sheet cells also calculate the dry hole cost of the well, the percentage of contingency included, and the daily operating cost.

Fig. 2-2 Summary Sheet Example
AFE Well Cost Estimate
By: Steve Devereux

BUDGET TITLE	Example Exploration Well	
Notes		1. Assumes well logged, tested, and abandoned.
DATE	12-Apr-97	2. Mob/demob costs are assumed split between 4 wells.
		3. Site survey costs assume mob/demob split between 2 wells.
		4. This well is designed with a conventional casing scheme.
VALUE of this estimate	$ 7,596,377	
Dry hole cost of well	$ 5,916,929	($1,679,447 less than total)
Contingency included	$ 851,099	(Equivalent to 11% of the total estimate)
Accuracy of estimate	15%	Budget cost should be $8,735,833

1) SUMMARY

This well cost estimate covers the drilling, logging, testing, and abandonment of the first maximum concession commitment well.

2) ESTIMATES

Well time estimate:		
Rig move, jack-up, preload	at 50 m	2.0 days
Drive 30 in conductor	at 135 m	2.0 days
NU diverter		1.0 days
Drill 26 in hole	at 1000 m	2.0 days
20 in surface casing		1.0 days
17.5 in hole	at 1700 m	2.0 days
13 3/8 in casing		1.5 days

12.25 in hole	at 2450 m	4.0 days
9 5/8 in casing		2.0 days
8 1/2 in hole	at 2900 m	4.0 days
TD logging		3.1 days
Testing		20.0 days
7 in liner		2.0 days
Abandon well		3.0 days
Release rig		1.0 days
Weather downtime 5%		2.5 days
Rig downtime 5%		2.5 days
TOTAL:		55.6 days

WELL COST ESTIMATE(see attached for details) :

Rig Move:	$	910,565
Drill, Log, & Suspend:	$	4,155,266
Compl. & Test:	$	1,679,447
TOTAL Base Estimate:	$	6,745,278
Contingency:	$	851,099
GRAND TOTAL:	$	7,596,377
Daily Operational Cost:	$	79,063

Costs broken down by cost code. The next level of detail gives a cost breakdown by code. Times and depths relevant to the estimate are carried forward from the front sheet. Some totals are calculated by this sheet (such as rig rate x days), and some of the figures come from the next sheet in the workbook.

The sheet is set out in three sections: the left column shows a code and description for each line, the middle set of columns are used to enter time or depth-dependent rates, and the right set of columns either multiplies time or depth-related costs by the relevant figure, *or* takes input from the next sheet. Most departments that require cost estimates will be happy with the summary and costs by code sheets and will not need further levels of detail. (See Fig. 2-3.)

Fig. 2-3 Costs by Code Sheet Example
AFE Well Cost Estimate

Well: Example Exploration Well

Reporter: Steve Devereux

Date :   12-4-97

A/C no.	Description	Cost Rates			Cost Estimate			
		Rig Move	Drill/Suspd	Compl, Test	Rig Moves 3.0 days	Drill/Suspd 30.6 days 2,900 m	Compl/Test 20.0 days 2,900 m	Total 53.6 days
TIME DEPENDENT ($/day)								
87201	Rig rate	45,000	47,500	45,000	135,000	1,451,363	900,000	2,486,363
87415	Vessels	8,000	8,000	8,000	24,000	244,440	160,000	428,440
87230	Additional (catering, etc.)		400	600		12,222	12,000	24,222
87530	Cement serv. & pers.		508	508		15,522	10,160	25,682
87524	Mud logging		600			18,333		18,333
87555	Conductor driving eq		2,000			10,000		10,000
87665	Dock fees & base overheads	4,400	4,400	4,400	13,200	134,442		147,642
87551	Rental tools		700			21,389		21,389
87554	Consultants on rig	1,400	1,400	1,400	4,200	42,777		46,977
87551	Anderdrift survey tool		500			15,278		15,278
87506	ROV mob; drill 26 in, set 20 in	3,000	3,000	3,000	9,000	15,000		24,000
87554	Water	5	5	5	15	153	100	268
87554	Fuel (rig and vessels)	500	500	500	1,500	15,278	10,000	26,778
	TOTAL		69,513		186,915	1,996,195	1,092,260	3,275,370

DEPTH DEPENDENT ($/m)								
87545	Deviation surveys			7		20,300		20,300
87315	Mud and chemicals					229,029		229,029
87330	Solids control consum's			3		8,700		8,700
87320	Cement and chemicals				44	29,688	1,467	31,155
87301	Bits					107,060		107,060
86201	Casing and accessories					835,478	62,120	897,598
86201	Completion						127,600	127,600
	TOTAL					1,230,255	191,187	1,421,442
FIXED COSTS ($)								
87110	Site survey	100,000			100,000			100,000
87220	Rig positioning	25,000			25,000			25,000
87220	Rig mob/demob	500,000			500,000			500,000
87220	Boats mob/demob							
87548	Casing crew & eq		20,000	5,000		20,000	5,000	25,000
87509	Electric logging		498,886			498,886		498,886
87509	Cased hole log & perf.		57,880			57,880		57,880
87554	Well testing			10,000			200,000	200,000
86211	Wellhead		50,000			50,000		50,000
87554	Insurance	20,000			20,000			20,000
88222	Fish & abandon services		10,250			10,250		10,250
88322	Well planning	50,000			50,000			50,000
	TOTAL				695,000	637,016	205,000	1,537,016

[2.1.5] Well Programming

SUPPORT COSTS ($/day)							
87554	Office overhead	2,500	2,500	2,500	7,500	76,388	50,000	133,888
89000	Communications	1,000	1,000	1,000	3,000	30,555	20,000	53,555
87665	Office sup't consultant	1,000	1,000	1,000	3,000	30,555	20,000	53,555
87554	Other drilling expenses	50	50	50	150	1,528	1,000	2,678
87420	Air transport	5,000	5,000	5,000	15,000	152,775	100,000	267,775
	TOTAL	9,550	9,550	9,550	28,650	291,800	191,000	511,450
	GENERAL TOTAL				910,565	4,155,266	1,679,447	6,745,278

Fig. 2-4 Details Sheet Example

AFE Well Cost Estimate

Well : Example Exploration Well

Reporter : Steve Devereux   Date :   13-Apr-97

DRILLING FLUID

Hole section	Interval	Time	Volumes Hole Vol.	Losses	Surface	Total	Mud Type	Unit Cost	Mud Cost
m	Days	bbls	bbls	bbls	bbls			$/bbl	
26.00 hole	865	2	1,863	3,726	500	6,089	Hivis PHB	6	$ 36,532
17.50 hole	700	2	683	1,024	0	1,707	KCl Glycol	42	$ 71,712
12.25 hole	750	2	359	538	1,500	2,396	KCl Glycol	45	$ 107,838
8.50 hole	450	4	104	155	0	259	KCl Glycol	50	$ 12,947
TOTAL	2,765	10	3,008	5,443	2,000	10,451			$ 229,029

CASING MATERIAL   (add 20% contingency)

Casing	Length m	Unit Cost $/m	Total $	Accessories $	Total $
30 in x 1.5 in WT J55	162	694	112,428	30,000	142,428
20 in #106.5 J55	1200	165.75	198,900	20,000	218,900
13-3/8 in, #72, N80, BTC	2,040	115.34	235,294	15,000	250,294
9-5/8 in, #47, N80, BTC	2,940	72.06	211,856	12,000	223,856
7 in, #29, N80, BTC	780	54.00	42,120	20,000	62,120
TOTAL	6,960				$ 897,598
Cost/meter					$129/m

CEMENT MATERIAL

Casing	Hole Size	Cement Interval		Cost/bbl	Excess	Total
	m	btm m	top m	$/bbl	%	$
20.00	26.00	1,200	50	35	100	21,574
13.38	17.50	2,040	1,000	35	40	6,302
9.63	12.25	2,450	1,800	40	25	1,812
7.00	8.50	2,900	2,250	50	100	1,467
TOTAL						$ 31,155
Cost/Meter						$9/m

Details. The third sheet shows how costs are arrived at for more complex items. This shows what assumptions have been made for the estimate. For instance, the cost per barrel of mud is matched against the hole volume, assumed losses, and required surface volume to be built. This is useful because if for instance it is decided to TD the well at a different depth, the depth on the Summary sheet is amended and then the totals are automatically recalculated. Thus, it is very quick and easy to revise the cost estimate when things change. New cost estimates can be made by revising previous cost estimates by using a different cost mud, inputting that cost, and updating the estimate.

In Figure 2-4, only the first few items are shown. The actual spreadsheet shows logging, abandonment, and other costs.

Contingencies. This is the last part of the workbook. When calculating what contingency to add to a base case well cost, it is reasonable to see what could possibly go wrong (use the offsets as a guide), estimate the cost of this event occurring, and multiply by a probability of occurrence to get a contingency cost. This is much easier to justify than adding a standard contingency amount or percentage to the base cost. The contingency amount is added to the summary sheet (see Fig. 2-5).

Fig. 2-5 Contingencies Sheet Example

AFE Well Cost Estimate

Well: Example Exploration Well

Reporter: Steve Devereux   Date :   13-Apr-97

Contingencies:		Contingency cost = $ 851,099 Total =			$ 2,463,764
Problem Probability,	%	Rig Days	Other Costs	Total Cost	Contingency Cost
Problems running 20 in casing into hole	25	2		$ 158,126	$ 39,532
Hole instability in 17-1/2 in. hole	25	5	$ -	$ 395,315	$ 98,829
Hole instability in 12-1/4 in hole	25	5	$ -	$ 395,315	$ 98,829
Kick in 12-1/4 in section	25	3	$ 50,000	$ 287,189	$ 71,797
Kick in 8-1/2 in section	25	3	$ 50,000	$ 287,189	$ 71,797
Losses in 6 in section	50	3	$ 50,000	$ 287,189	$ 143,595
Set 5 in liner & TD in 4-1/2 in	50	7	$ 100,000	$ 653,441	$ 326,721

$$\begin{bmatrix} 2.2 \end{bmatrix}$$

Well Control

The main objectives of the BOP program are to ensure that if primary well control (hydrostatic) is lost, personnel and equipment are not endangered and the well can be brought back under primary control. In the case of drilling with a diverter, the well cannot be closed in and the diverter system is for safe rig evacuation during an extended flow rather than for bringing the well back under control. If a shallow gas kick occurs on a nonfloating rig, nonessential personnel should be evacuated until the situation is brought under control and preparations are made to evacuate the remainder if necessary.

This section will consider well control from a drilling programming viewpoint. Practical well control aspects are discussed in Section 3, "Practical Wellsite Operations and Reporting."

2.2.1. Shallow Gas

The most complex and dangerous well control situation is a shallow gas blowout, occurring before BOPs can be installed. A shallow gas kick can occur with little warning, it may involve H_2S and presents a high probability of equipment failure. The risk of shallow gas should be assumed to be present in pre-BOP drilling, even for the last well on a multislot platform.

In top hole, the conductor shoe strength is too low to close in on a kick. If closed in, the flow could fracture formation from the shoe to the surface. A jack-up or platform could lose support. If the well is brought under control after such a blowout to surface, the well would probably have to be abandoned. A well can be drilled quite deep on a diverter system; any gas accumulation encountered before BOPs are nippled up is classified as "shallow gas."

A shallow gas kick may occur by drilling into an overpressured gas-bearing zone, by drilling into a producing well, or into an artificially charged zone due to a poorly cemented annulus on a nearby well. It may be high or low volume, it may or may not deplete quickly, and it may or may not bridge itself off before it depletes. It could also be normally pressured and kicks due to swabbing, losses, not filling the hole, gas from cuttings reducing hydrostatic, or other mechanisms that reduce hydrostatic below normal.

Advance detection. There are techniques that will aid in pre-spud detection of shallow gas. They include:

- Shallow seismic surveys (gives indications of gas accumulations but is no guarantee either way)
- Pilot hole drilling using a soil-sampling vessel. This allows optimizing of the conductor setting depth before spudding a well offshore and can help determine likely spud can penetration.
- Offset well reports
- Platform piling reports

Assess the likelihood of shallow gas, then decisions can be made regarding rig selection, drilling techniques, and contingency planning.

Drilling riserless from a floating rig is the best situation offshore if shallow gas is hit. The rig can be quickly moved off location, and the sea will inhibit flow and fill the hole once it has depleted or bridged off. Jack-ups normally cannot move off quickly enough and could capsize if the seabed fractures.

Planning and training are essential preparation to handle shallow gas.

Riserless drilling. When drilling from a floating rig, the riser cannot be connected until surface casing has been set. A jack-up will often drive conductor pipe extended to surface and set a diverter on top. If drilled and cemented rather than driven, it will be drilled riserless since there is nothing to attach a diverter to. If shallow gas blows out during riserless drilling, the well will flow to the sea until the gas accumulation is exhausted or until the well bridges itself off.

When drilling from a floating rig, riserless drilling is preferred to using a surface diverter. The slip joint is especially vulnerable to leaks and damage under diverted flow conditions, jeopardizing the rig and personnel.

A small diameter pilot hole will reduce the gas flow and will be more likely to bridge off due to produced formation. If the gas flow rate causes instability or fire concerns, a floater can move off location. The deeper the water, the more that the gas will disperse as it rises and any current will take the gas bubbles away from the rig. It is unlikely that rig stability will be seriously compromised except if there is a very strong flow in shallow water.

Santa Fe in Kalimantan has used a technique for diverterless drilling of a surface pilot hole for a jack-up rig. The rig is lightly pinned in place on the seabed and the towboats left connected. With zero air gap a small diameter bit is run in and the pilot hole drilled. If shallow gas is encountered, the string is dropped, the legs lifted, and the rig pulled off location by the towboats. This is only applicable where the bottom is fairly hard and gives little spud can penetration. See references in Section 2.2.5, "References for Well Control—Shallow Gas."

Note that shallow gas can occur at any depth. Therefore, it is possible that shallow gas can be met while conductor driving.

Diverter drilling. If the shoe strength is insufficient to close in a kick, flow must be diverted away from the rig—preferably downwind. Two lines are needed to prevent diverting upwind, whichever way the wind is blowing. A windsock should be visible to the driller. The diverter lines must be large diameter (at least 12 inches ID), straight, with no restrictions, and securely attached to the rig. Large pieces of formation may be produced with the gas. Erosion from produced sand can cut steel lines on bends. The line valve mechanisms must open fully and quickly, or bursting disks can be used. The line has to be kept clear of cuttings and debris.

If the equipment is available offshore, consider subsea diverter lines and valves. It is better to avoid bringing shallow gas up to the rig.

Run the casing with the diverter in place; do not nipple down until the cement is hard. Flush cement from below the wellhead with small diameter lines. A well can kick as cement sets and hydrostatic is lost.

Losses below the conductor shoe are common when drilling in divert mode offshore, due to the extra hydrostatic of bringing returns above sea level. LCM will not stop these losses. This will reduce the integrity of the shoe and, on a jack-up, could affect spud can integrity. It is risky to drill ahead with losses while diverter drilling; serious losses should be cured with cement prior to drilling ahead. Refer to "Suggested procedure for curing total losses with cement."

Other precautions when shallow gas is likely. A small diameter pilot hole should be drilled and opened up later. This will reduce the flow rate if shallow gas blows out and it will make bridging off more likely in the early stages, though it can enlarge fairly quickly. Care must still be taken when opening hole, since primary control can still be lost by swabbing, etc.

Logs can be run after drilling the pilot hole. If gas is seen, plug and abandon the well rather than risk opening the hole and running casing. If a kick occurs during either of these operations, the situation is dangerous. Alternatively a cement plug can be set from bottom to above the gas and casing set higher, if the formation strengths allow a BOP to be nippled up afterwards.

If the wellbore penetrates shallow gas, the drilling fluid may control the pressure. The well could kick later by swabbing, not filling the well on trips, lost circulation, cement gelling, etc. The driller must maintain good practices. The cement and mud programs should aim to minimize the chances of problems. Fast ROP is usual in top hole and if gas-bearing cuttings are produced, liberated gas from the cuttings could reduce hydrostatic enough for the well to kick, so restrict ROP.

Look at reaction times. Large bag preventers can take over a minute to close. Use the shortest length of large diameter hydraulic control lines on both sides (both open and close lines). The diverter line valve has to open the valve fully before the diverter fully closes, or a pressure surge could break down the shoe. A low-pressure bursting disk could be used in place of a valve. Ideally the divert sequence is started by the driller pushing one button, causing the overboard lines to open and diverter to close. If the return flowline shares any part of the diverter vent system, the flowline outlet must be closed, otherwise gas will reach the shale shaker area. Regular diverter drills should be held.

Planning points (equipment).

■ If the proposed well is offshore and there is thought to be some risk of shallow gas, consider using a floating rig if another location free from shallow gas indications cannot be used. Alternatively for a jack-up rig with a hard seabed, a pilot hole can be drilled with zero airgap and the towboat can be left connected ready to pull off location as described above.

■ Diverter equipment should be fast acting, preferably on an automated sequence that only needs one action to initiate. This should also automatically sound an alarm to alert the rest of the rig to the danger and warn the assigned personnel to close ventilation and stop unessential equipment.

■ Prepare a tank of heavy, viscous mud and be ready to switch to it as soon as the divert alarm has sounded.

■ Diverter lines should be large, straight, well-secured, and full-opening (ball valves or bursting disks).

■ Use large diameter lines (and short if possible) to the bag preventer to reduce closing time.

■ Have sufficient breathing escape sets available for the drill crew to evacuate to the muster point in an H_2S atmosphere.

■ Ensure all monitoring equipment is in perfect order: trip tank, flow meters, gas detectors, pump strokes per minute/pressure instruments, etc.

■ Ensure that the standby boats are fully briefed on what to expect if the well blows out, and avoid being downwind of the rig during top hole. Vessel crews should be H_2S aware.

Planning points (procedures).

■ Control ROP to prevent overloading the annulus, reduce the effect of gas in cuttings, and reduce penetration into a kick zone.

■ Use a small diameter pilot hole and *always* place a float in the drill string. Use large or open nozzles to reduce pump pressures and to facilitate pumping cement through the bit.

- Avoid swabbing completely, pumping out of the hole if necessary. If swabbing is suspected, run back to bottom and circulate again.
- Circulate clean and carefully flow check before pulling out. Consider displacing to slightly heavier mud before tripping.
- If using a closed circulation system, ensure good mud rheology to minimize surge and swab pressures.
- Investigate any anomalies while drilling, such as a slight reduction in pump pressure, gas cutting, drilling break, etc.
- Leave the diverter system ready to use and monitor the well until cement has set after running casing.
- Cure serious losses with cement before drilling ahead. See the procedure for drilling problems and advisory actions in Section 3.3, "Drilling Problems."
- Use a closed system with only one active suction pit and closely monitor volumes while circulating.
- Keep enough spare volume on surface to keep hole full if moderate losses are taken. If severe losses occur, keep pumping seawater if necessary to keep the hole full while preparing to cement.
- Monitor annulus pressures on adjacent wells for any increase in pressure.
- If additions to the active system are necessary, stop drilling/circulating and monitor the well while adding.
- If a kick is detected, first activate the diverters. If the well has not yet blown out, pump at maximum rate until the well unloads; the ECD may just be enough to prevent further influx. Switch to the tank of heavy, viscous mud.
- Apply the same precautions to opening hole.
- Conductor pipe should be set deep enough to give reasonable shoe integrity, to reduce the risk of losses and breaking formation while diverting.
- Where several conductors are batch driven before drilling, spot 30-50 ft of cement inside each before spudding the first well. This will prevent losses or kicks communicating to closely adjacent conductors. If the well blew out through an adjacent conductor, there would be a serious threat to life and the safety of the rig.

Planning points (contingency planning).

- Hold pre-spud meetings onshore and offshore, ensuring everyone is aware of actions to be taken in various situations.

- Be prepared to evacuate nonessential personnel if a kick is taken and for complete evacuation if necessary.
- Hold regular drills before and during drilling top hole. This should include function testing of the diverter system.
- Be prepared to plug and abandon the well if shallow gas is identified, even if it does not kick straight away.
- Ensure that written procedures are in place and accessible to the rig crews. These should define actions to be taken, move-off procedure (floater), etc.
- If the well kicks, men should be already assigned to shut down nonessential equipment, ventilators, etc.

2.2.2. Drilling with a BOP Stack

When the casing has a competent shoe, the blowout preventers can be nippled up. The setup will depend on company policies, type of rig, available equipment, etc. It should be possible to connect the choke and kill lines above the bottom rams and have spare, valved outlets for the choke and kill below the bottom pipe rams. If the choke and kill lines are below the bottom rams and a leak occurs at the ram side outlet, the flow could not be controlled.

The blind rams should not be the bottom set. This allows the pipe to land on the bottom pipe rams, be backed off (or sheared) higher up, and circulation to be established. On a surface stack the blinds should also not be the top set of rams, so that in the event of leaking upper pipe rams, the blinds could be closed allowing the top pipe rams to be changed out. On some BOPs (e.g., Shaffer), normal pipe rams are not designed to support the weight of the drillstring. Check with the BOP operating manual. Shaffer rams for supporting pipe weight have a special metal insert on the top edge where the tool joint will land.

Variable bore pipe rams (VBR) can close on a range of pipe sizes. VBRs should not be run as the bottom set; a drillstring cannot be landed on closed VBRs. Using VBRs can avoid having to spring the bonnets open and change rams for certain mixed drillstring configurations.

When drilling with BOPs, the casing shoe should be tested after drilling out to determine the formation fracture pressure and the maximum allowable annular surface pressure (MAASP). This establishes the point at which the formation will just start to take fluid and must

151

not break down the formation. This is a leakoff test. The test may be stopped at a predetermined limit, which is a procedure known as a limit test. Both of these tests are known as formation integrity tests. If weak or fractured formation is exposed, a limit test is usually performed. The recommended procedure is described in Appendix 2, "Formation Integrity Test Recommended Procedure".

BOP equipment rating. The rating of the BOP equipment has to take into account the degree of exposure to risk and the degree of protection needed.

IADC classifications require that the minimum BOP pressure rating shall be defined by the *least* of the following:

- the burst pressure of the casing to which the BOP stack is connected
- the formation breakdown pressure at the shoe minus the hydrostatic pressure of the casing filled with gas. (You can assume gas gradient to be 0.1 psi/ft.)
- the maximum anticipated surface pressure to which the equipment may be exposed

Sometimes a section of the well will be drilled with a certain mud weight when field experience indicates that a kick is possible in one of the formations in that section; one example would be potentially drilling into an overpressured raft within massive salt. A casing seat will be chosen so that the shoe is close to the potential kick zone. Some operators under these circumstances prefer to keep a tank of heavy kill mud standing by, ready to pump quickly. There are some reasons against doing this:

1. If the surface to bit volume is more than the bit to shoe volume, then heavy kill mud will enter the wellbore after the influx has risen into the casing. Therefore, pressures on the shoe and open hole will not be reduced. If this is the case then having kill mud ready serves no advantage at all. It may even have to be dumped afterwards if not used (water-based mud), wasting money.
2. It is unlikely that the gradient of the kill mud will be exactly what is needed and so would not be pumped straight down the well anyway; if it were pumped down this would complicate the kill and the annulus pressures would be more difficult to calculate. Rather than pumping it "neat," it could save some time if it is mixed into the active system to reduce the time needed to add barite.

3. One of the mud tanks is tied up for the time that the kill mud is kept standing by.

The circumstances may, however, dictate that it would be more prudent to have a tank of heavy mud standing by; if it is decided after discussion that this should be done then the drilling program needs to specify this.

2.2.3. High-Pressure, High-Temperature Wells (HPHT)

A well is said to be high pressure if the wellhead pressure could equal or exceed 10,000 psi with a full column of gas above the zone of highest pressure. A high temperature well could reach a wellhead temperature that equals or exceeds 150°C under conditions of a blowout from the zone of highest pressure through the back pressure manifold (BPM). High-pressure, high-temperature wells require special planning (for the relevant hole sections) to ensure that well control is maintained.

Operational planning and procedures. The following considerations can be made at the planning stage.

1. Provision should be made for running an additional casing string if a kick is encountered (or if the mud weight otherwise has to be raised to the point where the kick tolerance becomes smaller than acceptable).
2. Good drilling practices must be maintained, and close involvement of the rig supervisors is needed to ensure this. The supervisors (including contractor's toolpushers/drillers) should be fully briefed on the well plan. Consider organizing specific HPHT well control training courses off the rig.
3. Include in the well plan specific contingency instructions on actions to be taken where those actions are not the standard procedures for the envisaged situation. For instance, if oil mud was to be displaced with water mud.
4. If oil mud is in use, a gas kick may go undetected with a flow check until gas gets close to surface (due to gas solubility in oil) unless the formation is considerably underbalanced by the mud. If an

unexpected drilling break occurs it may be wise to have a policy of closing in and circulating bottoms up across open chokes before proceeding further.

5. Water-based muds should be used, if possible. This will reduce well control problems caused by gas dissolving in the mud system. If water based mud cannot be used for drilling then consider the (logistical) possibility of killing the well by replacing the drilling oil mud with water-based kill mud. This may involve having a supply boat on location for the duration of the HPHT hole section with a supply of mud on board. The gas-bearing oil mud returns may be hard or impossible to treat on the rig; if this is the case then it cannot be pumped back down the well for killing and has to be disposed of. If no other alternative is available, it may be possible to burn it off through flare booms.

   The main drawback to using water muds is the increased likelihood of hydrates forming in the surface equipment as the gas expands and cools. However, provisions should be made against this, even if using oil muds.

   This question has to be looked at from all possible angles and be thoroughly discussed with the drilling contractor (as should all the proposed HPHT plans and precautions).

6. Do not core long intervals in single runs. If the cores liberate gas while pulling out of the hole, a kick could be precipitated while a long way off bottom, which is harder to control.

7. Running a float sub in the drilling assembly will provide worthwhile safety in the event of a kick. A circulating sub run above the BHA will improve the position if the nozzles plug. In a jet bit, do not try to maximize HHP, but run large nozzles to help avoid plugging if LCM has to be pumped.

8. If swabbing is suspected at any time while tripping, run back to bottom and circulate bottoms up. Pump out of the hole if necessary, or even as a routine precaution. Calculate swab and surge pressures and maintain good rheology to minimize these occurrences.

9. If there are no concerns over formation fracture gradients, consider bullheading an influx back into the kicking formation prior to circulating kill mud around. This would be particularly applicable

if a large-volume kick causes concern over the casing shoe strength while killing.

Equipment considerations.

1. The entire BOP system must be rated to handle potential kicks. Elastomer seals (in ram and bag preventers, valves, hoses, etc.) have to work at the highest temperatures and pressures (for the duration of the section) likely to be encountered. The system should also be suitable for H_2S service.
2. The mud-gas separators should be able to handle the large volumes of gas at surface. Large-diameter vent lines will be needed (10 inches or more).
3. High-pressure flarelines should be connected to the back pressure manifold, rated to at least 5000 psi flowing pressure. The valves to the flowlines on the back pressure manifold (BPM) should be operable from the remote control panel as should those to the separators.
4. A facility should be available to measure the temperatures of the wellhead and also the BPM downstream of the chokes. This will ensure that temperature ratings are not exceeded and will give advanced warning of conditions for hydrate formation.
5. Methanol injection under wellhead pressures into the BPM must be possible to avoid hydrate formation.

2.2.4. Well Control in High-Angle and Horizontal Wells

High-angle and horizontal wells involve extra considerations in maintaining primary well control and in regaining it if a kick is taken. The normal kick calculation principles do apply, however, the calculations are more complicated for maintaining a constant bottom hole pressure during a balanced method kill. In some circumstances, it is preferable to use a drillers method kill. Refer to Section 3.1.4, "Well Killing in a High-Angle Well," for the calculations for killing a kick in a high-angle well.

Well planning considerations. With the well profile and casing points decided, write the well plan to reduce the chances of taking a kick. Normally in a horizontal well the pressures will be known and a kick caused by drilling into an unexpected, overpressured zone is unlikely. Hole cleaning is a much greater problem in horizontal wells and this can increase swabbing. If possible, keep hole angle slightly below 90°.

If the angle exceeds 90°, it may be difficult or impossible to circulate out a gas influx. A swabbed influx may stay in the horizontal wellbore without migrating, causing a kick when the new BHA is run back in and displaces out the influx.

Precautionary measures. The following precautions may be considered for inclusion in the well plan:

- Pump out whenever moving pipe out in horizontal hole, on connections, or when tripping out
- Flow check at the shoe and before the BHA when pulling out, also at the shoe and halfway in the horizontal section when tripping in (if a long horizontal section is open)
- Be particularly alert for a flow and "trip gas" after resuming drilling
- If pressures and conditions allow, bullhead the influx back into the producing formation
- If the inclination is over 90° then bullheading, if possible, is preferable

2.2.5. References for Well Control

IADC Drilling Manual, eleventh edition. Published by the International Association of Drilling Contractors.

Practical Well Planning and Drilling Manual on CD, http://www.drillers.com.

"Unocal, Santa Fe Devise 'Diverterless' Drilling Method for Jackups," article in *Drilling Contractor* magazine, published by IADC, July 1996.

$$\begin{bmatrix} 2.3 \end{bmatrix}$$

Directional Planning

In Section 1, "Well Design," the various options for planning the wellpath were covered. It was necessary to do this at the well design stage rather than the drilling program stage since it impacts the casing design. Having already defined the surface location, targets, and preferred wellpath in this section, the drilling program must show how the wellpath will be drilled to within an acceptable level of positional accuracy. Practical directional considerations will be covered in Section 3, "Practical Wellsite Operations and Reporting."

2.3.1. Downhole Tools Affecting Directional Control

The fundamental factor controlling directional performance is the design of the bottom hole assembly. Design in the context of directional drilling means the configuration and spacing of the bit, drill collars, full-gauge and undergauge supporting tools (stabilizers and roller reamers), sideforce generating tools (rebel tool and dynamic stabilizer), and downhole motors including turbines that can affect the direction that the bit drills. Other tools that do not affect the directional performance directly but give us controlling information include wireline run and/or retrieved survey tools, universal bottom hole orientating (UBHO) subs, measurement while drilling subs, inclination only measurement subs ('Anderdrift'), and wireline deployed gyro tools.

In terms of directional performance, the influence at the bit of a particular tool is greater the closer it is to the bit. For instance if the nearbit stabilizer goes undergauge by 1/4 in, it will affect directional performance far more than if the next stabilizer at, for example, 35 ft up goes undergauge by the same amount. It is also the case that only the bottom 90 ft of the BHA or the first three stabilizers (whichever is shorter) actually affects directional performance. Undergauge stabilizers are run higher up above the bottom 90 ft to minimize dynamic buckling and hold the drill collars off the wall for differential sticking concerns.

Following are the individual components and their effect on directional performance.

Drill bits. Some drill bit features can generate forces that produce deflecting or stabilizing forces. A roller cone bit which is dull or damaged will increase the natural tendency of rotary assemblies with roller-cone bits to make the bit walk to the right.

Other features will help the bit to resist deflecting forces or allow the bit to react more strongly to these forces. Long-gauge sections on fixed-cutter bits or gauge pads on the shirttails of roller cone bits will resist deflection forces. Parabolic or strongly convex fixed cutter bit profiles stabilize the bit whereas a flatter bottom profile will have weaker or no resistance to side forces. Directional fixed cutter bits sometimes incorporate cutters around the gauge which help cut hole to the side of the bit, helping to give stronger dogleg severities for a given bottom hole assembly and drilling parameters.

Some bits are designed to cut an overgauge hole; a bi-center bit has the center of the cutting structure offset from the centerline of the BHA for drilling in salts and very plastic shales. If the bit is cutting overgauge then the gauge area has no rock to work against and would be expected to react more strongly to deflection forces.

Stabilizers and roller reamers. A nearbit stabilizer or roller reamer is run immediately above the bit, as its name implies. The gauge of this stabilizer is critical to build or drop performance. An undergauge NB stabilizer will give a dropping tendency to a tangent or build assembly. Nearbit stabilizers or roller reamers have a box up and box down and the bit screws directly into the box down. Normally the box up is bored out so that a float valve can be placed inside. Nearbit stabilizers and roller reamers at full gauge are used in tangent and build assemblies.

Roller reamers may be used instead of stabilizers in abrasive hole or if high torque is a problem. In crooked hole country, where spiral holes can be drilled even with highly stabilized assemblies, the nearbit stabilizer can generate a lot of torque as weight on bit is applied. I experienced this some years ago; we drilled a 12¼ in hole section with a fully stabilized assembly and every time more than about 20 k was applied to the bit, the rotary table torqued up and stalled out. Back on surface the nearbit stabilizer and first string stabilizer were worn balloon shaped—full gauge in the center and up to ¼ in undergauge at the top and bottom edges. This had me extremely puzzled for a while, until I realized that the only way to wear a stabilizer like that was in a severely spiral hole. In the same interval on the next well we ran roller reamers in the first two positions, no torque problems at all and everything came out full gauge. In this case the final well report (FWR) of the first well made the recommendation to try the roller reamers on the next well, illustrating the central role of the FWR in improving future performance.

Stabilizers come in several flavors. Integral blade types have the whole tool machined from one block of steel. Sleeve types have replaceable blades that screw on to a body with a thread machined in the middle. The sleeve type is more versatile because a stock of blades can be kept on the rig and changed as necessary, and replacing the sleeve when it is too worn is much cheaper than replacing an entire integral blade type. Some people have reported problems with the sleeve type unscrewing downhole; I have never seen this and I consider that as long as it is properly installed and torqued up (and is not damaged in some way), unscrewing is very unlikely to occur. To make and break the sleeve safely a breaker plate should be used; check that this is available on the rig otherwise tongs will be used that may damage the sleeve and/or the tong.

In an 8½ in or smaller hole, integral blade stabilizers should be used rather than sleeve types because the flowby area becomes too small on sleeve stabs.

Stabilizers are also available that allow downhole gauge adjustment. The first adjustable stabilizer was the Andergauge downhole adjustable stabilizer, which is now in widespread use around the world. The benefits of a nearbit or string stabilizer, which allows downhole size adjustment, can lead to significant cost savings since trips are not needed to change the build or drop characteristics of the BHA. The Andergauge is available as string or nearbit types. They can be adjusted downhole between full-gauge and one undergauge size and (at present) are available for 12¹/₄ in and 8¹/₂ in hole sizes. These are used on rotary and steerable assemblies to control the build/hold/drop performance.

Some downhole motors can have stabilizer blades clamped on to the body of the motor to adjust the directional characteristics. Blades can be straight (aligned with the centerline) or spiral (wrap around in a spiral). In very soft formations, the straight blade may tend to dig into the side of the hole more readily than the spiral type. Spiral blades are most common in all but the largest sizes, which is strange considering that the largest ones are most likely to be run against soft formations!

Stabilizers may be made of steel or Monel. Monel is an alloy which does not distort the local magnetic field and so Drill collars and Stabilizers made of monel may be used to allow magnetic surveys to be taken while drilling. Refer to the following section on drill collars for positioning Monel stabilizers.

Drill collars. Drill collars are thick-walled pipes that allow weight to be applied to the bit. Drill collars are very stiff (especially compared to drillpipe) and have a relatively high resistance to buckling. The force required to buckle a drill collar is determined by the material, ID, OD, shape of the cross-section, and unsupported length. Larger ODs and shorter distances between stabilizers in gauge hole will both increase the compressive force required to buckle the collar.

Sometimes you want to make the collars buckle slightly because it will tilt the bit and impose a deflection force to change direction. With a build assembly, a nearbit stabilizer is run and a string stabilizer from 60 ft to 90 ft higher. Gravity tends to make the collars sag (assuming the hole is already inclined) and increasing weight on the bit will increase this deflection. Therefore in this configuration, placing more weight on the bit will increase the build rate. The further apart the nearbit and first string stabilizers are, the faster it will build. Unfortunately, this increased distance also reduces azimuth control and the bit will also turn left or right faster due to formation forces and formation-stabilizer interaction. As the hole angle increases, so too will the build rate if everything else stays the same.

This deflection, which is due to the weight of the collars, is also used to reduce hole angle. A pendulum assembly will have a bit, drill collar(s), and the first stabilizer between 45 ft and 90 ft higher. The weight of the lowest section of collars pushes the bit against the bottom of the hole and this will make the bit drill lower angles. Pendulum assemblies also tend to walk sideways and/or drill spiral holes; weight on bit needs to be kept low, otherwise directional performance can become completely unpredictable and so rate of penetration suffers with pendulums. Once a drop trend is established, increasing the weight slightly can help increase the speed of drop. As the hole becomes more vertical, the rate of drop decreases. It is very hard to drop to within a degree or two of vertical using rotary pendulum assemblies. A less severe pendulum could be set up like a 60 ft build, but use an undergauge nearbit stabilizer to drop the angle.

If uncontrolled drill collar buckling takes place, fatigue failures are likely to occur—usually a few threads down from the box-end shoulder. Apart from failure of the steel, this dynamic buckling is likely to lead to contact between the wall and the drill collar OD. This will lead to "stick-slip" oscillations where the string stops downhole (rotary table torques up and slows down) then the sticking is released, and the BHA spins very fast (rotary table speeds up; the string may overtake the rotary and the kelly bushing will bang against the drive holes). Stick-slip oscillations are seriously detrimental to bit performance and life. Sufficient stabilizers must be run so that dynamic buckling does not occur. A computerized analysis can be done or refer to the general BHA configurations at the start of this section for rule-of-thumb BHA design.

Drill collars are available in a square cross section. They are very stiff mechanically (i.e., resist bending or buckling better than round) and may be called for in severe crooked hole country.

The outer surface of round drill collars can be plain but mostly have spiral grooves machined in them. This minimizes the surface area in contact with filter cake and should reduce differential sticking tendency.

If magnetic survey tools will be used (magnetic single shot or magnetic multishot) then a section of the BHA has to be of a metal that does not distort the Earth's magnetic field. Monel drill collars and stabilizers can be run such that at the depth of the survey tool, magnetic interference from the rest of the BHA is acceptably low. The length of monel to be run depends on several factors, such as latitude of the drilling location, inclination of the wellbore, and azimuth (see Fig. 2.6).

Sideforce generating tools. The rebel tool is rarely run today. It is positioned just above the bit and can be set up for a left- or right-hand turn. It gives a gradual turn in the order of 1°/100 ft as long as the hole is in gauge. The tool has two arms or "paddles" linked by a steel rod. The length of the paddles determines the direction of walk. It should not be run where wellbore instability or balling is likely, since solids will pack in around the paddles.

The principle of operation is that when long paddles are used, the highest paddle hits the low side of the hole. The weight of the BHA sitting on this paddle turns it in toward the center; this force is transmitted along the rod to the paddle nearest the bit. When the upper paddle is on the low side taking weight, the lower paddle pushes against the right hand edge of the hole, exerting a sideforce to the left.

With the short paddles, when the upper paddle is on the low side the lower paddle pushes against the left-hand edge of the hole, exerting a sideforce to the right. Remember, *Long* paddles for *Left* turns.

A recent innovation from Baker Hughes Inteq, yet to be field proven at the time of this writing, is a dynamically adjustable stabilizer. The AutoTrak™ system uses an orientating sleeve that rotates relative to the BHA but does not rotate relative to the hole. Expanding stabilizer pads apply forces to the hole to continuously steer the tool. The system can be controlled from the surface and can apparently be set up to automatically geosteer within the reservoir—that is, using LWD measurements to keep the borehole within a certain part of the reservoir.

Downhole positive displacement mud motors. Power applied hydraulically downhole to the bit allows fine control of inclination and azimuth when combined with steerable systems or bent subs. On some tools the angle of the bend (and hence the dogleg severity) can be adjusted downhole. Using an Andergauge adjustable stabilizer above a straight motor also allows adjustment of build or drop performance.

Using downhole motors also requires careful attention to hydraulics and bit selection. The tool will give a specified power output and rotational speed for a given flow rate. This must be matched with the possible ranges of pump outputs for the total system pressure loss (including the motor) and the desired flow rate for hole cleaning. With high-speed motors the turn speed should be compared to the maximum recommended rotary speed for the bit, if a roller cone bit is used.

Motors incorporate a bearing pack above the bit sub. Normally a small amount of mud is diverted past the bearing pack to cool it. The seals across the motor will have a maximum pressure rating (i.e., pressure differential inside and outside the motor), and the bit pressure drop will provide most of this drop. The bit nozzles must therefore be chosen to give a pressure drop within limits at the anticipated drilling flow rates. (For specific details on a motor, consult the manufacturer's literature.)

2.3.2. Directional Measurement and Surveying

Survey intervals. The survey tools are your eyes. They tell you where the wellbore is relative to the planned position, and they tell you how much to correct the wellbore path if it strays enough to place the wellbore outside of the required targets.

How often surveys need to be taken may be dictated by government regulations or company policy. Otherwise, the following general advice can be followed.

Directional plan for the interval	> 81/2 in hole size	≤ 81/2 in hole size
Vertical (<5°)	200 m / 600 ft	100 m / 300 ft
Tangent section	100 m / 300 ft	50 m / 150 ft
Build/drop/turn section	20 m / 60 ft	20 m / 60 ft

Surveys can be taken on wireline as drilling progresses; either Totco surveys (which only give inclination) or magnetic single shot (which require monel drill collars and stabilizers to be run). These wireline surveys take time, which, apart from the cost, can be a problem if wellbore conditions are not good, and produce a risk of the wireline breaking and problems with the wireline winch.

UBHO sub. The universal bottom hole orientating sub is like a short sub of the same connections (pin x box) as the drill collars. Inside the sub is a sleeve with a key in it. The orientation of this key is set up and locked on surface so that it is aligned with the toolface azimuth, such as the inside of the bend on a bent sub or bent housing.

When a survey barrel is run or dropped into the UBHO sub with an orientating muleshoe, the tool rotates as it enters and aligns itself with the key. This is how the toolface azimuth can be determined.

Inclination-only surface readout sub (vertical wells). When drilling a vertical well, surveys still have to be taken to ensure that downhole events have not forced the bit to build angle. Circumstances could include setting high bitweights on insufficiently stabilized assemblies, faults, changes in dip, or boulders.

The Anderdrift Inclination Indicator is a sub run in the drill collars. It is mechanical only and it gives a set of pressure pulses, visible on the standpipe, depending on the inclination. A survey takes a couple of minutes (almost independently of depth), and the tool eliminates the risks associated with wireline surveys. It can be very cost effective for drilling vertical holes, especially with a high daily rate operation. A Totco ring can be placed above it so that Totco surveys (or single shots, if monels are run) can still be run on wireline.

Totco surveys (vertical wells). A Totco tool consists of an outer barrel that houses the working parts of the tool. The measurement part consists of a timer unit, an inclinometer, a punching mechanism, and a disk holder.

When a measurement is taken, the inclinometer reading is punched through a marked paper disk with a pin. The paper disk is then rotated 180° and another pinhole is made.

After recovering the tool, the paper disk is removed from the holder and examined. The two pinholes should both give the same reading and be 180° apart; if they do not give the same reading or are not directly opposite, then either the survey barrel was moving during the survey period or the tool itself is faulty. It is easy to check the tool on surface; load it, set the timer for a couple of minutes, and stand it almost upright somewhere so it will not move. Check the paper disk after the survey is taken.

The Totco tool can be run and recovered on wireline. Alternatively, it can be dropped down the string and recovered later, either by wireline or when the BHA reaches surface. It is a good idea to drop the tool whenever the bit is tripped out because this gives a survey with virtually no time impact. There is room past the tool for circulation so circulation is possible. However, if there are any potential problems when tripping out, it is better not to drop it for later recovery.

Since the Totco tool only gives an inclination reading, the wellbore position cannot be calculated. Also, it does not detect a low-angle spiral hole where inclination stays about the same but azimuth constantly changes. Therefore, the tool is only used in vertical wells to ensure that the well stays within a certain limit of verticality. A gyro or multishot survey will have to be run later if a definitive wellbore path and bottom hole location are needed.

Magnetic single shot (MSS) surveys. A MSS tool consists of an outer barrel that houses the working parts of the tool. The measurement part consists of a camera unit with timer mechanism, combined compass, and inclinometer.

There are three different camera units that use either a timer set at surface, a motion sensor, or a Monel detector that fires one minute after it detects the presence of a Monel drill collar. If the survey will be run on wireline, the motion sensor or Monel detector is preferable if the hole is not sticky, since it will fire one minute after reaching bottom. If the survey barrel is to be dropped, either to be fished on wireline or left in the string during a trip out, then use the timer unit.

The camera takes a picture of the compass/inclinometer unit onto a disk of film which is developed on surface. Inclinometer units come in different angle unit ranges, generally 0°-10°, 0°-30°, and 0°-90°. Choose the next range up from the maximum anticipated inclination.

The MSS can also give toolface azimuth if it is dropped into an orientated holder, such as a UBHO sub. On the film, the TFA appears as a line on the compass rose.

As the MSS gives inclination and azimuth, the wellpath and bottom hole location can be calculated as long as you know the location of the previous casing shoe. Generally, a gyro or magnetic multishot survey will have been run at the previous casing point to give a definitive shoe position. Some wireline logging tools also give deviation surveys and most, if not all, modern MWD tools give surveys of sufficient accuracy to be definitive.

For running the MSS to give azimuth, nonmagnetic or monel drill collars need to be run over the depth of the survey tool. Refer to "Monel drill collars and stabilizers—selection and use" later in this section.

Magnetic or gyro multishot surveys. At the end of a hole section, if monel drill collars have been run in the BHA, a magnetic multishot tool can be dropped. This produces a strip of film with shots taken at regular timing intervals. While tripping out, the surveyor takes accurate times and depths so that shots can be matched on the film with the survey depths. This slows down the trip out since the string must be held still for longer than it generally takes to rack a stand to ensure that a survey is taken.

Gyro multishots produce a similar strip of film that has to be matched to times and depths. The advantage of a GMS is that it can be run inside steel, either in casing before drilling out or inside the drill-string (e.g., after tagging the cement plugs and before drilling out). The main disadvantage of a GMS is that the quality control checks must be done properly to ensure that the gyro is correctly aligned at the surface, drift checks are done regularly to track the rate of drift, and the alignment is again checked on surface. All this takes time, so a GMS generally takes more time than a MMS. However, the GMS is often run in casing on wireline while flanging up the BOP and so the actual time impact may not be too much.

If a wireline unit is used for running a multishot survey, the depth counter must be accurate and the wire in good condition. A faulty depth counter will throw out your whole survey without you knowing about it. This should be part of the quality control checks that are witnessed by the drilling supervisor.

Even in a vertical well it is important to have a reasonably accurate picture of the wellpath. Apart from anything else, if disaster strikes and a blowout results, a relief well can then be drilled to allow the blowing well to be killed. I advise that a MMS or GMS be routinely run in the casing set before the first section which may encounter hydrocarbons, to provide a definitive wellbore position at that point.

Wireline deployed surface readout gyro (SRG). When kicking off the well in close proximity to steel interference in the Earth's magnetic field, a gyro tool can be run on wireline to give a constant surface readout of directional information. Normally a working stand is made up with a special gooseneck on the top, which allows mud to be pumped down while the top of the gooseneck seals on the wire. A mud motor is used for the kickoff so the working stand is not rotated.

The working stand is made up to the drillstring, the SRG is run on wireline until it lands in the UBHO. The seal is pumped closed and the mud pumps are kicked in. Drilling continues with a full surface data readout until the stand is drilled down. Once the stand is drilled down, it is pulled back, the pumps are stopped, and the SRG pulled back to inside the working stand. The stand is broken and racked back in the derrick, a stand of pipe made up and run in. Again the working stand is picked up and the process is repeated.

Alternatively a side entry sub can be placed higher up in the drill-string and the wire passes from inside the string to the outside, where it follows the annulus up to the surface. The major disadvantage is that if the BOP has to be closed, you now have a wire in the way.

Measurement and logging while drilling (MWD and LWD) tools. MWD tools started to become commonly used in the early to mid-1980s. At that time you had to have two engineers on the rig just to run the tool and it was generally not too reliable. Modern MWD tools are normally run by the directional driller and can usually stay for hundreds of hours downhole.

Most MWD tools send signals to the surface using pressure pulses in the mud stream travelling inside the drillstring. A very sensitive pressure transducer on the standpipe detects these pressure pulses and converts them into data, which can be presented to the driller. Real-time (delayed by a few seconds to a minute) data on tool inclination, azimuth, and toolface azimuth is displayed on the drillfloor allowing corrections to be made quickly as required.

There are several methods of generating pressure pulses. Negative pulses are caused by opening a small valve, momentarily allowing mud to bypass the rest of the BHA and bit. Positive pulses are caused by actuating some kind of restriction inside the BHA, which causes a positive pressure wave to travel up the drillstring. A carrier wave can also be generated that gives a continuously cycling pressure variation, and this carrier wave can then be modulated to give a data stream to surface in the order of 5 to 12 Hz.

More recent developments combine MWD and LWD tools to give real-time readouts of directional data, sonic, resistivity, gamma ray, and other formation parameters while drilling. The LWD package generally transmits its data electronically to the MWD transmitter package and this then transmits all the data up the hole. The modulated carrier wave transmission method allows a higher rate of transmission for large amounts of data.

LWD tools transmitting real-time formation data to the surface allow the drillers to actively steer the BHA within the reservoir. For instance, say you wanted to drill a high angle well in a sandstone reservoir. The plan is to drill 3-4 m below the shale cap rock. By running a gamma ray or resistivity tool, you would know when you approached closer to the cap rock because GR would increase and resistivity would decrease. Orientating your steerable assembly downwards for a distance would then bring the wellbore back to the desired position.

One major consideration with geosteering high-angle wells is that the LWD sensors can be some distance behind the bit, especially if a motor is used. By the time the sensor tells you a correction is needed, you could have drilled 60 ft or more. The closer the sensor is to the bit, the closer you can control the wellpath.

The Anadrill Resistivity at the Bit (RAB™) is a resistivity LWD tool that actually measures formation resistivity at the bit as well as resistivities at different depths of investigation a few feet behind the bit. This allows real-time decisions while drilling so that corrections can be made immediately and the results can be monitored.

LWD tools and applications are developing very quickly. If your well might benefit from this kind of measurement, check to see what tools are currently available. Also, check on prices, especially because, if they are lost in a hole, some of these tools can cost up to a half a million dollars.

MWD tools in horizontal wells. Most horizontal wells are drilled to intersect large vertical fractures (such as in the Austin Chalk, Texas). This can lead to losses. The larger the losses, the weaker the signal from the MWD tool, which you will be relying on to monitor and control progress. Baker-Hughes has an MWD tool for horizontal drilling that uses special techniques to get survey data to surface, even with total losses. It normally sends a conventional mud pulse signal along the mud column in the string, but once an hour it will send the special signal to surface. Unfortunately, if you have total losses, a signal once an hour (while better than no signal) is of limited use if you need to align the string to drill ahead.

There is a cable-run survey tool that can be used in horizontal holes. This is run after tripping part of the way in and latching into a special UBHO sub. The wire is cut and a wet connector is made up and suspended in the tool joint in a basket. Tripping in continues and to take a survey, the cable is run with the mating connector and latched into the bottom cable. Thus, the cable only has to be run in the top part of the hole. The connector can be a problem in conductive muds (saline water based). An alternative to this is using a side entry sub, leaving the downhole probe permanently on-line. The main problem is if the BOPs have to be closed because the wire will be outside the drill-string and the BOP may not seal around it. The wire is also quite likely to be damaged if the BOP is closed.

Directional surveys from wireline logging tools. Logging tools that are run at the end of a hole section can give directional surveys. If one of these tools is to be run anyway, it may be possible to save the time/cost of running a multishot survey. A wireline logging tool will also give very short intervals between stations (the survey is continuous), which should allow for a more accurate determination of wellbore position.

Dipmeter tools require an accurate wellbore location to give actual magnitude and direction of dip from the measurements that the tool takes while logging. A directional plot can therefore be obtained from a dipmeter run. This survey will have to be "tied back" or attached to a previous survey that gives a known wellbore location, inclination, and azimuth at or closely above the top of the surveyed interval. Schlumberger FMS and FMI logs also give directional surveys.

Before relying on these tools to avoid running a separate gyro multishot, check to ensure that the logging tool surveys can be used as definitive surveys per local government regulations and your own company policy.

Schlumberger also has a north-seeking gyro survey tool. It locates its own position on the Earth but unfortunately has to be left stationary for a long time in the rotary table to do this. Calibrating the tool prior to the run can take a very long time and probably for this reason does not seem to be run that often.

Monel drill collars and stabilizers—selection and use. The steel contained in the drill string components can become so strongly magnetized that it creates large errors in magnetic compasses. Monel drill-string components are not magnetic and are run to give a sufficient distance between the surveying instrument and the magnetic fields of the steel drill collars. The distance necessary to give sufficiently low errors can be calculated; it varies by the inclination and azimuth of the well-bore at the point of measurement. Refer to Figure 2-6.

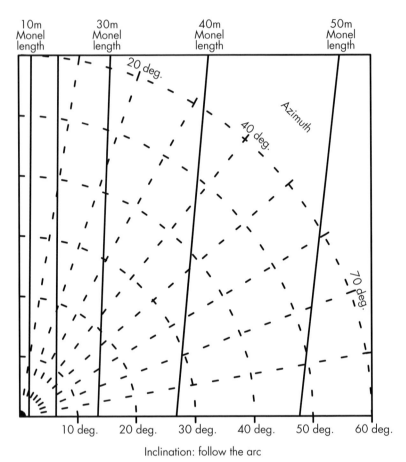

Inclination: follow the arc

Fig. 2-6 Monel Drill collar Length Required

For azimuths between 0° and 90°, use azimuth as shown in Figure 2-6.

For azimuths between 90° and 180°, subtract azimuth from 180° (e.g., for 125° use 55°).

For azimuths between 180° and 270°, subtract 180° from azimuth (e.g., for 205° use 25°).

For azimuths between 270° and 360°, subtract azimuth from 360° (e.g., for 295° use 65°).

The magnetic compass instrument should be placed 40% up from the bottom of the Monels.

Note that Monel lengths are minimum for wells drilled north or south and maximum for wells drilled east or west.

For example, for a wellbore inclination of 30° and a wellbore magnetic azimuth of 70°, start at the 30° arc (bottom axis). Follow the arc until it intersects the 70° radial. Read off the minimum length of monels required, in this case 40 m (131 ft). The compass should be placed about 16 m (53 ft) from the bottom.

Potential sources of survey errors. It is important to understand the potential errors that can arise in the various surveying instruments and methods of surveying. The main sources of error are summarized below.

1. *Long intervals between surveys.* The calculations give some sort of averaging between surveys and so if the interval is long, the averaged path may be significantly different from the actual path. The more the wellbore path may have changed during the interval, the more significant the potential error. A deviated well would give rise to more inaccuracy than a vertical well for the same depth interval between surveys.

2. *Inaccurate measured depths for the survey locations.* This error can arise from a drillstring tally mistake, faulty depth counters on wireline units, or a simple error in writing down the information.

3. *Reading errors.* When surveys are recorded on photographic film (single and multi-shot surveys), there is always the chance that the surveyor may make a mistake. It is good practice for two people to independently read the film and compare their readings.

4. *Instrument calibration errors.* All measuring instruments should be calibrated regularly and a record made of the calibration date and results. The drilling program could include a cautionary note to check the calibration certificate, which should be kept with the tool. If the tool has not been recently calibrated or if the data is not available, the tool should be replaced.

5. *Forgetting to account for magnetic variation.* At most places on the Earth's surface there will be a difference between true and magnetic north. Most survey calculations are done relative to true north,

which does not change. Magnetic north changes with time. To compensate for this, look at a recent aeronautical or marine chart or other suitable chart containing lines that join points of equal magnetic variation, called isogonals. The chart will also give a date of publication and state how quickly the variation changes in the area. It is therefore possible to assess what the current variation for your location will be. Variation can be east or west and this will be stated on the chart.

If variation is west then the variation should be subtracted from the compass survey reading; if east it should be added. There is a rhyme which can help you remember this: "Variation west, compass best. Variation east, compass least."

6. *Magnetic interference.* Even if the correct configuration of Monels is run, magnetic interference may come from various places. Monels occasionally develop magnetic "hot spots"; these can be checked by running a compass along the Monel collar. These hot spots tend to be near the connections. Also, if you run a magnetic single shot on sandline or wireline, the line itself can develop a strong enough magnetic field to affect the survey tool. To avoid significant error due to the line, position the survey tool over 4 m away for a line diameter up to 5 mm and 8 m away for greater sizes. The survey tool could also be dropped using a timer and then fished on wireline so that the line is not attached to the tool when the survey is taken.

Close proximity to other wells, fish that have been sidetracked past, or casings are also likely to cause interference.

7. *Gyro tools do not suffer from magnetic interference.* However, gyro tools have several potential sources of error. First, the tool has to be aligned on surface with a fixed reference that is at a known direction from the center of the rotary table. Incorrect initial alignment will throw out the entire survey. It may be possible to detect this by checking the recorded azimuths against a previous survey at the same depths.

Gyroscopes also "precess"; that is, the gyro will slowly wander away from its initial alignment. The rate of precession depends on several factors, such as friction from the bearings and gimbals, the rotation of the Earth, and small imperfections in the gyro. The rate of precession is measured by doing regular drift checks on surface and while surveying. A drift check simply holds the tool on depth for 10 to 15 minutes so that several shots are taken. Examination of these shots will show the rate of precession so that the survey readings can be corrected for the total drift at the time of the survey. Laser ring gyros are more accurate than mechanical gyros.

2.3.3. Kicking Off the Well

Initial separation from adjacent wells. When a series of wells are drilled from a single surface location (e.g., multiwell platform), the first requirement is that wellpaths do not interfere with each other. It may not seem possible, but a drill bit can cut through an adjacent casing string fairly easily. Magnetic interference high up will restrict your use of survey tools until some separation is obtained, and in this case a gyro surveying tool may be needed until magnetic interference no longer occurs.

If the kickoff point is planned somewhat deeper, wells with close, adjacent wells may be "nudged" in an appropriate direction after spudding below the conductor. This gives some initial separation. The nudge can be achieved by jetting, or with a motor or a whipstock.

Conductors can also be driven with a directional drive shoe. If it works then the well will start off aligned away from the adjacent wells. It is also possible that while driving, the conductor turns, and once this starts it is impossible to stop. In the worst case, the conductor may be unusable due to the high dogleg required to miss the nearest well.

Kickoff and build. Once the planned kickoff point is reached, the well has to be drilled to build angle towards a particular azimuth. If the well is a simple J profile then the kickoff will end when the well is generally pointing towards the target at an appropriate inclination.

Rotary tangent drilling assemblies often have a slight tendency to turn to the right ($1/4°$ to $1/2°$ /100 ft). This can be compensated for by finishing your build section slightly to the left of the line and pointing towards the left-hand edge of the target. Increasing the rotary speed generally decreases this tendency and in some cases can actually reverse it.

Further, with a tangent BHA, using high weights to maximize ROP may give a slight build tendency, as does drilling into formations with dips below 40°. A couple of degrees below the line at the end of the kickoff before locking up would compensate for this. The more weight run on the bit, the stronger the tendency to build. If the assembly is fully locked up, full bit parameters can be used for maximum rate of penetration and the build tendency can be quite small.

One other thing to remember when planning to finish the build a degree or two low is that if you build too much up to the direct line to the target (or above it) and later have to drop angle, you will compromise ROP with the low weights needed with drop-off assemblies. If, however, you need to build, a short-build assembly needs weight to work so your ROP will not be compromised. Pendulum assemblies lose lateral control and may wander off azimuth. Pendulums are also prone to drilling a spiral hole as angle decreases towards the vertical and this is detrimental to wellbore stability, hole drags, sticking tendency, and casing wear. Therefore, you should err on the side of being slightly too low rather than too high.

If offset wells indicate that certain formations have definite tendencies to deviate the wellpath (such as giving a slow build with a rotary locked assembly), then the plan can incorporate these measured tendencies to modify the kickoff slightly. If you expect a long slow build in the tangent section then use this information to improve your directional planning.

Kicking off by jetting or badgering. The fastest way to deviate a well in suitable (soft) sandstone formations is by jetting. One large nozzle is run with two small nozzles so that most of the flow goes to one side of the bit. By setting down weight on the bit and pumping fast, the formation is washed on one side and deviates the wellbore. Only the first few feet of each kelly are jetted and the rest is rotary drilled to the next connection. Ream once or twice before making the connection. The BHA will normally be set up as a build assembly with about 60 ft between the NB and first string stabilizer. If the formation can be drilled fast with a tricone bit (> 50 ft/hour) then jetting should be possible.

Smith Tool produces a special bit for jetting. This is like a tricone bit with one cone missing and a large circulation slot where the third cone would normally be. This bit has a right-hand walk tendency.

MWD tools are not normally used when jetting due to the large flow rates necessary. Single shot surveys are quite adequate and much cheaper than MWDs; string orientation is accurate from the table as there will be little difference between bit and kelly direction due to the shallow depth. Surveys during kickoff will generally be taken every single. Jetting can produce very high dogleg severities and must be monitored as the kickoff progresses.

Since drilling will be fairly fast and the annulus will get loaded up with cuttings, a float in the string is necessary. Otherwise, you will have to spend time circulating before every survey to prevent strong backflow.

Kicking off with a downhole motor. A motor may be configured with a bent sub above it or it may have a bent housing. Where the housing bend is close to the bit, the motor may be used in orientated (sliding) mode or may be rotated while drilling ahead. Such an arrangement is a steerable system, so called because it can deviate the well by sliding or drill straight by rotating the drillstring.

If the kickoff is fairly shallow, a UBHO sub can be run above the motor or bent sub, orientated with the bend. This allows a single shot survey to be run on wireline with an orientating muleshoe so that the inclination, azimuth, and toolface azimuth are shown on the survey. By determining TFA after running the survey, the driller can mark the drillstring and drill orientated with the drillstring locked in rotation. Deeper kickoffs call for a measurement while drilling tool is run. This is due to the time taken to run wireline surveys (and the risk involved of breaking the wire) and as reactive torque generated at the bit will have a marked effect on toolface orientation once on bottom with a longer drillstring. The MWD tool gives a constant readout of inclination, azimuth, and TFA so that reactive torque can be compensated for.

Whichever system is used (bent sub or bent housing), deviating the wellbore is achieved by generating a side force at the bit that makes it cut sideways as well as ahead.

Kicking off on a cement plug (direction not controlled). In a vertical well, it may be necessary to sidetrack around a fish and re-establish a vertical well to target depth. It is possible to set a cement plug and kickoff from it with a flexible drilling assembly, as long as the cement plug has a higher compressive strength than the formation.

Once the cement plug is set, run in with a 60 ft pendulum BHA and a mill tooth bit. Start to drill on the top of cement, building up to a high WOB/lower rotary speed. The bottom drill collars should start to buckle dynamically, pushing the bit sideways. If the cement plug is harder than the formation, the easiest route is for the bit to come off the plug. Monitor the returns for formation cuttings; if progress is too fast to monitor progress properly then drill around 10 ft and circulate for samples before continuing. An inclination survey (Totco or Anderdrift) can be compared to the original inclination at that depth to confirm that the well was kicked off.

Once all the returns are formation cuttings with little or no cement, drill ahead with reduced WOB and reaming two or three times before making a connection to allow the well to start dropping angle towards vertical. When the inclination is satisfactory a locked assembly can be run, carefully reamed through the kickoff depth and drilling ahead continued.

This technique would only be viable where it did not matter on which side the well kicked off. If the downhole target was large or close enough then this could be considered. Since no downhole motor or MWD is required, this could be a low-cost option in the right circumstances. See "Kickoff plugs" in Section 2.7.5, "Cementing Design for Cement Plugs and Squeezes."

Kicking off in casing from a window (direction controlled). A well may be kicked off from inside casing for various reasons, such as to reuse a nonproductive well by drilling to a new target, collapsed casing lower down, etc. A whipstock in casing method is described below, but a window can also be milled in the casing and the well can be kicked off in a cement plug set across the window.

Run a casing mill to the desired depth. The length of window required will vary with the casing size and the expected dogleg severity that can be obtained with the motor and bent sub or bent housing.

Mill a window of the desired length. Set a cement plug from about 100 ft below the window to about 20 ft above.

Run in with the directional motor assembly and tag the top of the cement carefully while circulating and rotating slowly. Drill off the cement to the top of the window. Orientate the motor in the desired direction and commence the kickoff.

Kicking off in casing with a whipstock. A whipstock is basically a long wedge-shaped metal guide that is run in attached to the BHA and set on bottom, oriented in the direction of desired kickoff. The whipstock is held in place on the milling assembly by a shear bolt; once oriented (checked by a single shot survey run to a UBHO sub) setting down weight forces steel teeth into the casing to anchor the whipstock, setting down more weight shears the bolt. The mill follows the concave face of the whipstock and drills into the side of the hole. After drilling for one or two singles (sufficient for a single shot survey to be taken in open hole away from the steel whipstock), the mill is pulled out, and the whipstock is pulled by the BHA. A full sized bit is then run on a suitable drilling assembly and the pilot hole is drilled out. The process can be repeated if necessary.

A combined whipstock and mill can also be run inside casing; the whipstock is oriented and set permanently in the casing. The whipstock is released by setting down weight to shear a bolt, the mill (guided by the whipstock) cuts a window in the side of the casing, and is pulled out. A balloon mill is run to smooth off the window and then a bit may be run to drill ahead as desired.

2.3.4. Drilling the Tangent Section

The tangent section of the well is designed to be drilled (mostly) straight. As mentioned above, the wellbore is almost never truly straight but will be influenced by the bit, BHA, drilling parameters, and formation characteristics.

With a rotary assembly rotating clockwise, the forces acting on the BHA normally generate a slight side force at the bit tending to turn it to the right. This tendency is increased at lower rotary speeds, also as the bit dulls or if it is crippled. Older directional hands will tell you stories about cutting one cone off a bit to cripple it and make it turn to the right for correcting the wellpath. In this case, they are not kidding! With a sharp bit and high rotary speed, it may be possible to reverse the normal trend so that a left-hand turn is initiated.

The tangent section generally needs to be drilled quickly, with minimum dogleg and not spiral or overgauge so that logs and casing can be run without problems. If any other rotary bottom hole assembly is run than a fully stabilized one, the above objectives will be compromised to a greater or lesser extent.

Tangent section in vertical wells. In many operating companies, vertical wells are routinely drilled with 60 ft pendulum assemblies. This is a mistake. Of all the possible assemblies to use, this is about the worst—with the exceptions of a 90 ft pendulum or a completely slick assembly. The problems with using a pendulum assembly in a vertical well include drilling a spiral hole (seriously bad for increasing drags, problems running logs and casing, spiral communication paths in cement outside casing, casing wear, etc.), inadvertently kicking off in streaky formations or on boulders, and seriously restricting the maximum weight on bit.

If the well is pointing in the right direction, whether it is vertical or deviated, then lock it up and save yourself a lot of problems. If you have concerns about those stabilizers causing excessive drags then use one or two roller reamers to replace the NB and first string stabilizers. In fact, unless you are likely to drill a spiral hole with your locked BHA, or if you have seriously mobile formations, you should not see excessive drags and torques with the packed BHA. In either case, the roller reamers should handle the problem.

If balling is your concern then revise your drilling fluids program. Modern water-based muds can largely eliminate balling due to shale hydration. Stabilizers above the first three can be run a little bit under-gauge to minimize dynamic buckling and differential sticking without a large contact area with the formation.

2.3.5. Dropping Hole Angle

Of course there will be occasions when you need to drop angle. In this case, either a pendulum assembly or some kind of motor assembly is required.

In most cases, a rotary assembly will be preferred. The problem with a motor is that when the toolface is aligned to the low side of the hole, the tool will tend to flip over once reactive torque is applied as the bit contacts the bottom. It can be quite tricky to hold the motor in this position if the hole is deep and at a high angle. A steerable assembly is preferred to a motor-bent sub combination since the steerable has a smaller bend, close to the bit.

Drilling parameters are also important with a pendulum assembly. Until a drop trend is established, very low bit weights will be required. Once the drop trend is established (that is, once the second stabilizer is in the section of hole that is dropping), it should be possible to bring

WOB up somewhat. This will improve ROP and may also increase the drop rate a little as it forces the drill collars to buckle outward due to the bend in the hole between the bit and second stabilizer.

Reaming once or twice before each connection can also help in soft to medium hardness formations.

If using a fixed cutter bit, choose one with a short-gauge length, flatter face profile, and side cutters. With a roller cone bit, avoid pads welded on to the shirttails. In abrasive formations with insert bits, use diamond-enhanced heel and gauge cutters but again avoid shirttail pads that effectively increase the gauge length.

Maximizing hydraulic impact force may help to erode formation at the bit face that could help to drop angle.

There are only two occasions in general when a pendulum assembly is justified; when you actually want to drop angle and when drilling large diameter surface hole with large, heavy drill collars ($17^1/_2$ in hole with $9^1/_2$ in drill collars). Do not use a pendulum assembly if already vertical and you intend to drill a vertical well. Any sideforce at the bit (e.g., a hard streak) will cause the bit to kickoff, causing a potential problem later on with a ledge, apart from the effect on directional performance, which may cause you to miss the target.

$$\begin{bmatrix} 2.4 \end{bmatrix}$$

Drill Bit Selection, Parameters and Hydraulics

Field data in this section was allowed into the public domain with the kind permission of the Badr Petroleum Company, Egypt (BAPETCO).

2.4.1. Overview of the Structured Approach

One of the two major elements in efficient drilling is maximizing drilling rate. This is intimately related to selecting the correct drill bits and using correct drilling parameters to maximize drill rate and bit life.

For well planning, it is necessary to construct a good overview of field drilling history. Where several wells have been drilled, the amount of data that can be usefully analyzed is large. Often such offset data is not analyzed properly, which leads to less effective planning and higher drilling costs.

A structured approach is shown here to make the best bit selection based on all available data. Actual selection on the rig may be different but if it is, it should be justified by noting all the factors which went into that decision and why the selected bit would improve performance over the programmed bit.

2.4.1 Overview of a Structured Approach

1	2	3	4	5
Summarize all relevant offset data	Identify the best two bit runs in each interval or formation	Determine how to improve on the previous best bit run	Select bit, BHA, parameters, and procedures. Predict expected performance.	Compare actual vs. expected performance and make recommendations
■ Update the BITREX bits database. ■ Update field information notes. ■ Examine mud logs. ■ Evaluate wireline logs. ■ Update hole section summary.	■ Examine the hole section summary and mark the best and second best runs in each interval.	■ Look at the dull bit gradings. ■ Determine likely causes of major dull features and how to improve bit selection, BHA, or drilling practices in order to avoid them on the next run. ■ From all records in the same formation, should a softer formation bit be considered?	■ Remember this is an *economic* decision. ■ Which combination of bit, BHA, parameters, and practices should give the lowest cost per foot? ■ Should a different bit type (e.g., PDC) be considered? ■ What overall ROP and distance should be expected from the best combination?	■ After the well, compare actual vs. expected performance and CPF. ■ If performance is below expectations, explain possible reasons. ■ Note ideas for further improvement on the next well.

The elements of this approach are explained individually in this section. It works well in practice, but the usefulness is relative to the amount of effort and expertise put into it.

There are two (paper) methods that summarize and present a large amount of data in a way that is useful for well planning and drilling operations. These are described under Preliminary Work for Well Design. In addition, a computerized bit database that is good for analyzing the bit runs and for fast access to a large amount of data in the office or field is used.

The central data summary tool is the *hole section summary*. This document can display a lot of information in a way that allows fast and meaningful comparison of data from multiple wells. It can be constructed on quite a basic level using only information from offset bit records and daily reports. However, by incorporating other information from mud logs, electric logs, the bit database analysis, final well reports, etc., it allows a more precise analysis and optimum bit selection. Refer to "Hole section summaries" in Section 1.1.2, "Data Acquisition and Analysis."

The other data summary tool is *field information notes*. These give a detailed (drilling) description of each formation to be encountered and show problem areas, recommended bits, etc. It is particularly good for drilling personnel new to an area if it incorporates the current state of knowledge within the company on those formations. If a well is to be planned where offset data is available but no in-depth analysis has been done, it is extremely useful to take the time to work through each formation and write these notes for reference. These can be appended to the drilling program. Refer to "Field operational notes" in Section 1.1.2, "Data Acquisition and Analysis."

Both the field information notes and the hole section summaries should be updated after each well is drilled.

2.4.2. Evaluating Offset Well Drilling Data

Comparing bit records using the BITREX database. When comparing offset runs, it is important to compare a shorter, faster run with a longer run at lower ROP and say which is the better run in the same formations.

BITREX is a database application, created using Microsoft Access that runs under Windows. It can be downloaded free from the web site at *http://www.drillers.com*. In addition to recording data from the bit run, it can calculate a performance index value to allow comparisons of offset bit runs.

The Normalized Bit Performance Index (NBPI) takes the total distance drilled and divides it by the total time for the bit run (including an assumed trip time based on depths with bit cost expressed in rig hours). In that respect, it is a type of cost per foot calculation. However, it also takes into account the time taken by the next bit in to ream to bottom, *where this is caused by the bit coming out undergauge.* This is important because if the next bit in has to ream undergauge hole for (say) eight hours to bottom, the penalty for this should go against the undergauge bit when comparing performance. A judgment may have to be made to decide what proportion of next bit reaming time to use, for instance, if the bit came out undergauge but a stiffer bottom hole assembly was used on the next run. Fishing time could also be put against a bit if it came out in more than one piece.

A direct comparison of the NBPI for several bits run in the same interval in different wells is only *completely* valid if both the following conditions are met:

■ The bits under comparison were pulled at the end of their economic life. This would normally be on ROP (e.g., cost per foot calculations) or on indications of bearing failure.
■ The bits under comparison were run using parameters for best performance, normally on fully stabilized rotary assemblies where no constraints on WOB or RPM are applied, except the manufacturer's recommended maximums.

However, even if both these conditions are not met, the NBPI still provides guidance on how successful or otherwise the bit run was.

The NBPI is calculated as follows:

$$\text{Normalized Bit Performance Index} = \frac{\text{Bit footage}}{\text{Drilling Hours} + \text{Normalized Tripping Hours} + \left(\dfrac{\text{bit cost}}{\text{rig hourly cost}}\right) + \text{Next Bit Ream hours}}$$

- Normalized trip time assumes 1 hour per 1000 ft.
- Next bit ream time is the time spent reaming back to bottom *as a direct consequence of this bit coming out undergauge.*

Therefore, by entering all offset bit runs into the BITREX database, a printout can be generated that shows the NBPI for each bit which can then be put onto the hole section summary against each bit run.

The database allows sets of bits to be selected from the database for generating reports. The bit selection can be done manually (by "tagging" each record of interest) or it can be done based on various criteria (such as field and range of depths in, well and bit size, bit serial number, etc.). Selection by criteria adds to the current selection. Thus bit runs of interest can be quickly examined and compared.

Once a decision is made as to which bit is to be used for a certain interval, another report can be generated showing all runs in the database with that bit type and size. This can give an indication of expected bit life and performance with various parameters.

Comparison of bits run in the 12¼ in Obaiyed hole sections. The following analysis was made from the BITREX database comparison of the 12¼ in hole section bits run on the Obaiyed field in Egypt. It compares the best performing bit in each formation with the second best bit. The percentage of improvement shown is taken by comparing the Normalized Bit Performance Indices.

- In the Abu Roash, the (best) Hughes ATM05 (417X) outperformed the Smith F1 (517X) by 20%. (The Smith bit was third best after the two ATM05 bits.)
- In Lower Abu Roash and Bahariya, the Hughes ATM11HG (437X) outperformed the Hughes ATM22GD (517X) by 15%.
- In the Kharita, the Security SS84FD (517X) outperformed the Smith F2OD (517X) by 14%.
- In the Dahab and Alamein, the Reed EHP51HD (517X) outperformed the Hughes ATM22 (517X) by 20%.
- Alam el Bueib did not perform enough complete runs for analysis.

It can be seen immediately that in no case did a harder bit outperform a softer bit. This might imply that we can probably make better progress with slightly softer bits than the previous best in some cases; it would help to examine all the bits run in a particular formation, not just the best two.

This quick evaluation *could* form the basis for our bit selection on the next Obaiyed well by simply using the best performing bit to date in each formation. However, what we should do is try to establish which bit will *improve* on the previous best run. For that, we have to do further analysis.

Writing field information notes for bit runs. Daily drilling reports often leave a lot of relevant information unrecorded. Drilling programs rarely give sufficient information to the drilling supervisor about the anticipated formations. Both these concerns can be overcome by writing and updating field information notes.

As operations proceed, the on-site drilling supervisors should make detailed notes of operational events. At the end of a hole section, the drilling supervisors should use these notes to write a detailed report, which can be incorporated into the final well report. An example is shown in Section 1.1, "Preliminary Work for the Well Design." This information should also be used to add to or amend the field information notes. The information that is required in particular is:

- Details and comments on each bit run
- Any problems that were experienced and how they were solved
- Suggestions for improvements to the bit program for drilling the next well

All the available data relating to each formation can then be summarized for future reference when planning and drilling. As experience is gained, this data should be kept up to date. It can be seen that there is much useful information to aid in bit selection and use.

Mud log. If a logging unit is on-site, it is worth getting the loggers to record drilling data every foot or meter in a format that can be imported into a spreadsheet. This allows the drilling engineer to make very detailed analyses. As shown on the hole section summary, the ROP data with changes of bit or formation that came from the mud logger's data (recorded every meter). By correlating with wireline logs, a comprehensive picture can be built. The data recorded in the spreadsheet was:

- Depth in 1 m intervals
- Drilling rate in both minutes per meter and meters per hour
- Weight on bit

- Rotary speed
- Rotary torque
- Pump output
- Pump pressure
- Mud density in
- Mud density out
- Mud temperature in
- Mud temperature out
- Gas readings

In addition, the mud loggers take samples from the shakers and describe lithology, cuttings appearance, and cavings percentage and appearance.

2.4.3. Drilling Hydraulics

Much of the power produced by the mud pumps is lost in the circulating system through the surface lines, drillstring, and annulus. These are parasitic losses that produce no direct benefit to the drill bit performance. The power that is left can be used in different ways to help clean the bit/hole bottom, aid ROP by the direct effect at the bit face, and drive downhole motors or turbines.

The calculations involved in drilling hydraulics are best done with computer programs and these are available free from various bit companies (e.g., Reed or Smith). Using a hydraulics program allows different scenarios to be compared quickly and easily; the entered data can usually be saved and modified later as required by different mud properties or pump capabilities. Calculating these pressures by calculator is tedious since they have to be repeated for each change in the flowpath size, consuming valuable time if you are playing "what if" with different mud properties or flow rates.

There are two current theories for optimum hydraulics. One gives the total nozzle area to maximize hydraulic horsepower. The other calculates for maximum hydraulic impact force on bottom. Of the two methods, maximizing HHP gives greater pump pressure and lesser flow rate.

To maximize bit hydraulic horsepower, the pressure drop across the bit should be 65% of the total pressure loss in the system. If the nozzles were sized smaller than this, less HHP would result on bottom because the flow rate would have to be reduced to keep the same surface pressure. If the nozzles were larger, the same flow rate with less bit pressure drop would also reduce the HHP expended at the bit.

To maximize hydraulic impact force on the bottom, the pressure drop across the bit should be 48% of the total pressure loss in the system.

The actual percentage loss across the bit will decrease as drilling progresses due to increased pressure loss in the system (adding more drillpipe). The calculations can be made for the middle of the expected bit run interval, unless the expected bit run is very long and the pumps will be running at or near to maximum pressure; in this case, calculate for the expected end of the bit run.

Hydraulic horsepower expended at the bit is often expressed as horsepower per square inch or HSI. Using HSI above 5 is likely to lead to significant bit erosion except on very short bit runs. Erosion on the bit is acceptable as long as it does not lead to premature bit failure (such as PDC cutters dropping out of the body).

Following are some advantages to optimizing for impact force rather than hydraulic horsepower:

1. Larger nozzles will reduce nozzle plugging and will be better for pumping LCM.
2. Lower pressures give less pump wear, reducing downtime due to pump failure.
3. Higher AVs give better hole cleaning.
4. Erosion of the wellbore due to annular velocity, except in very unconsolidated formations or fractured formations with a nonsealing mud system, probably does not occur. It is a commonly held opinion that high AVs lead to washouts. What is more likely is that the high hydraulic energies at the bit lead to hole enlargement or washouts and optimizing impact force may help reduce this effect.

Maximizing bit HHP is not guaranteed to lead to improved ROP through better bottom hole cleaning (especially in harder formations) at normal pressures. Some success has been reported with experiments using ultra high pressures (up to 30,000 psi) in hard rocks, however, this is not in widespread use (see references in Section 2.4.9, "References for Drill Bit Selection").

If maximizing HHP does not improve ROP, then either maximize impact force or choose nozzles for the flow rate required at a pressure below maximum at section TD. Calculating exact nozzles and flow rate for optimum hole cleaning under the actual conditions at the time cannot be done due to the large number of variables, some of which are unknown.

Note that it will not always be possible to fully optimize bit hydraulics since other considerations may take precedence. For instance, if the use of LCM is anticipated, then use nozzles of at least $16/32$ in size, even if the optimization calculations suggest smaller nozzles. Pressure drops through downhole motors and motor seals may restrict the minimum bit pressure drop.

Apart from calculating the nozzle total flow area (TFA), the type and arrangement of nozzles can make a significant difference to bit/bottom cleaning and therefore ROP. In a rotationally symmetrical bit (such as a conventional tricone drill bit) without a center jet, placing three equally sized nozzles may lead to a dead zone of no flow at the center of the bit. Asymmetric nozzle sizing (i.e., using one large and two smaller nozzles for the desired TFA) can cause a cross-flow effect across the bit face, ensuring that no dead zone exists. Experience and lab results indicate improved ROP from cross-flow configuration. The large nozzle also creates turbulent pressure fluctuations which helps lift rock chips off the bottom, reducing redrilling of cuttings and effectively increasing bit efficiency and ROP. Some operators also use one blank nozzle to give a cross-flow effect from the two remaining nozzles and this has demonstrated some ROP improvements. Laboratory experiments indicate ROP improvements from asymmetric nozzle configurations of around 20% (see references in Section 2.4.9, "References for Drill Bit Selection").

Extended or mini-extended nozzles can also be used (in combination with asymmetric sizing if required). By decreasing the distance between the nozzle and formation, less dissipation of energy takes place before the jet impinges on bottom. In softer formations this can give a worthwhile increase in ROP, probably due to a combination of better bottom cleaning and some erosion of the formation. Extended nozzles have to be handled carefully; if the bit breaker is worn, then makeup torque can be transmitted between the nozzles and the bit breaker, which may lead to downhole loss of the nozzle.

A relatively recent nozzle design works in theory by creating both positive and negative pressure regimes (i.e., below mud hydrostatic) on bottom. The Vortexx nozzle has an asymmetric inside profile which also produces a directional flow; on a PDC bit these nozzles are aligned in such a way as to produce cross-flow effects along the blades by using a template which has been worked out for a particular bit design. To allow freedom in alignment, the nozzles have to be threadlocked in place. A set of these nozzles was run in a PDC bit in the Nile Delta. While there were no directly comparable runs to draw technical conclusions against, the run was very impressive with the bit drilling at controlled ROPs over 160 fph with 1 to 3 kips on the bit and 80-100 RPM. It was determined that the nozzles had contributed significantly to this performance.

Whatever is decided about hydraulics at the bit, a rule of thumb is to use an annular velocity over 100 ft/minute to ensure that cuttings can be lifted up the annulus. However, this is only a general guide. Higher angle wells may need turbulent flow to prevent cuttings beds from building up (if the mud rheology does not allow turbulent flow, then thin, turbulent slugs may be needed). Washouts are particularly detrimental in lowering AV and allowing cuttings beds buildup. If the flow rate is restricted by losses, the AV should never fall below what is needed to lift cuttings up to the loss zone, even if total losses are occurring and you are drilling blind. You should use 50 ft/min as the minimum AV.

Mud rheology becomes very complex in non-Newtonian fluids. There are many different theories and methodologies to relate mud shear rates to shear stress. Polymer fluids can exhibit complex behavior. In addition to shear thinning (exhibiting lower viscosity at higher flow rates), they may be time-dependent and exhibit different velocities at a particular shear rate depending on whether the pump speed is increasing or decreasing! Temperature, and to a lesser extent pressure, will also make a significant difference so that modeling the actual effect on bottom of particular flow rates can get tricky.

If these complex fluids are in use and it is important to calculate accurate pressure drops (such as in deep and/or slimhole wells), the fluid itself needs to be tested and a suitable model needs to be used for the circumstances.

2.4.4. Using Log Data to Aid in Bit Selection

Sonic, lithology, and porosity logs can be interpreted to give various rock mechanical properties, such as Bulk Compressive Strength, Shear Modulus, Young's Modulus of Elasticity, and Poisson's Ratio. Some of the major bit vendors have done work on relating bit selection to these properties. (Note: The definitions of these mechanical properties are covered in Section 1.4, "Casing Design.")

At the time of writing, one of the best evaluation tools is the Dipole Shear Sonic Imager (DSI) from Schlumberger. This tool gives compressive, shear, and Stoneley wave data, which can be interpreted with other log data to yield mechanical properties. These results can be displayed in a log format. (Note: monopole sonic tools only yield useful information when sonic velocities are faster in formation than in mud, i.e., hard rock; dipole tools can be used in softer rock too.) Correct interpretation of the results takes considerable skill and area knowledge and is therefore not something that drilling engineers would undertake themselves. In fact, there are many useful interpretations for drilling purposes that the DSI tool allows and these are covered in more detail in Section 1.4.7, "Mechanical Properties of Steel."

Application of the DSI tool to bit selection is in its infancy. Some general conclusions can be drawn at present, but the complexity of all the various interrelated factors leaves plenty of room for future research. There are quite a few papers on these topics (see the listing of references at the end of this section) which can be followed up for further study.

Following are guidelines that relate various log interpretations to the basic suitable bit type.

Sonic transit time vs. PDC cutter type. *(Courtesy of Smith International Inc.)*

μsec/ft	110	100	90	80	70	60	50

3/4 in (19 mm) cutter ----------------------------------|
3/4 in (19 mm) dome cutter   ----------|
5/8 in (16 mm) dome cutter --|
1/2 in (13 mm) cutter   ---|
1/2 in (13 mm) dome cutter   ------------|
3/8 in (9 mm) dome cutter   ---------------------------|
Natural diamond 1-2 stones/carat   -----|
Natural diamond 3-4 stones/carat   --------------------|
Natural diamond 5-6 stones/carat   →

Cation exchange capacity vs. PDC profile. *(See Reference 1)* This relates to the potential for bit balling in water-based muds.

CEC	40	35	30	25	20	15	10	5

Fish tail design --|
Large cutter applications   ---|
Conventional cutter sizes   ------------------------------|

(Cation exchange capacity is approximately API gamma ray units ÷ 4)

The ultimate objective is to be able to describe a formation purely using sets of numbers for the various properties that can then be manipulated mathematically by an expert system to give optimum bit features to drill the desired interval length at the lowest cost. Given sufficiently detailed offset bit runs to analyze, bit performance could perhaps be predicted with enough accuracy to recognize how worn the bit is while drilling. Therefore, even in nonhomogenous formations, the bit could be pulled when its efficiency (specific energy required to destroy a certain volume of a particular rock, ideal for that bit ÷ actual) drops to a level that justifies a round trip. Changes in drillability due to increasing pore pressure could also give better kick prediction.

The major bit vendors will draw up bit proposals based on log information, which they interpret for the rock mechanical and other properties, together with offset run data. Examining proposals from several vendors for a well can give you a good feeling for the requirements of the best overall bit to use.

2.4.5. Types of Drillbits

Mill tooth bits ("**rock bits**"). Mill tooth bits are most useful in soft formations, usually the top sections of the hole. One large bit may drill top hole for several wells. It is also possible to have the cutting structure built up again (by a specialist firm) and extend the bit life, as long as the bearings are still okay.

Some operators use mill tooth bits to drill the shoetrack prior to running a PDC bit, although PDC bits are routinely used to drill out plugs. (For a plug to be PDC drillable, it should contain no aluminum that tends to wrap around the cutters, which then fail through overheating. The nonrotating type of plug is also preferable.)

Mill tooth bits are also used in formations that contain harder nodules (e.g., chert and conglomerates), which may cause TCI bit insert or PDC cutter breakage due to shock loading. Running a shock sub will be helpful here. Some mill tooth bits are designed with bearings and cutting structures to be used with mud motors at higher speeds.

Certain strong, "elastic" shales can be drilled better with a mill tooth bit than a PDC or insert.

Mill tooth bits are the only ones capable of drilling on small pieces of steel junk. Large amounts of junk will of course cause serious damage to the bits and may create even more junk by knocking off cones if they are not run carefully.

Tungsten carbide insert (TCI) bits. Bit teeth made of sintered tungsten carbide revolutionized tricone bit drilling since they were introduced. The materials and techniques used today have made these cutters very long lasting. Diamond coatings can be applied to make them even more wear resistant, which is especially useful for gauge protection in abrasive sands. These bits are more expensive than mill tooth bits using the same bearing structure and are far more durable and certainly more popular in medium, hard, and very hard formations. The tungsten carbide inserts, while extremely hard, are also brittle and break under shock loading.

When using tricone bits in softer, sticky formations (shales) which may ball the bit, center jets are often helpful. Extended jets that take the flow closer to the bottom also help bottom cleaning. In a bit with three nozzles, a "dead zone" of little fluid flow in the bit center can result from running three equally sized nozzles. Running one large and two smaller nozzles of a different size gives a cross-flow effect on bottom, which may improve cleaning, especially when drilling fast. Some operators blank off one nozzle completely to give a very strong cross-flow effect. This does not seem to have an adverse effect on bearing life (due to overheating) as might be expected; the blanked off area will experience strong flow upwards from the two nozzles.

Polycrystalline diamond compact (PDC) bits. PDC bits use a thin wafer of diamond mounted on a stud. This cuts efficiently like a lathe tool. They are good in plastic formations (e.g., medium shales and salt) and can give fast ROP over long intervals. Early PDC bits used in water-based muds tended to ball the cutters, then overheat, which delaminated the diamond. Better water mud technology has overcome this.

PDC bits use either a steel or a matrix body. The manufacturer can refurbish some steel-bodied bits as long as the bodies are in good shape, giving you "new" bits for a reduced cost.

PDC bits have a wide choice of cutting structures. Large cutters work aggressively in softer formations, removing large cuttings at high rates of penetration. Bits for harder formations will be heavier set with a greater number of smaller cutters. Gauge protection is often with natural diamonds. They are not suitable for very hard or abrasive formations.

PDCs do not give a good D exponent trend. When it is important to recognize a pressure transition zone, it may be better to run a tricone bit.

PDC bits are usually not suitable for formations containing hard chert or other nodules, though this is possible with some bits and careful optimization of parameters. By using a large flowby area around the cutter pegs, maximum flow rate, minimum RPM, and moderate WOB, small hard nodules can sometimes be washed away before damage is done to the cutters. The Hughes B11M or Eastman "Eggbeater" types have demonstrated this.

Natural diamond bits. Natural diamond bits incorporate diamonds directly into the bit matrix. These bits work by abrading the formation, producing very fine cuttings or "rock flour." A diamond bit will drill any hard formation, but at low ROP. Diamond bits, like PDCs, need to have the cutters effectively cleaned and cooled for a full working life.

Due to their construction, diamond bits are available in a wide variety of profiles and cutting actions. For sidetracking, bits with a good side cutting action may be ordered. Parabolic shapes allow a greater concentration of working cutters on bottom and tend to be directionally stable.

2.4.6. Defining Recommended Bits

In order to define which is likely to be the best bit to use for each part of the hole, a set of questions can be asked. The first question is "what are the best two offset bits?" The second question is how can we improve on the previous best bit run?"

Examine the two best runs. Look at the bit gradings as well as the run details. Differences between the best two may "point the way"; if the best bit was a slightly softer type than the second best, is it likely to help if an even softer formation bit is used? It is clear that drilling supervisors and engineers often choose bits that are too hard a type for the formation being drilled and therefore end up sacrificing performance.

If the best bit was pulled at the end of its economic life, what caused the bit to be pulled? For instance, if there were an excessive number of broken teeth, was that caused by bad run practices, drillstring vibration, or formation conditions? Can this be avoided on the next run by a change of practices, BHA configuration, or correctly set up shock sub?

If the bit came out undergauge, did this cost time on the next bit in to ream to bottom? Would the cost of using premium gauge protection (such as diamond-coated heel teeth) be repaid in better drilling performance and less reaming?

Were optimum hydraulics used? If not, is this likely to have caused a lower ROP? Are there any constraints on optimizing hydraulics (such as having to use larger nozzles for possible LCM)?

Economics is the bottom line to the decision. Which bit will drill the most cost-effectively? PDC bits in larger hole sizes tend not to be economic on low-cost rigs, but this changes if: the hole sizes decrease (there is less extra PDC cost and a higher chance of losing cones off tricone bits), the hole deepens (trip time becomes more significant), or the rig day rate increases. As wells are drilled slimmer than in the past and as PDC bits improve in their cost and performance, PDC options should be examined where a potential application exists.

By examining all aspects of the bit runs and gradings and by considering all restrictions imposed by the BHA, the drilling engineer should be able to see what bit features may be changed to obtain a better run next time.

Particular bit features and how they relate to bit selection. Certain downhole conditions can include or preclude particular bit features at the well planning stage (see Table 2-1). Later in Section 3 when practical operations are discussed, dull bit features are used to identify possible downhole conditions that may lead to a modified bit selection.

Using IADC codes to identify a general class of suitable bit.

IADC tricone bit classification. Each bit manufacturer, having their own way of naming their bits, made identifying which particular tricone bits to use for different formations more difficult. The IADC recognized this problem some time ago and devised a code system with three digits for identifying principal bit features. This is a useful comparison of different makers' bits since the IADC code is usually given as well as the bit name.

The first code digit is formation hardness series and can be 1 to 8. Low numbers relate to softer formations; 1 to 3 are for mill tooth bits and 4 to 8 are for TCI bits.

Table 2-1 Bit Features

Condition	Helpful Bit Features	Detrimental Bit Features
Abrasive sands and mobile or very permeable zones leading to under-gauge hole (reaming in).	Diamond-enhanced gauge protection and roller reamers.	PDC bits are generally not suitable.
Highly plastic shales.	Mill tooth cutters.	Tungsten carbide inserts sometimes do not perform well in very plastic formations.
High-compressive strength.	TCI bits and sealed journal bearings.	Mill tooth bits will wear quickly and PDC bits are generally limited to applications below about 20,000 psi. Security is researching PDC designs for formations up to 50,000 psi.
Very high-compressive strength.	Natural diamond bits.	
Reactive shales in water-based mud.	Bit balling may be expected; center jets on tricone bits. Large flowby areas on PDC bits (bladed designs).	
Motor applications.	High-speed sealed roller bearings and fixed cutter bits.	Journal bearings tend to overheat at very high rotary speeds.
Deep, hot, high pressure.	Metal to metal sealed bearings or fixed cutter bits.	Elastomer seals are more likely to fail early in harsh conditions.
Directional wells with high build rates.	Bits with side-cutting action and small gauge lengths.	Parabolic fixed cutter bits and/or bits with long gauge areas.
Hard nodules in formation (cherts).	Mill tooth bits are generally best, but TCI may be applicable.	PDC cutters may shatter when hard nodules hit.

The second code digit is type and it is a subclassification of hardness. The code can be 1 to 4 for bits that increase hardness within the formation hardness series.

The third code digit is feature classification and can be 1 to 9 as follows:

1 - Standard, nonsealed bearing
2 - Air lubricated, for air drilling
3 - Standard, nonsealed bearing with insert gauge protection
4 - Sealed roller bearing
5 - Sealed roller bearing with insert gauge protection
6 - Sealed journal bearing
7 - Sealed journal bearing with insert gauge protection
8 - Directional
9 - Special application

For example, a bit type IADC 5-4-7 could be a Hughes ATJ33, Reed HP54, Smith F37, or Security S88F. Each of these bits has broadly comparable application as journal bearing insert bits for medium formations of high-compressive strength. Note that the bit manufacturer and not the IADC gives a bit its IADC classification.

The system also defines 16 special codes that can be placed after the IADC bit classification to show particular features.

IADC special feature code and description

A   Air drilling application
L   Lug pads
B   Special bearing seal
M   Motor application bit
C   Center jet
S   Standard steel tooth model
D   Deviation control
T   Two-cone bit
E   Extended jet (full length)
W   Enhanced cutting structure
G   Additional gauge and body protection
X   Predominantly chisel tooth inserts

H Horizontal steering application
Y Predominantly conical tooth inserts
J Jet deflection bit
Z Other shape inserts

For instance, a security bit type ERASS84FD could be described by IADC code 517WMBXC as a soft to medium sealed bearing insert bit with enhanced cutters, motor application, special seals, predominantly chisel-shaped teeth, and a center jet.

IADC fixed cutter bit classification. As with roller cone bits, IADC have defined a classification system for describing fixed cutter type bits. There are four digits that classify the type of body material, cutter density, size/type, and body profile. Following is a summary of how the system works.

First digit: body material
M for matrix or S for steel.

Second digit: cutter density
Ranges 1 to 4 for PDC bits and 6 to 8 for surface set (e.g., diamond) bits. The PDC bit number relates to the cutter count; there is a higher number for a heavier set. The surface set number relates to diamond size.

0 Reserved for future use
1 Less than 31 x 0.5 in diameter PDC cutters
2 30 to 40 x 0.5 in diameter PDC cutters
3 40 to 50 x 0.5 in diameter PDC cutters
4 50 or more x 0.5 in diameter PDC cutters
5 Reserved for future use
6 Larger than 3 stones/carat
7 Between 7 and 3 stones/carat
8 Smaller than 7 stones/carat
9 Reserved for future use

Note: For larger or smaller diameter cutters, use the ratio of diameters. So 50 x 0.25 in cutters would equal 25 x 0.5 in cutters. This does not include extra cutters, e.g., gauge trimmers and side-cutting cutters (sidetrack bit).

Third digit: size (PDC) or type (diamond)

PDC	Surface set
1 >24 mm diameter	Natural diamonds
2 14-24 mm diameter	Thermally set polycrystalline (TSP)
3 8-14 mm diameter	Mixed (e.g., natural and TSP)
4 <8 mm diameter	Impregnated diamond

Fourth digit: body profile

1 Flat face surface set or "fishtail" type PDC
2 Almost flat profile
3 Medium length profile (e.g., round or short parabolic)
4 Long profile (e.g., long flanked turbine bit)

Mud motors, steerable systems, and turbines. When planning to run a downhole motor, the following other considerations apply:

1. If the string is rotated while drilling with a motor, will the maximum bit RPM be exceeded?
2. Will the flow through the motor clean the hole at the envisaged flow rates?
3. Are there any limitations on the bit pressure drop imposed by the motor?
4. Is the proposed bit suitable to use on the type of motor to be used?
5. Can LCM be pumped through the motor and, if not, should a circulating sub be run above? Is LCM needed while drilling with the motor?
6. What sizes of liners are needed in the pump for the necessary flow rates and pressures?
7. Are there any problems with the mud properties (e.g., chemical compatibility with seals, sand content, etc.)? Check with the motor supplier.

What is the plan for before and after the motor run? For instance, if running in for straight hole turbodrilling with a PDC bit, the previous assembly should be fairly stiff to avoid reaming in with the turbine. Any junk in the hole would require a junk run first, or if a steerable system run is going to terminate at casing point, then a wiper trip with a rotary assembly should be made to ream to bottom and reduce the chance of mechanically stuck casing by reducing doglegs and ledges.

2.4.7. BHA Considerations Related to Bits

The BHA has an impact on bit performance and durability. Logically it also has an impact on bit choice. For optimum bit performance, the preferred rotary BHA is fully stabilized, giving several advantages:

1. If the well is already aligned in the desired direction, a fully stabilized BHA is less likely to go off course and generate directional work to correct the wellpath.
2. Any inadequately stabilized rotary assembly will drill a slightly spiral hole. This leads to high drags, reduced formation stability, and problems running logs and casing.
3. As WOB is applied, the drill collars will tend to buckle. If they buckle sufficiently, they may contact the wall, causing stick-slip oscillations. Whether they may buckle under dynamic loading can be calculated, but the significant factors are the length of unsupported collars (i.e., between stabilizers), the mechanical stiffness of the collars (related to OD, ID, and material), and the compressive force exerted on the column. Buckling can also lead to fatigue failure and twistoffs. The BHA should be designed with stabilization that will avoid dynamic buckling under the maximum planned WOB.
4. If the hole is enlarged, a fully stabilized BHA will still tend to buckle less than an unstabilized one. The situation is improved if larger drill collars are run above the bit (in the area under greatest compression); for instance, if 9½ in drill collars can be run instead of 8 in collars in a 12¼ in hole.

For directional work, the bit will have to work at the parameters imposed by that BHA. For instance, a downhole motor will dictate ranges of possible RPM, WOB, and flow rates. If a drop in hole angle is required using a rotary BHA, then light weights will have to be used which may make it advantageous to run a softer formation (or PDC) bit than might otherwise be the case.

If the assembly is not fully stabilized, then make a list of the available ranges of parameters that the driller can apply to the bit. Next, check to see if the "ideal" bit choice can work within those parameters. If not, the choice should be revised.

2.4.8. Drilling Program: Bit Selection and Drilling Parameters

Having arrived at a carefully considered position on which bit is likely to be best for each formation, the following information should be included in the drilling program:

1. Recommended bits for each hole size, showing in each case the best offset bit and why the recommended bit differs (if it does)
2. Anticipated BHAs for each part of the well
3. Recommended ranges of drilling parameters for each bit
4. Any other information directly relevant to selecting or running the bits
5. Expected performance of each recommended bit (e.g., footage to drill and average ROP)

2.4.9. References for Drillbit Selection

Atkin, et al. "New Nozzle Hydraulics Increase ROP for PDC and Rock Bits," *SPE* 37578. (Refers to the Vortexx nozzle design.)

Brandon, B.D., et al, "Development of a New IADC Fixed Cutter Drill Bit Classification System," *SPE/IADC* 23940. Presented at the IADC/SPE Drilling Conference, New Orleans, February 18-21, 1992.

Gault, et al. "PDC Applications in the Gulf of Mexico with Water Based Drilling Fluids," *SPE* 15614, June 1988.

McGehee, D.Y. et al. "The IADC Roller Bit Classification System," *SPE/IADC* 23937. Presented at the IADC/SPE Drilling Conference, New Orleans, February 18-21, 1992.

Veenhuizen, et al. "Ultra-High Pressure Jet Assist of Mechanical Drilling," *SPE* 37579.

Wells, M.R. (Amoco), and R.C. Pessier (Hughes Christensen). "Asymmetric Nozzle Sizing Increases ROP," *SPE* 25738. Presented at the SPE/IADC Drilling Conference, Amsterdam, February 23-25, 1993.

Drilling Fluids Program

(Most of this section contributed by Stuart Smith, drilling fluids consultant)

Mud has a great influence on the stability, safety, cost, and production potential of the well. Mud properties must be engineered to minimize potential hole problems (for instance, as seen on offset wells) and meet the other objectives.

Casing points are often designed to separate pressure regimes. Several potential problems can occur in the same hole interval, which may call for conflicting mud properties. An example might be the tertiary section of a northern North Sea well. The 17½ in interval may comprise the complete tertiary sequence with a pressure transition zone in the Paleocene, which requires the setting of a 13⅜ in casing. The 17½ in section can comprise young, "gumbo" type, reactive shales which overlie the mature, brittle shales of the lower tertiary. The optimum water-based fluid to deal with the reactive shale may not stabilize the lower shales.

A mud program should outline the concise approach for each hole interval to address problems. It should define the range in which individual relevant fluid parameters will be controlled and explain how the fluid will be maintained within this range.

The daily mud report should include all those properties that were defined in the mud program. This allows you to ensure that the properties are maintained as specified. You also need to ensure that performance indicators are reported, such as low gravity solids levels, sand content, etc. The following topics look at some basic principles and then go on to examine various mud systems from a well planning viewpoint.

2.5.1. Reaction of Clays to Water: General Principles

When a well penetrates a rock, the equilibrium that has developed over geological time can be immediately disturbed. With clays, the dominant effects of diagenesis are compaction of clay structure and dehydration. Some clays will hydrate and expand when they are penetrated and supplied with water (from the drilling mud). A process that has taken millions of years can, in some cases, be reversed in less than an hour.

The essence of drilling clay formations is to provide a drilling mud that will inhibit the tendency of the clays to reverse the process of diagenesis. The potential reactivity of a clay formation will depend on the types of clays present and the physical environment. Some clays are more likely to hydrate and expand than others.

One highly reactive clay mineral in the presence of supplied water is montmorillonite. The montmorillonite crystal structure comprises large, flat sheets of alternating octahedral and tetrahedral layers. For this reason, it is described as a "mixed layer" clay. Other types of mixed layer clay also occur.

Water that is allowed to enter the crystal structure can cause the crystal lattice to expand because of changes in electrostatic forces. This expansion is described as dispersion. The more polar in nature water can become, the more it will cause clay dispersion. The polar nature of water can be increased by the addition of alkalis, such as the monovalent bases, sodium hydroxide, or potassium hydroxide. The theoretical surface area of fully dispersed montmorillonite is around 800 m^2/g. Fully dispersed, a clay such as montmorillonite will have its clay platelets completely separated and held apart by negative charges on the faces of the platelets.

Mechanical forces also affect clay formation stability. For instance, because the well is exposed to surge/swab pressures while tripping, the clays can become unstable much faster than they would otherwise. Starting and stopping the mud pumps too fast can exert destabilizing forces. Good drilling practices help minimize or avoid such problems. This is covered in more detail in Section 3.3.7.

2.5.2. Dispersion and Flocculation of Clays in Water

The ability of montmorillonite to expand due to hydration is exploited in muds. When expanded, it provides a large ionic charge and in a polar medium such as water, ionic forces will cause the clay platelets to repel each other. The resultant slurry will have developed viscosity. As it is commonly known, bentonite (montmorillonite) is widely used as a viscosifier in certain fluids. The platelet nature of the clay also produces filtration control.

In some cases, clay platelet dispersion is disturbed by an ionic charge imbalance at certain parts of the clay structure. The clay platelets will not evenly separate from each other but they may have some random and loose attachment. The edge of one clay platelet may attach itself to the edge or the middle of another clay platelet. A clay in such a condition is said to be flocculated. The presence of divalent cations, metal ions such as calcium or magnesium, or anions such as sulfate, chloride, or carbonate/bicarbonate can flocculate a clay. These ions can come from the formation drilled, the makeup water (especially if it is hard water or seawater), or the addition of a base such as calcium hydroxide or magnesium oxide.

If a clay has not been dispersed in water but is put in such a water environment where it would have flocculated if it were dispersed, the clay remains in a nonexpanded or aggregated state. This is one method of stabilizing formation clays by providing such an environment in the mud. This would not be conducive to dispersing bentonite in order to obtain viscosity. Consequently, such an environment is avoided when mixing bentonite by treating out ions that would cause flocculation, e.g., calcium.

Various chemicals can negate the imbalance of ionic charges that cause flocculation. These are known as deflocculants. The more powerful deflocculants will put the clay in such a condition that it will fully disperse. Hence these deflocculants are sometimes described as dispersing agents.

Not all dispersants are deflocculants. Sodium hydroxide in water is a powerful dispersant but has no deflocculating effect. The various lignosulfonates and lignites can be strong deflocculants that result in full dispersion. Examples of other deflocculants are some polyacrylates.

2.5.3. Mud Types Available

There are three mediums in use today as makeup fluids for drilling mud: water, oils (hydrocarbon liquids), and air/gas.

The most commonly used medium is water and the resultant drilling mud may be described as water-based mud. There is a substantial variety of water-based fluids that may be used in the drilling industry and these types may be classified in many different ways. One common classification of water-based mud types is to divide them into two groups, dependent on the state of the clay that is present in the mud (see the previous topic). Thus, a drilling mud can be described as either dispersed or nondispersed (see the subsequent topics). Oil, air, and miscellaneous systems will be discussed later.

There are a substantial variety of mud types and special applications available in drilling. It is beyond the scope of this book to cover them all. However, outlined below are some mud types that have been (or are being) commonly used. Although they have been classified on the basis of dispersed and nondispersed systems, attention is drawn to the varying ways in which inhibition of shales is achieved; e.g., potassium systems, calcium systems, silicate systems, etc. The mechanism of inhibition varies with each. The type of mechanism to use can best be selected on the basis of the type of shale being drilled. In many cases, the idea that there are closely defined mud systems (particularly with water-based muds) should be dispelled. As can be seen from some of the examples shown below, one type of mud blends into another type of mud in the same way that one type of shale might vary into another.

2.5.4. Dispersed Water-Based Muds

An uninhibited dispersed mud is where any clay present has been allowed to hydrate and expand (disperse), or has had a chemical such as a lignosulfonate added to it which has defocculated it and ultimately allowed it to disperse. Fully dispersed, a clay may be at its most expanded condition and it will not be able to impart any more viscosity than it already has to the mud. This can give some temporary stability to the mud in respect of its rheology, but the resultant mud may be capable of causing further dispersion of formation clays. As such, it would be considered to have very little inhibition to clay dispersion.

Some degree of inhibition can be provided to a dispersed mud. The addition of some form of calcium (from lime or gypsum for example) to a fluid whose makeup clays has been deflocculated and dispersed will cause drilled clays to remain in the aggregated condition or, at the least, limit their dispersion. Lime or gypsum muds are examples of such inhibited dispersed systems. Nondispersed variants of lime or gypsum muds now exist but the amount of clay in them is minimized.

A characteristic difference between a dispersed and a nondispersed mud can be seen in their typical rheologies. A dispersed system will usually have a low yield point value and a high value for plastic viscosity. The plastic viscosities of dispersed muds will be higher than those of nondispersed muds. Usually improved drilling performance, better hole conditions, and less formation impairment can be achieved if a nondispersed mud can be used (as compared with a dispersed system). In the early days, in many cases, only dispersed systems could be used. It was the improvement in efficiency of rig solids control equipment that allowed nondispersed systems to be successful. Following are some dispersed mud systems.

Spud mud. Often in top hole drilling, a fluid based upon bentonite in water is used to provide viscosity and filter cake. This fluid is often described as "spud mud." Such mud would be described as dispersed mud when the bentonite was in a dispersed condition. It would become flocculated mud when it was mixed in with seawater or had lime (calcium hydroxide) added to it. The flocculation of a prehydrated bentonite (dispersed in fresh water) with lime, in the case of spud mud, is done on purpose to obtain additional viscosity and gel strengths. The gel structure of flocculated mud is invariably stronger than in the dispersed state.

Caustic lignosulfonate mud. This is a very simple form of dispersed mud. It will cope with varying quantities of clay present in the mud. However, in using a caustic lignosulfonate mud those varying quantities of clay will tend to increase. That is because this type of mud is a highly dispersive mud and has no inhibitive properties. This mud can cause substantial formation damage if there are clay minerals present in the reservoir sand. The problem for the formation is that the lignosulfonate does not know when to stop its dispersive action. If it is necessary to run a dispersed mud, it is always better to use one (such as a gyp mud) that has some inhibition.

Normally lignosulfonate is added until the rheological properties and gels are brought to the required levels. Usually 2 to 3 ppb of lignosulfonate is used. The pH of a caustic lignosulfonate is normally maintained above 9.3 or whatever level is required for the lignosulfonate in use to solubilize. Filtration control is normally obtained with CMC or starch. The use of any other polymer for filtration control such as PAC is a waste of money since any special benefits that might be seen in a nondispersed system will be completely suppressed in a caustic lignosulfonate mud. Caustic lignosulfonate muds are only relevant for nonreactive shales and sands where formation damage is not important. In such formations, it may be used where inferior rig solids control equipment may cause a solids buildup. Quite often a nondispersed polymer system that fails will end up, at least temporarily, as a caustic lignosulfonate mud.

Gypsum lignosulfonate mud. Originally, gypsum or gyp lignosulfonate muds were developed from other bentonite-based systems for drilling through gypsum ($CaSO_4.2H_2O$) and anhydrite ($CaSO_4$) zones. Substantial contamination of the bentonite muds was experienced and they became highly flocculated. It was decided to contaminate these muds on surface with gypsum to stabilize their properties. In particular, they had to be treated with a high concentration of lignosulfonate to ensure that they remained deflocculated and that their rheology remained the same. A correctly formulated gyp mud would therefore suffer no further contamination and deterioration of properties when a gypsum or anhydrite sequence was drilled.

The use of gyp lignosulfonate mud has been taken a lot further than its original role in drilling gypsum and anhydrite formations (e.g., the Guadelupe Mountains of the United States). Because of the inhibitive properties of the presence of substantial filtrate calcium, it can stabilize moderately reactive shales. Of all the available dispersed mud systems, gyp lignosulfonate is probably the most inhibitive. The mechanism of inhibition is that the divalent calcium ion attempts to maintain drilled clays in an aggregated state or at least suppress clay dispersion. This is similar to attempting to disperse bentonite in hard or brackish water.

Gyp lignosulfonate muds would not provide the stability to very reactive shales as would, for example, a potassium chloride nondispersed polymer system (which uses a different mechanism of inhibition). However, gypsum lignosulfonate systems can still be relevant against some reactive shales. Engineering a gyp lignosulfonate mud is similar to engineering an invert emulsion oil-based mud. Every component of the mud has its particular function. The concentration range of each component chemical is closely defined and, if correctly maintained, can result in a very stable mud. Each chemical interacts with each other.

The basis of gyp mud is 15 to 18 ppb of bentonite prehydrated and fully dispersed in fresh water; 6 ppb of gypsum is added to the system and 5 to 6 ppb of lignosulfonate deflocculant with caustic soda is added to counter the flocculating effect of the gypsum on the bentonite. In adding 6 ppb of gypsum, it is intended that 2 ppb of gypsum will solubilize in the mud and 4 ppb will remain in suspension as a "reservoir" for the supply of calcium ions. The pH will control how much gypsum will go into solution and how much will remain in excess (in reserve). As mentioned above, the means of inhibition is the presence of soluble calcium ion and the objective is to get as much into solution as possible. At a pH range of 9.5 to 10.0, it should be possible to achieve a filtrate calcium concentration of 1200 to 1500 ppm. The higher the pH, the less soluble calcium will be present.

The lower pH range for gyp mud is defined as that which is necessary to allow lignosulfonate to solubilize. Normally, this is around a pH of 9.5. If you used another deflocculant that would be active at a lower pH and you used a lower pH, more calcium would be in solution. At the pH range, however, of 9.5 to 10.0, 2 ppb of gypsum will be solubilized and 4 ppb will be excess gypsum.

The amount of caustic soda (sodium hydroxide) that has to be added to a gyp lignosulfonate mud to obtain a pH of 9.5 is quite substantial (usually around 3.5 ppb). This is because of the acidic nature of lignosulfonate and, to a lesser extent, gypsum. Filtration control in gypsum mud is normally achieved with the addition of 5 ppb of CMC

LV, but starch can also be used. There is little point in using a filtration control agent such as PAC (which is more expensive than CMC). In an environment of 5 ppb of lignosulfonate, any encapsulating effect from PAC can be forgotten.

Gypsum muds are stable up to 250°F depending on the polymer used for filtration control. Always add new volume to the system by whole mud prepared to the specific formulation. Gypsum muds have also been formulated in the nondispersed condition.

Lime lignosulfonate mud. The principles of low lime lignosulfonate mud are very similar to that of gypsum lignosulfonate mud. The means of inhibition is the presence of filtrate calcium. In this case, the filtrate calcium is derived from the addition of calcium hydroxide.

The pH of low lime lignosulfonate mud would normally be in the range of 11.5 to 12.0. Excess lime is normally 1 to 2.5 ppb. Care should be taken in breaking over to a lime mud because a viscosity hump will occur. The order in which you add the chemicals is very important.

A high lime variant of a lime lignosulfonate mud exists. The excess lime in such a case would be controlled in the range of 5 to 15 ppb. Low lime lignosulfonate muds can withstand substantial carbonate contamination. Low lime lignosulfonate muds are not commonly used these days but some mud companies have their propriety variants. They have been supplanted by lime polymer systems or gypsum lignosulfonate muds.

2.5.5. Nondispersed or Polymer Water-Based Muds

A nondispersed mud is where the hydration and dispersion of a drilled clay is minimized. There are a number of ways to achieve this. The most common is to limit the amount of water that reacts with the clay by encapsulating the clay with polymer as quickly as possible to prevent further access of water to the clay. Such mud systems are described as encapsulating polymer muds.

Nondispersed polymer systems in the drilling industry are mostly based upon anionic and nonionic polymers. There are also some systems that have been developed which are based upon cationic and nonionic polymers. A nondispersed system should have sufficient anionic polymer present for encapsulation of clays and other low gravity solids and their removal by the solids control system.

If the amount of clay in the mud, the cation exchange capacity (MBT), or the amount of low gravity solids are not controlled, the rheology of the fluid can become uncontrollable in the nondispersed state. The polymer makeup should be able to provide encapsulation and good low shear rate rheology but, most importantly, there should be sufficient encapsulating polymer in the system at all times. If there is not, the MBT will go out of control first and this will be followed by a lack of control of the rheology. It should be remembered that the encapsulating polymer concentration will deplete all the time in the system as it is removed with the clays over the shakers.

In some cases, the use of nondispersed polymer muds is attempted and fails. The muds fail because the low gravity solids are not kept under control. This may be due to poor solids control systems. They may fail because insufficient encapsulating polymer is maintained in the system or because filtrate chemistry is not correctly controlled. Failure of a nondispersed polymer system will invariably result in an inability to control viscosity. The resultant action is to disperse the system. However, when the system is dispersed it usually, and at least temporarily, acts as a dispersed mud without any substantial inhibition. In planning a mud program, it is better to evaluate the ability to maintain a nondispersed polymer system. If the conclusion is that this cannot be successfully achieved, it is better to consider the application of a gyp lignosulfonate mud.

Types of polymer. A polymer consists of molecules (monomers) that have been processed to be chemically joined (polymerized) to form a chemical chain (polymer). The monomers are usually organic (based on carbon chemistry), but they may also be based on the chemistry of other atoms, e.g., silicon or phosphorous.

Polymers may be specified in a number of ways. Their molecular weight will give an idea of how long their polymer chains are. Their degree of polymerization (or number of repeated units) will also demonstrate this. Their degree of substitution (of charged side chains) will give an idea of the net charge on the polymer and their charge density (charge per mole) will also show this.

Polymers can be classified according to the ionic charge they carry. There are three main types of polymers in this classification:

- *Anionic polymers.* These carry a net anionic (negative) charge
- *Cationic polymers.* These carry a net cationic (positive) charge
- *Nonionic polymers.* These carry no net charge

There is a fourth group known as amphoteric polymers. The charge nature will vary depending on the environment in which they exist. They are not commonly used in drilling muds as "dry powder" polymers, but they may be present as surfactants.

Polymers can also be classified according to their origin and manufacturer. First, there are natural polymers. These are derived from vegetable material and they are used in their extracted form without any alteration. Guar gum would be an example of a natural polymer. The guar polymer is extracted from the soybean. Cornstarch or raw potato starch would be another example. Second, there is the class of polymers known as semisynthetic polymers. These are natural polymers, often derivatives of cellulose, which have been altered by chemical groups substituted onto their molecules or processed to undergo further polymerization.

Finally, there is the group known as synthetic polymers. These are polymers that have been synthesized from chemical derivatives such as acrylic acids, acrylates, various acetates, etc.

Polymer usage. Polymers may be classified by their usage. Listed below are some uses of polymers in drilling muds.

Flocculants. High molecular weight synthetic polymers whose molecules are long enough to cause clays and ultra fine drilled solids to be joined together.

Deflocculants (or thinners). These negatively charged polymers work by adsorbing onto clays to negate positive charges and provide a net negative charge to the clay platelet. Examples are lignosulfonates, lignins, polyphosphates, and low molecular weight polyacrylates.

Viscosifiers. Normally these are high molecular weight polymers that in water do not "unravel" very well because their polymer chains are so big. In water, there is a strong interaction between the polymer chains that demonstrates itself as viscosity. Shear decreases these polymer interactions and thus a polymer can provide a viscosity that can be described as pseudo-plastic or shear thinning. The best known polymer viscosifier is xanthan gum, which will provide substantial viscosity at low shear but is shear thinning. Other examples are various polyacrylamides, guars and semi-synthetic guars, polyanionic, or hydroxy ethyl celluloses (PAC and HEC).

Xanthan gums are exceedingly long molecules with molecular weights in many millions. The digestion of cellulose by the bacteria, xanthomonas campestris, synthesizes them. They are also commonly known as XC polymers, after the name of the bacteria.

Filtration control additives. These act in a number of ways. They may impart filtration control by improving the nature of the filter cake of mud by their deflocculating action. They may increase the viscosity of the filtrate and make it more difficult for it to pass through the filter cake. They may also provide colloidal-size particles, which will bridge off in the filter cake if the polymers are not completely water-soluble. Examples are carboxy methyl cellulose (CMC), which in their low viscosity forms will provide a deflocculating effect. High viscosity forms will provide viscosity to the filtrate. Some starches will provide colloidal particles. Examples of other filtration polymers are asphalts, lignites, resins, and some polyacrylates

Other uses of polymers are as surfactants, shale stabilizers such as glycols, emulsifiers, de-emulsifiers, foaming agents, lubricants, and corrosion inhibitors.

Anionic polymers: PHPA and PAC. Two types of anionic polymers are available to provide encapsulation. These are partially hydrolyzed polyacrylamide (PHPA) and polyanionic cellulose (PAC).

PHPA provides far more anionic charge than PAC because its molecular weight is much greater and its charge density is a lot higher. If encapsulation is required, PHPA is the best polymer to use if the chemistry of the drilling mud allows it to be used. It is only stable at a pH maximum of 10.0 and a filtrate calcium of 300 to 400 ppm. Sometimes it cannot be used, such as where the mud needs a high pH and the presence of high-filtrate calcium. An example would be if substantial and persistent CO_2 contamination was experienced that would require a high pH and the addition of some form of calcium. However, encapsulation can also be achieved by using PAC, but the concentration of PAC required will be a lot higher than that of PHPA. Normally with PHPA, 1 ppb is kept in excess in the system, which would necessitate a makeup concentration of 1.2 to 1.3 ppb. To get some amount of satisfactory encapsulation with PAC, a 2.5 ppb minimum excess would be required.

PAC is available in various forms dependent on the amount of viscosity that it will impart. PAC polymers that impart viscosity are longer molecules than those that provide less viscosity. PHPA, which provides some viscosity, will provide more anionic charge than PAC, which gives less viscosity. This is because the degree of polymerization or molecular weight is different and the longer molecules carry more ionic charge. It should always be remembered in using PACs that they can break down at about 1200 ppm calcium with a pH above 9.5, or at higher levels of calcium if the pH is more neutral. PHPA will provide encapsulation and viscosity to a fluid but not filtration control. The amount of viscosity that it will provide will be very substantial at a concentration of around 0.5 ppb. This is called a "viscosity hump." The level of PHPA in the mud should not be run at this concentration. Some drilling fluid companies have some specialty variants of PHPA, and you should consult with them on the specifications for their performance and stability.

PAC will provide encapsulation and filtration control to the mud. It will also provide viscosity, dependent on the type of PAC.

Cationic polymer systems. The majority of polymer systems that are commonly used are based upon anionic polymers and nonionic polymers. There are, however, systems available that are based upon cationic and nonionic polymers. Some of the cationic polymers are cationic polyacrylamides. These systems are particularly proprietary to the drilling fluids companies that have developed them. Consequently, reference should be made directly to these companies. However, some general comments can be made.

Cationic polymers adsorb onto clays by attachment to anionic sites on the clay structure. This electrostatic bonding is far more powerful than the Van der Waal forces that cause an anionic polymer to adsorb onto clay. The question arises as to whether there is any benefit from this stronger form of bonding. Properly engineered anionic systems have stabilized many extremely reactive shales and have demonstrated that the form of adsorption was sufficient.

Cationic systems are more expensive than anionic systems for two reasons. First, cationic polymers are overall more expensive to manufacture and second, the polymers are proprietary. There is a reason as to why they should remain proprietary. Major drilling fluids companies have completed substantial research in the applications of polymers of all types. In the case of the anionic polymers, these have become more or less commodity products. The mud company does not always recover the benefit of its research. The field of cationic systems allows, at the present, a mud company to market its (cationic) polymers as proprietary products.

Cationic polymer systems have mostly been applied for clay stabilization. They do have the benefit of being more stable in make up waters with high concentrations of divalent cations being present. However, the cases where such high concentrations of cations are encountered are not in the drilling of reactive shale but more in respect of evaporite sequences. It is the case that many of these evaporite sequences have been successfully drilled using mixed salt systems with nonionic polymers.

Engineering of a polymer system. In putting a polymer system together, the viscosity provided by the polymers has to be reconciled. XC polymer may be required to provide low shear rate rheology, but there could not be enough room for it in respect of the apparent viscosity of the total fluid if the polymer selection is not optimized. PHPA will provide viscosity in itself. It also depends on which mud company is used because the PHPAs will vary with company.

Potassium chloride is often used in conjunction with encapsulating polymer systems because it provides clay stabilization by base ion exchange. KCl or any other inhibiting electrolyte such as NaCl is beneficial to the performance of an encapsulating polymer system because it affects the adsorption characteristics of PHPA. A clay when it is "free" in water is not immediately encapsulated by PHPA. It will tend to expand all the time until the encapsulation is complete. Expansion causes increase of surface area to be encapsulated. Potassium ions will slow the rate of expansion.

KCl will also inhibit the development of viscosity provided by the PHPA. It will minimize the extent to which the molecule will "unravel" in water and thus provide a higher charge density per surface area of polymer. For polymer encapsulation, the best results are seen when the polymer has the highest charge per surface area and is adsorbing onto a clay whose surface area has been minimized.

The performance of the polymer system must be continuously monitored and any potential problem must be addressed proactively. Emphasis must be put on looking at all the parameters and not on any one alone.

The indicators for performance of the system will be:

1. The buildup rate of the cation exchange capacity. It is necessary to keep the CEC below 10. The figure of 15 ppb equivalent clay is an absolute maximum because with such a CEC, the rheology will be difficult to control (and the hole could be washed out). The build rate of the CEC must be addressed long before the CEC ever gets to 15.
2. The condition of the cuttings at the shaker.
3. If PHPA is used, the concentration of PHPA can be measured. It could show trends but could not accurately provide a definitive concentration. It is a difficult test to do as it involves measuring the rate of ammonia generation in a gas train. It is easier just to make sure all polymer additions are by measured premix.
4. The buildup of low gravity solids.
5. The lagged in and out rheologies in the lower shear rates.
6. The pH of the fluid would normally be run at around 9.0 to 9.5.

Some considerations to make in respect of selection of the best pH to run would be as follows. A pH of 9.3 will have removed most of the magnesium ion in the system if seawater is used as the makeup fluid. The pH must be run within a range that will allow for stability of the polymers present in the mud. Most water dispersible polymers disperse best in a pH of 9.0 to 9.5. Minimum dispersion of clays occurs at a neutral pH (when water has its least polar nature), but the KCl and the polymer, if run properly, will minimize dispersion at a higher pH.

The hardness must be controlled below 200 ppm if PHPA is in the system. If PHPA is not in the system, the hardness will be controlled below 500 ppm (a safe level for PAC) since there is no need to keep it low unless it needs to protect the PHPA. Allowing some calcium ion to be present in the filtrate is a good protection against carbonate contamination.

Gel CMC mud. In a way, this is not a true nondispersed polymer mud, as any clay present would tend to become dispersed due to the effect of pH. However, this type of mud is normally used for drilling top and intermediate hole sections containing long sand intervals where there is not much clay present. The formation will not contribute much clay to the mud.

In drilling these kinds of sand formations, there is not a lot of reactivity with the formation. The mud should provide viscosity and filtration control. Buildup of thick filter cake should be avoided. The commonest problems in drilling these formations are a tendency for solids buildup, substantial losses over the solids control system and, if the mud weight is not controlled, there is a potential for downhole losses. Substantial dilution is normally required.

A gel CMC mud should meet the requirements for viscosity and filtration control but at the same time be relatively inexpensive in view of the high dilution rate required. Normally, a gel CMC mud will contain at least 8 ppb of prehydrated bentonite in order to provide some viscosity and a filter cake for the CMC to act upon. CMC is used for filtration control and usually, as viscosity is required, it will be in the form of high viscosity or extra viscosity CMC. Additional viscosity, if required, can be obtained from xanthan gum.

The usage of anionic polymers to provide encapsulation is unnecessary, but PAC might be used to give viscosity and filtration control dependent on its price against the high viscosity forms of CMC.

Gypsum polymer mud. The principle behind this type of polymer mud is to create inhibition by providing an environment with as high a concentration of calcium ion as possible. See also "Gypsum lignosulfonate mud" in Section 2.5.4, "Dispersed Water-Based Muds."

Gypsum is the source of the calcium ion but, unlike a gypsum lignosulfonate mud, the pH is maintained at a neutral level. This allows substantial calcium to be soluble in the system (in excess of 2500 ppm calcium ion). PHPA cannot be used in this mud but PAC can. The stability of PAC to calcium ion is greatly increased due to the reduced pH.

Lime polymer mud. The principle means of inhibition of this system is the presence of soluble calcium and the use of the anionic polymer, PAC. This principle was discussed previously in "Gypsum polymer mud" and Gypsum lignosulfonate mud."

Lime (calcium hydroxide) will normally come to a base level at a pH of 12.4 when added to water. However, if a monovalent base such as sodium hydroxide is added to the calcium hydroxide, it is possible to control the amount of calcium ion that will go into the solution. For example, at a pH of 10.5, it will be possible to control the filtrate calcium at around 400-500 ppm depending on how much lime is held in excess (not in solution). Lime polymer muds are normally run at a pH of 10.3-11.0. PAC is the only encapsulating polymer that can be used. PHPA would break down in a lime polymer system due to the pH and the filtrate calcium that is present.

A lime polymer mud would not provide the inhibitive properties that a potassium chloride polymer system would against a highly reactive shale. It would be capable of stabilizing most moderately reactive shales. A lime polymer mud can be prepared from drillwater, hard water, or seawater.

The relevant principles previously described in this section will apply. It will be necessary to ensure a sufficient quantity of PAC polymer is kept in the system at all times. This may be in the form of PAC regular (which will provide viscosity) or a low viscosity PAC. Additional filtration control should be obtained from stabilized starch. If additional viscosity is required (to that provided by PAC regular), use xanthan gum or wellan gum.

The MBT should be controlled in the range of 6 to 12 ppb clay equivalents (15 ppb maximum). The lower range of MBT is to provide some filter cake; the upper range is where viscosity problems would occur in the nondispersed condition. It is unlikely that it will be necessary to add any prehydrated bentonite to the mud unless the MBT dropped below 6 ppb clay equivalent. This would only be likely after drilling a long clay-free section (e.g., a long chalk section).

A lime polymer mud would be a relevant mud to use against a moderately reactive shale that would be drilled with an aerated mud and require a high pH. It would be applicable for moderately reactive shales where a persistent carbonate contamination problem existed.

Mixed salt systems. Mixed salt mud is designed for drilling through salt sequences that include the most soluble ones, such as the Zechstein formation encountered in the North Sea and northwest Europe. Evaporite sequences occur as a result of seawater evaporating, leaving the soluble salts behind. There is a definite order of precipitation as the least soluble salts come out of solution first. If a saltwater lake evaporated without further influx of saltwater, the order of precipitation would be calcium carbonate ($CaCO_3$), dolomite ($CaMg[CO_3]_2$), gypsum ($CaSO_4.2H_2O$) (which is converted to anhydrite [CaSO$_4$] as heat and pressure remove the associated water molecules), halite (NaCl), and, finally, various rare potassium and magnesium salts. These last ones are very soluble and would only precipitate out if dehydration was almost complete. Any fresh influx of water would tend to dissolve these last rare salts again and a fresh cycle of precipitation could then occur, but these rare salts may not therefore be in the first sequence. The Zechstein, for instance, consists of four evaporation cycles; three of which contain the most soluble potassium and magnesium salts, which often leads to problems associated with large washouts.

Evaporites consisting only of the salts up to NaCl in the sequence can be drilled with common salt saturated mud. The more complex sequences present several problems, which are not solved with a conventional salt (NaCl) saturated mud. The highly soluble magnesium and potassium salts will dissolve in a sodium chloride saturated solution. This can lead to very large washouts with attendant problems of lost directional control, difficult cementing, keyseating, and difficult or impossible fishing operations if a fish gets lodged across a large washout.

As the solubility of salts increases with temperature, a solution that is saturated at surface or flowline temperature will not be saturated at bottom hole temperatures. Therefore, an excess of salt is required to be held in suspension at surface in order that the solution will still be saturated downhole. Potassium and sodium salts precipitate as fine crystals that are small enough to pass through the shale shakers and thus remain in the system. Magnesium salts, however, form large crystals that would be removed at the shakers; so to maintain a saturated magnesium system, you would need to continually add magnesium, which would be very expensive. Solubility of magnesium decreases as Na and K concentration increases and so it will usually be possible to inhibit washouts in Mg salts with moderate Mg concentration and high Na and K saturations. The actual concentrations needed for a particular hole section will be dependent on the salts present and the downhole temperature. Use desanders and desilters only as much as is necessary, otherwise large amounts of the fine salt crystals will be removed and have to be replaced.

Because of the substantial concentration of divalent cations that are present in a mixed salt mud system, only nonionic polymers should be used to control viscosity or fluid loss. The MBT should also be minimized. See the discussion on polymers in "Saturated salt polymer mud" following in this subsection. Yield point should be kept above 20 to keep the excess salt crystals in suspension. Since salts are impermeable unless fractured, fluid loss is not important as far as the evaporites go. However, salts are usually associated with carbonates that may be highly permeable, calling for an API fluid loss below 10 cc. The minimum practical density of a mixed salt system is about 0.57 psi/ft (11 ppg) and it can be weighted up with barite as normal.

Potassium chloride polymer mud. This type of mud system is one of the most commonly used throughout the world (where it is applicable and consequently, it is felt that it warrants more detail in discussion than other systems mentioned). The basis of this mud system is normally the anionic encapsulating polymer fluid. Potassium chloride is added to the mud to provide a source of potassium ions. The potassium ion will assist in stabilizing reactive clays—particularly mixed layer clays.

The mechanism of inhibition is described as base ion exchange. The potassium ion enters the clay structure and will replace calcium, magnesium, or sodium ions present in the clay crystal. Since the ionic radius of the potassium ion is smaller than the other ions, there will be a tendency for the clay structure to contract. Hydration and dispersion of a clay will tend to cause expansion of the clay structure. The tendency for contraction caused by base ion exchange by the potassium ion can be seen to negate the expansion effect that might be induced by the hydration of the clay.

The required amount of available potassium ion in the mud will depend on the types of clays or shales being drilled. Nonreactive shales that do not contain high proportions of mixed layer clays will not require a lot of potassium ion to be present. A potassium chloride concentration of 3% would normally be sufficient. However, a shale containing a substantial amount of mixed layer clay (this would be a gumbo type shale) would require a substantial amount of potassium ion to be available. In this case, as much as 10 to 15% or more potassium chloride may need to be added to the mud.

If seawater is used as the makeup fluid, more potassium chloride will be required than if drill water is used. This is because the potassium ion would be required to be present in such quantities as to compete with the sodium ion at exchange sites on the clay structure. When drilling a reactive shale, it should be remembered that there would be a tendency for the potassium ion to deplete in the system. This is because of its removal in the clay after it has taken up its "position" in the clay. It is possible in well planning to use information that might be derived from analysis of shale samples to determine how much potassium chloride must be added to the mud to satisfy the shales demand for potassium ion. The ion cation exchange capacity (CEC), or what is often called the shale factor of the shale sample, will give an idea of the amount of mixed layer clay present. An XRD analysis that gives a quantitative mineralogical composition of the shale is also of benefit. Sometimes such shale analysis is not available and it is necessary to proceed by only using the symptoms seen in drilling such a shale with a less inhibitive mud (e.g., mud rings, bit balling, hole washout).

Potassium chloride is not the only available source of potassium ions. Potassium polymer muds have also been formulated from other potassium salts such as potassium carbonate, potassium citrate, potassium acetate, and potassium carbonate. Each potassium salt has its own specific chemistry, which will allow or not allow various chemicals to be used or various parameters to be run. In using these other potassium salts, it is necessary to study the chemistry of the specific salt.

As mentioned earlier, the concentration of potassium ion in the mud should vary according to the clay mineralogy of the shale being drilled. Sometimes excessive amounts of potassium chloride in a system can destabilize a shale that is not particularly reactive. This type of shale will have very little swelling clay such as montmorillonite present. It will contain substantial illite and kaolinite (a clay with a very small clay structure), probably various micas and other minerals. The potassium ion can cause this type of shale to slough because it causes the kaolinite to disperse, and if any exchange of ions is made on the illite, the net effect would be contraction of the structure.

Drilling ahead with an excessive concentration of potassium ion for the shale being drilled will normally show no problems. However, two or three days later the shale may begin to "come in on you." Excessive potassium ion can be one cause of time destabilization of shale (but not the only cause). Examples of this destabilization are seen in the tertiary sequence of the North Sea. The lower shales in the Eocene and Paleocene can destabilize with time against the high concentration of potassium ion in the mud that is required to avoid "gumbo problems" at a shallower depth against the Pliocene. These problems (of "gumbo" and a time-destabilized shale in the same interval) could be avoided by using oil-based mud. Prior to that when using potassium chloride polymer muds, the most successful operators were those that "slammed" their 13³⁄₈ in casing into the ground when reaching TD on the transition zone in the Paleocene.

Other examples have been seen in the Middle East. In the Malay basin, better calipers and longer "time envelopes" have been achieved by controlling the potassium ion concentration to meet the needs of the shales. In some cases, the concentration of potassium chloride used has been half that previously employed.

If the clay mineralogy of the shale is available, this can be very helpful in selecting potassium ion levels. If this is not known, then the best approach is to start at a predetermined level and increase as the cuttings condition dictates (but also ensure that there is an excess of at least 1 ppb of PHPA in the mud because this can be another reason why the cuttings are soft and sticky). With subsequent wells, depletion levels of potassium chloride can be studied.

If the clays in the shale require a certain concentration of KCl, the clays will take potassium ion from the mud. To keep a steady concentration of potassium chloride in the mud, more KCl would have to be added than would be calculated for the KCl concentration being run (and adjusted for new volume). The depletion rate for KCl should confirm that the correct KCl concentration is being run.

Saturated salt polymer mud. In drilling a salt sequence, the mud downhole must be saturated in order to avoid leaching out of the salt. If the formation to be drilled comprises only halite (NaCl) and there is no excessive difference in temperature between surface and bottom hole temperature, the salt mud can be prepared by saturating it with sodium chloride. If, however, other salt minerals are present or the mud that may be saturated at surface will be below saturation down hole, a mixed salt system should be used. The makeup of a saturated sodium chloride mud or mixed salt system is more or less the same. The presence of clay in the mud should be minimized since it will be flocculated in the presence of the salts, requiring the use of a deflocculant. In some cases, this is unavoidable if the hole section comprises a shale section above the salt. The usage of anionic polymers such as

PHPA or PAC should be avoided, since they are not compatible with any divalent cations that will be present in the salt. Viscosity should be obtained from the use of a biopolymer such as xanthan or wellan gum. Some forms of starch can also provide viscosity. Starch can be used to provide filtration control as long as the bottom hole temperature is not much above 200°F. At higher temperatures, starch would require to be stabilized or a variant of starch that is more temperature stable should be used. At temperatures above 250°F the usage of CMHEC (carboxy methyl hydroxy ethyl cellulose) or other temperature stable polymers should be considered.

There is little point in running a pH in the system much above a neutral pH unless any deflocculants or any other material used in the system requires a higher pH. Increasing pH results in increased insolubility for most salts. Any salts that come into solution from the formation are best kept in solution. For corrosion control, the use of a film forming amine may be considered.

Seawater polymer mud. If shale is not particularly reactive, it may be adequately drilled with a seawater polymer system. All the principles of an encapsulating polymer fluid as previously described in this section will hold true except that no additional inhibitive chemical such as potassium chloride would be added to the system. Inhibition would be obtained from the encapsulating polymers and the chemistry of the seawater.

Using a seawater polymer system will provide for all the advantages of a nondispersed mud, e.g., the rheology and ability to control low-gravity solids.

It will be the case that for a seawater nondispersed polymer system, there will be more inhibition if it is made up on PAC and not on PHPA. The reasoning behind this is as follows:

In a potassium chloride polymer mud, PHPA is the best available polymer because of its encapsulating ability but it has to be protected or otherwise it will break down. The hardness has to be controlled at below 200 or 300 ppm depending on the pH and the pH must be kept low and definitely a maximum of 10. However, the potassium ion is providing substantial inhibition. In using a seawater polymer system as opposed to KCL PHPA mud, the chemical inhibition from the potassium ion will not be present. Second, any inhibitive effect from calcium that comes from the seawater will not be present because the calcium has to be treated out to protect the PHPA. This would not be necessary if PAC were used instead of PHPA.

Inhibitive KCl-Glycol mud. This system is based upon the potassium chloride polymer mud, but with the addition of a polyglycol. Certain glycols can impart many of the properties of diesel and mineral oils but without the toxicity. Most glycols, unlike diesel or mineral oils, are miscible in water at normal temperatures. There are several theories as to why glycol inhibits shale hydration; with a TAME mud (thermally activated mud emulsion) the glycol solution is designed so that glycol precipitates out of the filtrate when it heats up after entering the shale, which blocks further filtrate entry. This requires the mud to exhibit a cloud point behavior, whereby at a temperature between the bottom hole circulating temperature and the (higher) formation temperature, the glycol starts to de-emulsify. Glycol systems are also formulated without cloud point behavior.

Glycol coats steel with a hydrophobic film that repels water. This helps to prevent bit and BHA balling and gives good lubricity, reducing drags and torques. If balling is suspected, a drum or two of glycol + surfactant pumped downhole may remove the balling. Glycol can also be used in noninhibited water-based systems for lubrication and prevention of balling. This makes the use of PDC bits more effective in WBM, as any balling of PDCs causes rapid heat buildup that leads to cutter damage.

Glycol improves filter cake quality, reduces filtrate loss, and reduces cake thickness while improving filter cake lubricity, therefore reducing differential sticking tendencies. There is also some evidence to show that glycol improves tolerance to drilled solids, reducing the need for dilution.

Silicate muds—general. Refer to the discussion on wellbore stability in Section 2.9.1 for background information on reaction of shales with drilling fluids.

Silicate muds provide pore pressure isolation (see Section 2.5) and prevent hydration by blocking off the pore throats of the shales while drilling. Silicates are also effective at encapsulating cuttings (assisting removal at the surface and preventing solids content buildup) and preventing dispersion (of limestones as well as shales, due to the high pH). Silicate muds are currently one of the most effective water-based fluids for drilling shales, limestones, and sands.

Groups of silicate molecules are small enough to penetrate into the tiny pores in the shale surface. In the mud, filtrate pH is kept high—

usually around 10. As the filtrate enters the shale pores, the pH drops towards neutral. Silicate molecules form gels at neutral pH and solid silicates precipitate out when reacting with polyvalent ions, which are always present in shale pore fluids. These combined gels/precipitates block the pore throats against any further invasion. In addition, this causes a semi-permeable membrane at the shale surface so that osmotic forces can be crafted to further dehydrate the shales, which causes the shale to further harden.

The mechanism also seems to work well in microfractured shales. These types of shales often go unrecognized and cavings from fractured shales are made worse with increased mud density or swab pressures, which is why the standard answers of increased mud density and wiper trips are seriously detrimental when certain types of cavings are seen.

Shale samples left exposed to silicate formulations actually show an increase in hardness (measured by a penetrometer), probably due to these precipitates cementing the shale crystals together combined with shale dehydration.

Silicate muds were tried in the 1930s for shale inhibition with some success. The chemistry involved was not understood and the main problem was getting stable rheology. However, these problems have been solved and silicate muds have been successfully used in the field.

Due to the way they work, any use in reservoir rocks must be carefully planned; silicates in general are only suitable for the reservoir in development wells where the wells will be cased and perforated. Open hole completions may be a problem due to the presence of gels and precipitates giving an effective mechanical skin. Return permeability tests on core samples will indicate whether the silicate mud is suitable for the reservoir section.

Silicate muds are nontoxic, environmentally acceptable, noncorrosive, can accept all available polymers, and the constituents are readily available at a reasonable price. Clear fluids can be used since the silicates are completely soluble at high pH. Acid gas influxes (such as H_2S) are neutralized by the alkaline mud. In addition, the anionic silicate prevents buildup of shale cuttings on the drillstring (balling). However, silicate depletion rates can be high; the mud needs careful design and proper supervision to be successful. The high pH needed may cause problems with elastomers.

NaCl or KCl/polymer/silicate mud. The mud is formulated as a low-solids polymer using seawater, KCl, or NaCl with 20 to 40 ppb of a soluble anionic silicate network co-polymer. NaCl can be used for saturation or KCl up to 35 ppb. The inhibition works better with the addition of an alkaline metal oxide such as Na_2O. Several mechanisms can be used to further improve the basic inhibitive properties:

- Increase the ratio of SiO_2 to alkali metal oxide
- Modify the alkali metal blend
- Increase the SiO_2 concentration
- Vary the base fluid salt content
- Optimize the mud pH

Salt saturated silicate mud. A salt saturated silicate mud can be formulated. This would be effective where salt and shale sequences were present in the same hole section. Actual formulation (salts used) would depend on the composition of the anticipated evaporites.

2.5.6. Formation Damage with Water-Based Muds (and Cements)

This matter should always be a key consideration in the drilling fluids program for the reservoir sections of a well. The optimal fluid that will minimize formation damage and allow for maximized productivity will not necessarily have the same attributes as the mud that was used to drill the overlying sediments.

In some cases, the mud can be designed to drill the reservoir with minimal damage. At other times, a specially designed drill-in fluid should be used. In other cases, it may be only sufficient to contain the damage if it can be assured that the well can be perforated past the zone of damage. Some of the mechanisms in which water-based muds cause damage can also be present with cements and cement filtrates. The commentary made below confines itself to damage caused by muds (and cements). It is the case that in some instances formation damage can be caused by the reservoir fluid, and the act of producing the well can generate this damage. Refer to production operations manuals and other literature for more detailed study of formation damage.

Just as a water-based mud can be divided into two components (the insoluble solids and the liquid fraction containing solubles), the causes of formation damage can also be classified as those due to the insoluble solids and those due to the liquid fraction.

Formation damage due to insoluble solids. Solids from the drilling fluid may plug the face of the wellbore or the perforations, or migrate into the pore spaces of the formation. These solids may be weighting agents from the mud, low-gravity drilled solids and clays, mud makeup materials such as insoluble polymer, corrosion by-products and scale, cement solids, excess drillpipe dope, or insoluble hydrocarbon material from the reservoir.

The larger solids tend to bridge off against the wellbore face. In some cases, they can be removed by back flowing the well but this may not always be the case. If back flow only causes movement in a small part of the zone, there may not be enough differential pressure along the rest of the zone for the solids to be removed.

If particles are small enough, they may enter the formation pores and block them off. Blockage of pores inside of the reservoir does not only come from migration of fine solids from the fluid in the annulus. Disturbed interstitial material in the pores may become free to migrate further into a sand and plug off. Some reservoir liquids may provide precipitates when contacted by mud filtrate or when subjected to pressure change. These precipitates may become another source of plugging material.

Formation damage due to the liquid fraction. In some cases, the liquid fraction, usually in the form of filtrate (but in some cases as whole mud invasion) can contribute to formation damage. The fluid can cause hydration of interstitial clays. It may disperse or flocculate these clays which can move further into the reservoir and plug pores. It may dissolve matrix cement material, allowing fine particles to become mobile, and migrate further in to block pore constrictions. It may cause precipitates to be generated by reacting with connate water or formation crude. It can be appreciated that the liquid fraction of the mud, as with the solids phase, can set up solids blocking mechanisms.

The liquid fraction, however, can cause damage in other ways as well as initiating solids blockage. Substantial fluid invasion can cause an increase in water saturation, which will affect the relative permeability to oil. It will affect the flow of oil.

To really understand the effect of change in water saturation, the reader is directed to reservoir engineering and production operations manuals. An attempt is made here to condense and simplify the content (if such a thing is possible).

A water aquifer is a single phase reservoir and it is saturated with water. The term absolute permeability is a measure of the capacity of a rock to transmit a fluid under a pressure differential when 100% saturated with that fluid. It might be said that the permeability in the aquifer is the absolute permeability.

However, most hydrocarbon reservoirs are multiphase. There are at least two fluid phases (in some cases there are three phases): oil and water or gas and oil. In a study of a hydrocarbon reservoir, the permeability of each phase must be quantified. For each phase, there is an effective permeability that is a measure of each specific phases ability to flow, i.e., effective permeability, oil-effective permeability, gas-effective permeability, and water.

The most significant effect on the effective permeability of each phase is the saturation of each phase (amount of each phase present). The saturation of each phase is described as the residual oil saturation, residual water saturation, and residual gas saturation. To describe the effect of saturation on effective permeability, the term relative permeability is introduced. The relative permeability for each phase is the ratio of the effective permeability for that phase to the absolute permeability. (The absolute permeability is measured in the laboratory by using common mediums such as air, but also refined oils, synthetic brines, gas, nitrogen, etc. It is beyond the scope of this book to provide more detail.)

How does filtrate invasion cause damage by increasing water saturation? In most rocks, oil will cease to flow when the residual oil saturation decreases to the range of 20-30%. The residual saturation for oil where flow has ceased or the relative permeability to oil has decreased to zero is described as the irreducible oil saturation. Similarly, the effect is seen in the variance of water saturation. The irreducible water saturation is the point where water will not flow.

It can be appreciated that where a substantial increase in water saturation has been caused due to substantial filtrate invasion (it also will happen with water coning), the residual oil saturation will decrease and thus hinder the flow of oil. Eventually the irreducible oil saturation will be reached and the well will not flow oil. This condition is described as water blockage.

A further effect that the liquid fraction of the mud can have on productivity impairment is if there is a change in wettability (the adsorption of one phase onto the rock). In any two phase reservoir, there will always be a "wetting" phase and a "nonwetting" phase. In an oil/water system, water will normally be the "wetting" phase and in the case of a gas/oil system, it is oil. Changes in oil or water saturation can alter wettability of a reservoir. However, some surfactants in the filtrate of a drilling fluid can also cause changes in wettability. A change in wettability in a reservoir from water wet to oil wet can reduce the permeability to oil. This is particularly the case in low-permeability reservoirs.

The surfactants that will cause a change in wettability from water to oil are usually cationic, but some nonionic surfactants can also have this effect.

Proactive measures to be applied to drilling fluids to minimize formation damage. There are some reservoirs that are not at all susceptible to damage by most drilling fluids. The damage that occurs is easily cleaned up. Heavily fractured carbonates would be one example. However, listed below are some considerations for minimizing formation damage.

Whatever mud is used, minimize on all insoluble solids and particularly low gravity solids. The cation exchange capacity of the mud should be as low as possible. If necessary, consider displacing at the top of the reservoir (if there is a casing point) to a drill-in fluid. In this drill-in fluid, do not use any clay to get a filter cake. (In a polymer mud, around 6 ppb of clay is necessary to get filtration control irrespective of how much fluid loss control additives are added). Calcium carbonate bridging agent might be used to aid filtration control along with HEC for viscosity and appropriate fluid loss control additives.

The bridging agent's particle sizes should be designed around the pore throat diameter, if known. The best way to determine the optimum bridging agents is by return permeability tests on core samples. The fluid needs to be designed such that a thin filter cake is formed with low spurt loss invasion that will come off the formation when flow starts. This will require a combination of inert bridging solids (e.g., $CaCO_3$) of a size range from one-third pore size upwards, together with hydrocolloids such as starch. The coarser particles bridge first, then progressively finer particles, and finally the hydrocolloids progressively block the remaining spaces. Spurt loss takes place at this stage. Finally filtration occurs where all solids are filtered out by the wall cake and only clear fluid invades the formation.

One point about sized calcium carbonate bridging agent—it is often promoted because it is acid soluble. However, it can usually be cleaned off the wellbore by back flushing. If an acid treatment is required to remove it, acidizing should always be carried out underbalanced. If acidizing was overbalanced, there is a danger that the particle size of the bridging agent will become smaller as it dissolves. The smaller particles may invade the sand. Also, acid may not dissolve all of the remaining $CaCO_3$. It should also be remembered that the plastic viscosity of the mud will increase proportionately to the amount of calcium carbonate that is added.

Calcium carbonate is not the only bridging agent that might be used. Other forms of bridging agents are sized salts used in saturated solutions and oil soluble resins. The sized salts can be removed by dissolution with water. The oil soluble resin, which is not soluble in water, dissolves in the crude as the well flows. These techniques have been highly developed and particularly the use of sized salt, the accompanying polymer mix, and the methods of removing the polymers. It is recommended to review these techniques further with the mud companies or the specialty companies that produce these materials.

Control of solids and use of bridging agents (to bridge on the face of the wellbore) is important as it minimizes on solids invasion and also on fluid invasion of the reservoir. For the same latter reason, the filtration rates of the fluid should be as low as possible. The lowest safe differential pressures should for similar reason be applied.

Most drill-in fluids are in a nondispersed condition. This is because dispersed muds carry much more (damaging) low-gravity solids. At the same time, the use of dispersants such as lignosulfonates should be avoided. Lignosulfonates, as well as acting on formation clays, can disperse interstitial clays in the porosity of the reservoir. If these clays disperse, they may cause plugging or even weaken the matrix of the rock. It is common in nondispersed drill-in fluids to have 2 to 3% potassium chloride present to protect these interstitial clays.

Avoiding using dispersed muds can show some large increases in productivity, particularly in marginal fields. One field in Zaire, increased well productivity from 400 BOPD to 1100 BOPD because of changing from a lignosulfonate CMC mud to a lime nondispersed polymer mud. A field in northern central Siberia increased production per well from 40 tons per day to above 120 tons per day. The change was that, instead of the previously used mud with dispersants (lignin liquors such as cmad, kssb) and poor control of solids, a 3% KCl polymer system was used.

Another factor to consider is pH. High pH might also cause dispersion of interstitial clays. It may also cause emulsions dependent on the crude in the reservoir.

The occurrence of precipitates should be avoided. Potential to generate precipitate scales because of reaction of filtrate with connate water should be investigated. It should also be remembered that in some cases scaling can occur due to changes in wellbore pressure. If there is a possibility that scaling may occur, a scaling tendency for the fluid planned to be used should be made against the chemistry of the connate water. It should take into consideration chemical concentrations as well as temperatures and pressures. Most major mud companies can do this as well as operators and independent laboratories. Without going into complete detail on scaling, an idea of the potential causes of scaling can be derived from the type of scales experienced in the oilfield. In many cases (but not all) one ion will come from the filtrate and another from the connate water. The reader is directed to further reading in production operations manuals for further detail. Types of scales are:

- Carbonate scale; the carbonate or bicarbonate ion with the calcium ion
- Gypsum or anhydrite scale; the sulfate ion with the calcium ion
- Barium or strontium scale; the barium or strontium ion with the sulfate ion
- Iron scales; ferric or ferrous ions, usually as corrosion by-products with oxygen or sulfides
- Sodium chloride precipitates; changes in pressure and temperature

Carbonate ion or sulfate ion can cause scaling if present in sufficient or certain quantities in the connate water (or filtrate). Often minimizing the concentration of divalent cations such as calcium is enough to avoid this type of scaling. The use of monovalent brines such as sodium chloride, sodium bromide instead of calcium chloride, and calcium bromide might be considered.

These are some but not all first principles in avoiding formation damage. Some or most might be applied in exploration wells. To really be sure of minimizing the damage to a reservoir, nothing can replace knowing the reservoir in depth. The reservoir conditions, the connate water conditions, the mineralogy, the permeability, and the porosity. There is no standard answer to this and the solutions are highly reservoir specific.

2.5.7. Oil Muds

As described above, certain clay formations contain clay minerals, which will readily hydrate, expand, and disperse when supplied with a polar liquid such as water. The hydration of a clay is the reverse of the process of diagenesis.

Oils and many other (but not all) hydrocarbon liquids are nonpolar or have very low polarity. They will not have present dissociated hydroxyl or hydrogen ions which can interact with a clay mineral and cause it to destabilize. This fact is an attribute of the use of oil (or certain other hydrocarbon liquids) in certain drilling fluids. It is possible to make the definition that the term oil mud is directed to oils or other hydrocarbon liquids that are not miscible in water and (these oils) comprise the continuous phase. Glycols and glycerols when employed as a component of a drilling fluid could not be described in such a way since they are miscible in water.

The use of oil in drilling fluids can be divided into three categories: emulsion muds, invert emulsion oil muds, and oil-base mud. Only the latter two are properly described as oil muds.

Emulsion muds. In this case, oil is emulsified into water or a brine and the water will be the continuous phase. These types of muds are often called straight emulsions. They are really water-base muds with between 5 and 45% of oil emulsified through the water phase. Oil is sometimes added to increase lubricity or assist in lowering fluid loss.

Droplets of oil are dispersed through the continuous phase (water) and held in dispersion by anionic emulsifiers. In some cases in the past, lignosulfonates were effective in emulsifying oil in water if the amount of oil was not substantial. These muds are not at all common except for some amounts of oil emulsified in a water-base mud. Those with high oil contents are more likely to be used for specific cases such as reduced density (below 1.0 specific gravity).

Conventional wisdom says that adding small amounts of oil also reduces density and provides some shale inhibition. Adding some oil does initially decrease density but this reduces the efficiency of the hydrocyclones and can lead to an increase in solids content, affecting rheology and ultimately allowing density to increase again unless the oil percentage is high. As far as shale stabilization is concerned, there is no evidence to support the common assertion that an oil in water emulsion will have any stabilizing effect whatsoever. In addition, as water is the continuous phase the oil does not prevent clay dispersion.

Invert emulsion oil muds. In this case, water, or dominantly brine, is emulsified in oil (or other hydrocarbon liquids). This is the most common type of oil mud that is used in drilling. The brine is emulsified in the oil as minute droplets by the presence of cationic emulsifiers.

Oil-base mud. This term, according to convention, has covered fluids where there is no brine emulsified in the continuous phase of oil or where it is present in quantities of less than 10%. Another term that might be used is "all oil systems." Today, the terms oil mud, oil-base mud, etc., are interchangeable and really define fluids where the continuous phase is oil and if an emulsion is present, it is as an invert emulsion. The presence of a water phase in an oil mud allows for versatility in control of parameters such as rheology.

2.5.8. Components of Invert Oil Emulsion Muds

The oil (continuous) phase. This is the main component of the system and often called "base oil." It may be made up from (in historical order of usage):

Crude oil. This was the first continuous phase to be used but seldom now in use because of environmental concerns, potential elastomer solvency, high-inherent viscosities, and high flammability.

Diesel oil. This was most commonly used until the early 80s when low aromatic oils began to appear. It is still used in some cases, particularly when it is regulated that oil mud cuttings have to be processed at a disposal or treatment site. It is less expensive and often logistically more convenient than low-aromatic mineral oil. Because the kinematic viscosities of diesel (and crude oils) are higher than most low aromatic mineral oils, they will demonstrate, in the oil mud, correspondingly higher plastic viscosities.

Diesel normally contains approximately 26% total aromatics of which more than 5% of the total are polycyclic aromatics. Most polycyclic aromatics with three or more aromatic rings are toxic and carcinogenic. Low-aromatic mineral oils normally will contain less than 5% total aromatics of which less than 5% of the total aromatics are polycyclic aromatics. In some low-aromatic mineral oils, the concentration of hazardous polycyclic aromatics may be less than 5 ppm. The resultant toxicity of the formulated oil mud is less (than those based upon diesel).

The use of diesel as a base oil was outlawed in most areas by the Paris Convention. There are still some countries that permit its use, however, most of the big operators are so environmentally conscious that even in unregulated areas, they do not use diesel in mud.

Low-aromatic mineral oil. These oils superceded diesel in offshore operations because their toxicity was substantially lower (at least 10 times lower for most marine organisms tested).

The development of low-aromatic mineral oil systems was greatly advanced (particularly in the North Sea) in the early eighties as it allowed PDC turbine combinations to be used on offshore rigs which did not have cuttings washing systems. Early PDC bits required the use of oil muds as a flush.

Nonmineral oils and synthetic oils. These were developed to address the problem that a substantial amount of hydrocarbon material in a cuttings pile will cause an environment that is not conducive to a benthic population and where anaerobic conditions may be set up.

In this group of hydrocarbon liquids there are vegetable oils, stabilized esters, ether systems, and oils synthesized and formulated to provide for stable oil muds with low toxicity, biodegradability aerobically, and in some cases anaerobically.

Early work on these "oils" was directed to fish oils and vegetable oils. However, these oils showed very high kinematic viscosities with high plastic viscosities in the formulated muds. The resultant muds showed poor system stability. Stabilized ester systems later developed, followed by ether systems and liquids synthesized to specific specifications.

This group of oils is very diverse and it is recommended to investigate them further for specifications, performance, and environmental acceptance of local legislation through contact with the drilling fluids companies and legislative bodies.

Some very brief generalizations for this group are:

1. It is easier for anaerobic bacteria to break down stabilized esters than ethers because the chemical bonds are different.
2. Ether systems have higher temperature stability than ester systems.
3. Nonmineral oil and synthetic oil systems are substantially more expensive in varying degrees than oil mud systems based upon mineral oils. The economic viability of using such systems depends on the following factors, at least:
   a) An operation with a high total daily operational cost
   b) Depth/time curve will be greatly reduced by using an oil mud
   c) The use of any other form of oil mud would be banned
4. Although these systems are biodegradable in seawater, the oil mud system can be reused on multiwell programs.
5. In some systems, care must be taken to avoid hydrolysis caused by too alkaline or acid an environment.

The brine (or water) phase. This is the other liquid phase in an invert emulsion oil mud and will be discontinuous. Although it is usually present as a brine, it is described as the water phase as it is quantified by volume on the basis of the percentage of water present in the total volume of the oil mud. It consists of a salt, normally calcium chloride but in some cases sodium chloride, dissolved in water. This brine will be present in the form of droplets, sub-micron to a few microns in size and surrounded by emulsifier.

The brine phase of an invert oil emulsion mud has two functions. It allows for flexible control of certain fluid parameters. It also generates an osmotic force to initiate water transport from shales to the fluid or at the least prevent water penetration of the bore hole wall.

Properties of the brine phase—osmosis. The emulsified brine will not be completely isolated from the formation because the layer of emulsifiers around each droplet of brine can act as a semi-permeable osmotic membrane. If an osmotic gradient is set up between the brine emulsified in the oil and formation water, it is possible for water transport from the formation to occur. The osmotic gradient will depend on the concentration of the salt(s) in the brine. This concentration is known as the water phase salinity (WPS). It can be quantified by chemical titration of a sample of the water phase (obtained by breaking down the emulsion of a sample of the mud, diluting with water, and creating a straight emulsion for test purposes).

A measure of the osmotic gradient that is present can also be derived from the water activity (Aw) of the mud. This test measures the percentage relative humidity in air above a sample of mud in a closed space using a hydrometer.

Originally, it was considered that the osmotic gradient should be such that it be sufficient only to avoid a net movement of water from the water phase of the mud to the formation. This was described as a "balanced activity." However, the benefits of an osmotic imbalance in favor of water transport from a shale to the water phase of an oil mud can be seen in improved hole stability. In fact there are several other forces acting to create flows of water apart from the osmotic (chemical)—hydraulic (overbalance) and electrical potentials also play a part. It is the net of these forces that determines whether and in which direction flow will take place.

It is possible to define the required water phase salinities or water activities for any given matrix stress for any shale at any depth. To create the osmotic gradient in favor of water transport being directed to the water phase of the mud, there are two forces to be considered. First, there is the osmotic pressure generated by the water phase salinity of the mud. Acting in the opposite direction, there is the osmotic pressure of the shale. This is related to the matrix stress of the shale.

The osmotic pressure of a shale is generated in two ways. First, during diagenesis, water is forced out of the shale because of compaction (or generation of matrix stress). Diagenetic water leaving the shale has a lower salinity than water remaining in the pore spaces. The resultant shale salinity will be higher. Additionally, because of compaction, pore spaces are reduced. In drilling a shale, there will be a tendency for the rock to be released from stress at and near the bore hole wall. This release of stress will tend to cause expansion and resulting increase in pore volume. If there were to be an increase in pore volume, there would be a suction potential for water into the shale. In effect, this would be the opposite of diagenesis. If water can be prevented from entering the shale, the suction potential will provide stabilization by minimizing pore volume expansion.

A young "wet" shale (such as a gumbo shale) at a shallow depth, will require a lower water phase salinity than a more diagenetically mature shale (with higher shale salinity) at the same depth. The water phase salinity requirements for any particular shale type (with the same shale salinity) will increase with depth because matrix stress will increase with depth.

In Figure 2-7, a shale that has a matrix stress of 5074 psi and a shale salinity of approximately 100,000 ppm NaCl equivalents (intercepts at A) is shown to require a 265,000 ppm CaCl₂ water phase salinity (B) to obtain balance of osmotic pressure.

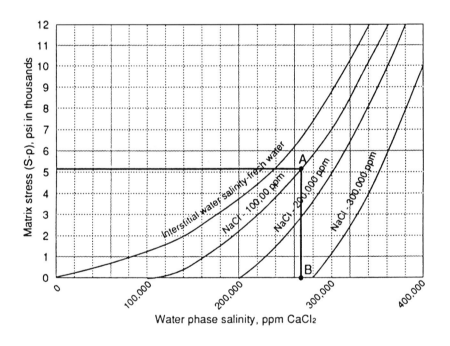

Fig. 2-7 Oil-Mud-Water Phase Salinity Requirements to Balance Matrix Stress of Shales (courtesy of Baroid)

How do you quantify the matrix stress at the rig site? How do you come up with the shale salinity? As accurately as you practically can. At the least, you may be able to quantify the minimum water phase salinity that you require.

Matrix stress. In a few cases, you may be running a downhole tool that will tell you the matrix stress. If this is not possible, you have to use the most accurate measurement of pore pressure and work on the rough model that matrix stress + pore pressure = overburden pressure (derived as having a gradient of 1 psi per foot unless it is known more accurately). For this example, this rough model was applied. The shale is at 12,000 feet and pore pressure was determined to be equivalent to an 11.1 ppg fluid or 6296 psi. The matrix stress is obtained by subtracting 6296 psi from 12,000 psi (the assumed overburden pressure at 12,000 feet).

Shale salinity. This may be harder to quantify at the rig unless the logging unit is measuring it. It can be measured in the laboratory with a certain degree of accuracy. You may have to make a best guess at shale salinity. Discussion with geologists may help in respect to shale type. A marl will have a higher shale salinity than a claystone for example. If you study the graph, you will see that at low matrix stresses (shallower depths), that variance in shale salinity is more important than at higher matrix stresses (deeper depths) except for shales with salinities of 200,000 to 300,000 ppm NaCl equivalent. These shales are not very common. At shallow depths, the shale salinity is more likely to be in the range of up to 100,000 ppm NaCl equivalent salinity.

It is worthwhile to study the graph, irrespective of whether you apply it or not, since it shows how the two parameters—matrix stress and shale salinity—affect the required water phase salinity.

Properties of the brine phase—emulsion. The water phase, because it is discontinuous and coated with emulsifier, will behave as a solid in the oil mud and demonstrate itself in the plastic viscosity dependent on the strength of the emulsion and the amount of brine phase present. A tight emulsion will provide for smaller sized droplets of emulsified brine and a lower plastic viscosity. It should be remembered that the base oil, the insoluble solids in the mud, and the oil wet condition of the insoluble solids also contribute to the plastic viscosity.

Properties of the brine phase—oil/water ratio. The size of the water phase in an invert oil emulsion has a significant effect on almost all the properties of the resultant mud. The size of the water phase, as well as being measured as the volume percentage of the total mud volume, can be quantified in what is the most fundamental of parameters of an invert emulsion oil mud—the oil/water ratio (OWR). This is defined as the ratio of the volume percentage of oil in the liquid phase to the volume percentage of water in the liquid phase.

Many of the components of an invert emulsion oil mud are present or perform in the oil phase. Consequently, if the oil phase is reduced because the oil/water ratio has been reduced (in favor of water) or because the solids phase has increased due to increase in density or solids, the performance of some of these components will be affected.

For example, the yield point and gels of the mud can be raised with the addition of viscosifiers. These viscosifiers are present in a whole mud concentration but only act in the oil phase. If the oil/water ratio is increased, the viscosity will decrease (unless additional viscosifier is added) because the viscosifier concentration in the oil phase will have been reduced. The viscosity will also have been reduced because the water phase (and its contribution to viscosity) will have been reduced.

This effect is an attribute of an invert oil emulsion mud. It allows for a stable control of viscosity by altering the oil/water ratio to counter the tendency for viscosity to increase as fluid density is increased. Normally a low-density oil mud will have a lower oil/water ratio than one with a higher mud weight.

Determination of oil/water ratio. Oil water ratios can been engineered to as low as 40/60 or as high as "all oil" (100/0) and still provide a stable mud. Very low oil/water ratios have sometimes been used to provide a mud that will give a low concentration of oil on cuttings. All oil systems have been used as drill-in fluids to study formation water without any contaminating effect of water. They are sometimes described as "native state coring" fluids.

Normally when using oil mud on a complete well, the shallower intervals with lower mud weights would be drilled with oil muds having 60/40 to 70/30 oil/water ratios. Deeper intervals with higher fluid densities would see oil water ratios in the range of 75/25 to 85/15.

Specific consideration of the oil/water ratio should be made in respect to any contaminating fluids entering the mud due to an influx. If this is liable to happen, it is better not to have too small a water phase. The contaminating influx will show more effect.

OWR also depends on density. The higher the density, the higher the OWR required. Table 2-2 provides the lowest OWR recommended for each density range.

The mud should be built with an OWR corresponding to the highest density that might be needed during drilling; otherwise, a density increase would require addition of oil. This would dilute the mud so more weighting material and chemicals would be needed. This would also increase the volume.

Table 2-2 Lowest OWR Recommended by Density Range *(courtesy of Baroid)*	
Mud weight, ppg	Lowest recommended OWR
7-9	60/40
9-11	65/35
11-14	70/30
14-16	75/25
16-19	80/20
19+	85/15 to 95/5

2.5.9. Environmental Aspects of Oil Muds

Seabed studies of multiwell sites in the North Sea by the mid-eighties (after three or more years of usage of low-aromatic mineral oils) showed cuttings piles below platforms where in some zones there were incomplete benthic populations. The problem was the vast amount of hydrocarbon material in the cuttings piles, and in some cases or some parts, anaerobic conditions persisted. Subsequently, legislation required that the amount of oil on cuttings discharged to the seabed be reduced. Permitted levels of oil on cuttings were persistently reduced. This required improvements in primary shaker equipment and eventually the reintroduction of cuttings wash systems. In some offshore areas (e.g., spawning grounds) the discharge of cuttings from oil muds based upon low-aromatic mineral oils was banned irrespective of how little the oil content was on the cuttings.

2.5.10. Oil Mud Additives

Emulsifier. The droplets of brine are held in an invert emulsion in the oil phase because they are coated by emulsifiers. Emulsifiers are surfactants that have an organophylic end(s) and a hydrophylic end to their molecule. Each end aligns itself to be in either the oil phase or the water phase. They bridge the oil-water interphase. The emulsifier, coupled with shear, forms the minute emulsion droplets.

Emulsifiers that consist of soaps will have their cationic ends aligned in the water phase. Monovalent cations such as sodium will form soaps that formulate straight emulsions. A typical example is sodium oleate.

However, divalent cations such as calcium can form soaps with two large organic groups attached that are organophylic. This "imbalance" between the hydrophylic (calcium) ends and the organophylic ends will create an invert emulsion.

The type of soaps used in oil muds are commonly formed from tall oil fatty acids. Tall oil is derived from pine trees. It is an oil synthesized by the tree and its quality is dependent on the climate and weather when it is produced. The quality of tall oil can be significant in the effectiveness of the emulsifier. "Tall" is the Norwegian word for "pine." Not all soaps are generated from tall oil fatty acids. An example of another type of soap used as an oil mud emulsifier would be soap derived from fish oil fatty acid.

Not all emulsifiers used in oil muds are soaps. Various amine based surfactants such as polyamines or polyamidoamines are also used. They also will have hydrophylic and organophylic ends to their molecules. In some cases these may be mixed in with the soaps in their packaging or may be supplied separately.

Initially, in early oil mud systems, the term "primary emulsifier" was applied to the soap type emulsifiers and "secondary emulsifier" to other types. This terminology is now very loose because many invert systems are based upon non-soap type emulsifiers with no soaps present.

Alkalinity control. Alkalinity control in most invert emulsion muds is obtained by the addition of common calcium hydroxide (lime), or in some cases calcium oxide. The use of a calcium base is particularly relevant when tall oil fatty acid soaps are in use.

Oil-base muds do not suffer from the problems of hydrogen sulfide or carbon dioxide contamination that can occur in a water-base fluid. However, such contamination in an oil-base mud will show itself as the increased addition rate of lime to maintain alkalinity.

Viscosifiers—organophylic clays. Hydrophilic clays (e.g., bentonite) that do not normally disperse in a nonpolar fluid such as an oil, can become oleophylic and dispersible (in oil) by treatment with amine salts. In this case, the cations of the bentonite clay are replaced by cationic groups from the amines. The amines are also adsorbed onto the clay surfaces. When this treated clay (organophylic clay) is present in an oil, it will disperse. The clay surfaces will be displaced apart because the cationic groups repel each other and because the amine salt will extend itself in the oil medium. This process is assisted by shear.

Dispersion of the clay can be further increased and the gel can be stabilized if an organic group such as an aromatic ring can enter between the clay surfaces. It is often the case that a stronger gel (higher yield point) per amount of organo-clay added is obtained in diesel rather than in a low-aromatic mineral oil mud. To obtain a sufficient yield point when using base oils that are low in aromatics, it is necessary to either add additional material, provide increased shear, add an effective oil wetting agent, or use a clay with an amine salt treatment that is optimum for the base oil in use.

Organophylic clays vary in a number of ways. When used in drilling fluids, there are three types of clays that are treated to become oil dispersible. The most common is bentonite and invariably, it will be a sodium bentonite (Wyoming). Bentone 38 is an example of such a type of clay but there are many types. Another clay that is converted to an organo-clay is hectorite. The hectorite clay is a layer clay, similar to bentonite but with a much larger clay structure. An example of such an organo-clay is Bentone 64. This type of clay has the advantage of producing good stable rheology at very high fluid temperatures. Attapulgite (seawater gel) can also be treated to make it organophylic. This type of clay can improve the suspension properties of the mud but will not assist in filtration control.

Another way that organo-clays will vary is the type of quaternary amine salt that is adducted to the clay. The performance of these "quats" will vary depending on the type of base oil in use. Some quats are more effective than others, and change in quat on a clay may provide as much as double the yield point.

The process by which the quat is added to a clay can also make a difference to the performance of the organo-clay. There are two processes involved. One is a dry process and the other is a wet process with subsequent drying. Some companies will make proprietary variants of organ-clays by adding other materials to the clay before it goes to the dryer.

As is the case with bentonite clay in a water-base mud, the presence of organo-clay in an oil mud will have an effect on filtration control of the mud. If the concentration of organo-clay becomes too low (e.g., less than 2.0 ppb), filtration control can become difficult. Such a low concentration of organo-clay is not very common in an oil mud but could be the case in an oil mud with an ultra low oil/water ratio.

Filtration control additives. It is possible in an invert emulsion oil-base fluid to obtain filtration control without any fluid loss additives if the emulsion is strong enough. This is because the water phase droplets can act as a solid and reduce filtration. Every time emulsifier is added to the mud, it will be noticed that the HTHP filtration rate is reduced to a certain extent.

It is, however, the case that filtration control additives can effectively improve HTHP filtration. These are particularly relevant in HTHP control at very high temperatures.

In some muds, asphalts, gilsonites, or various proprietary variants are effective filtration control agents. In using these materials, the softening point temperatures in whatever base oil is used should be considered. These materials are relatively inexpensive and at normal temperatures can perform effectively.

Organophylic lignites are used as filtration control agents, particularly in high-temperature formulations. They are more expensive than asphaltic or gilsonitic materials because they have been processed to be organophylic. They are produced by the addition of a quaternary amine salt as an adduct onto the lignite. This treatment, as in the case of organophylic clays, makes the lignite dispersible in an oil. The same factors that effect the dispersibility of the organo-clays also affect the organo-lignites.

In addition to the filtration control agents mentioned, some drilling fluid companies also provide liquid fluid loss agents. These materials are proprietary and the particular mud company should be contacted for further details.

Rheology modifiers. These are usually diamer or triamer salts which, when added to an oil mud, can alter the low-shear rate rheology or the thixotrophy (speed and strength of development of gel) of the oil mud. These are usually system specific and their performance will depend on the emulsifier package and the type of organophylic clay in use. Rheology modifiers are often used to avoid or minimize the problems of barite sag in oil muds. Their function is to provide as much thixotrophy as possible in a low rheology.

Oil-wetting agents. As well as emulsifiers that are present in invert emulsion muds, in some cases specific oil-wetting agents are used to ensure that drilled solids and barite are oil wet. Emulsifiers that are currently in use in oil mud systems are capable of oil-wetting barite and drilled solids. Specific oil-wetting agents are not usually required as compared with earlier use of oil muds. If an oil-wetting agent is added to an invert emulsion system with substantial barite concentration, a noticeable decrease in plastic viscosity can sometimes be achieved. This is due to a decrease in interfacial tension.

An example of an oil-wetting agent used in an oil mud is lecithin. This comprises a negatively charged phosphate group and a quaternary amine that is positively charged. The phosphate group will be dissolved in the oil phase and the quaternary amine is attracted to metal or mineral surfaces that are normally negatively charged.

As previously described in "Viscosifiers" and "Filtration control additives" within this section, oil-wetting agents are processed with clays and lignites to form organophylic clays or organophylic lignites.

In some cases, the use of oil-wetting agents is specified to be avoided to ensure that oil wetting of a reservoir sand is prevented. Oil wetting of sands, which is theoretically possible with an invert emulsion oil mud, is usually limited because of the low amount of filtrate (oil and some emulsifier but not water unless the emulsion is weak) that enters into the formation. The zone of damage is usually very low. (See Section 2.5.11, "Formation Damage with Oil Muds.")

Bridging agents. Materials such as calcium carbonate bridging agents can be added to the oil mud to prevent whole mud invasion of the formation. In some cases against sands with some degree of permeability, better return permeabilities have been achieved with bridging agent present in the mud.

2.5.11. Formation Damage with Oil Muds

Concern has often been shown regarding possible formation damage that may be caused by oil-based muds. In some cases, this concern has been justified where real cases of formation damage have been experienced. In general, formation damage caused by an oil mud is not at all common. Possible causes of formation damage specific to oil muds are described below.

Oil wetting. In theory, the emulsifiers and oil-wetting agents that will oil wet droplets of brine to form an invert oil emulsion will also oil wet a sandstone reservoir. Oil wetting will destroy the layer of water on the sand grains of the reservoir over which the crude oil flows. It will result in decreased relative permeability to oil. Consequently, mobility of crude oil flow is reduced. (See Section 2.5.6.)

This type of damage is usually very localized because of the low filtration rate of oil muds into the formation. Usually penetration of filtrate is a few inches or at least within the distance that can be perforated out. Obviously, running a low, tightly controlled filtration rate and not having high differential pressures will minimize filtrate invasion of the reservoir. The use of a bridging agent that can be back flushed will also be beneficial.

As well as the action of the emulsifier, wettability might also be changed by an increase in oil saturation if there is a substantial invasion of oil filtrate. Low-permeability reservoirs would be more likely to be affected. This type of damage is not so likely if the filtrate of the oil mud is kept very low (which is usually the case). However, in the past, before the advent of PDC bits, oil muds usually drilled slower than water-based muds. Engineering the oil mud to have a high filtration could enhance penetration rate with an oil mud. Oil wetting of the reservoir might have been more likely with this type of mud.

Oil muds with high filtration rates are no longer used. Advances in bit technology now provide all the penetration that is desired. A "relaxed" filtration rate will have the only effect of increasing dilution costs.

Emulsion blockage. The filtrate of an oil mud is normally oil with some dissolved emulsifiers. This filtrate if mixed with formation water might form an emulsion in the formation. Such an emulsion would have an increased viscosity and thus impair the mobility of crude oil.

Unreacted or partially soluble emulsifier. If emulsifier present in oil mud filtrate were to become insoluble, it could cause damage by blocking pore throats. This was more likely to occur with chemical change of soap-type emulsifiers. With advances in emulsifier chemistry, this type of damage is not likely in most cases.

Whole-mud invasion. If the permeability is sufficiently high and/or differential pressure is excessive, whole mud might be forced into the formation. This would also happen with a water base mud. All the damage that would occur with a water-base mud such as migration of fines, permeability blockage, etc., would occur with an oil mud. Also, there would be the possibility of emulsion blockage and oil wetting. Further, the addition of properly sized bridging agents will help to avoid whole mud invasion, as will minimizing differential pressure to the lowest safe level.

2.5.12. Air, Foamed, and Aerated Systems

Air can be used in several ways to improve rate of penetration and reduce or eliminate formation damage. However, the circumstances must be suitable for the system and extra equipment and personnel may be needed.

With all the techniques summarized here, specialist advice is necessary to evaluate the proposed solutions, costs, and equipment requirements.

Air drilling. This is only suitable where formations to be drilled contain no liquids. Large volumes of compressed air are used as the circulating medium. The areas of application are hard, dry formations: dry geothermal zones and dry gas production zones. However, air drilling cannot be used if any formations will slough or extrude into the wellbore in the absence of hydrostatic control (unconsolidated formations, plastic salts, etc.).

Where air drilling is used, several significant advantages are gained:

1. The rate of penetration may be increased by a factor of five or more
2. Formation damage due to hydrostatic, solids, or chemical reactions is eliminated
3. Lost circulation is eliminated
4. The cost of the alternative mud system is saved, which offsets the cost of the air drilling equipment and personnel
5. Bit life may be extended due to lack of erosion and seal damage from mud solids
6. The well is continuously tested in gas zones while drilling

Mist drilling. Similar to air drilling, mist drilling is an option when small amounts of water are produced by the well. Increased air volumes are needed and a mist pump is used to inject small quantities of a water foaming agent solution. The benefits of mist drilling are similar to air drilling, but at a higher cost.

Stable foam drilling. This is the most versatile of the air-assisted drilling techniques. The liquid is a continuous phase and contains encapsulated air bubbles within it. The percentage of liquid will vary between 2% and 15% by volume. An annular back pressure valve is used which, with a varying liquid fraction, allows a variable bottom hole gradient of between 0.026 psi/ft (0.5 ppg) and 0.312 psi/ft (6.0 ppg). The lifting capacity of stable foam is superior to that of drilling muds and it is possible to displace fluids from the hole using foam.

Advantages of foam are similar to those for air and mist drilling. Oil and salt water influxes are likely to destroy the foam stability, precluding the use of foam in those conditions.

Stiff foam drilling. Stiff foam is a very stable emulsion of air in mud. It is formed by injecting air into a thin mud slurry, with air-mud ratios of between 100:1 to 300:1. There are several formulations for the mud slurry, depending on what fluids will be encountered in the well-bore. The foam can be tailored to resist salt water or oil in small quantities. High volumes of water or oil produced will cause problems.

This type of system may be used where the air-assisted techniques mentioned above are not suitable. Again, specialist advice should be sought.

Particular applications that may be suitable are:

1. Large diameter holes where air requirements for dust or mist drilling would be excessive. Air requirements for a 26 in diameter hole using stiff foam are only 400 to 600 cubic feet per minute (CFM).
2. Drilling unconsolidated formations. Stiff foam can stabilize this kind of formation.
3. Preventing downhole fires, which sometimes happens when drilling oil zones with air.

Aerated fluid drilling. This is simply injecting standard drilling mud with air, effectively gas-cutting the returns and lightening the fluid column. The main advantages are:

1. Maintaining full circulation in loss zones.
2. Increased ROP by reducing chip hold-down.
3. Reduced incidence of differential sticking.
4. Reduced formation damage.

Air may be injected at the standpipe or by other means, at an appropriate rate in proportion to the mud circulation rate. Generally, the technique is limited to a maximum depth of about 2800 ft, since injection pressures become excessive at greater depths.

2.5.13. Tendering for Mud Services

Following are some points that might be considered. In many cases, operating companies do not pay enough attention to tendering for the mud services and suffer expensive wells as a result.

Lowest price is not always the lowest cost on the mud bill or the total operating cost. Studies of many wells where a comparison between mud companies performance can be made, can often show that the lowest bidder can be the most expensive. This is not always true in "high technology" areas, but is very common when working in areas where not so much control can be exerted or is applied. Some companies actively pursue a lowest price policy. The most alarming case of how this can be detrimental would be in respect to oil muds. If the lowest bidder's oil mud causes substantial skin damage (compared to others), it will not only have an effect on total operating cost but also on field economics. It might necessitate costly remedial treatments. At the worst, if wells do not flow because of damage, that damage becomes 100% damage. *The oil stays in the ground.* This scenario is not very common but it has happened on some occasions.

Direct mud service price may not be the only criteria for awarding a contract; also evaluate technical support, valued added service, area experience, and engineering quality criteria. When you award to a mud company, you may wish to take on their area experience or additional back-up services. With regard to engineering, often an inexperienced or unqualified engineer can make a poor job of a good mud program, and yet a good engineer can turn around a poor program or succeed even with a lack of appropriate or high quality materials.

Quality of materials is an important factor. Most reputable major drilling fluids companies have their own specifications that they apply in supplying material. The mud companies reputation is more important (in regards to repeat business) than any quick short-term gain from supply of inferior products. However, some mud vendors, especially in third world countries, can find plenty of ways to make extra profit at your expense. The basic problem at the tender stage is ensuring that you specify exactly what is acceptable to you. Many major oil companies have their own "in-house" specifications, which can make the tendering process easier because they can be included as specifications in the scope of work. The vendors know exactly the specification that they are bidding on. It is worthwhile to get a drilling fluids expert to look at the wells you are likely to drill and write out a tender to ensure that you have recourse against the vendor if inferior products are supplied. Following are a few, but by no means all, of the materials that may be specified:

All barite must meet at least API 13 A specification. The vendor should state the country of origin and other specifications that barite will pass. Barite can be tested to ascertain that it will meet the stated specifications.

Bulk bentonite should meet at least API 13 A specification for European bentonites if it is to be used in spud muds. Sack bentonite is recommended to meet API 13 A specification for Wyoming bentonites if it is used as a component of mud systems other than spud muds. The vendor should state the country of origin in each case.

Many polymers are supplied in liquid form (as a suspension in a carrier oil). The percentage activity or purity should be stated, which will allow you to compare the actual amount of active chemical per dollar.

- Polyacrylate deflocculant liquid; state % of activity
- PHPA, liquid; state % of activity
- HEC, liquid; state % of activity
- Salts (such as sodium chloride) should have the degree of purity stated

CMC and PAC polymers are supplied in *pure* grade (which is just what that implies) or in *technical* grade (which is where sodium chloride, a by-product in the manufacturing process, is not removed from the final product). Pure grade polymers should be specified in your tenders with the exception of using CMC in a sodium chloride brine or in seawater and it should be made clear to the vendors that tech grade PACs will not be accepted.

XCD polymer should be xanthan gum, coated for dispersion. No polymer blends are acceptable (such as guar gum and urea). This can be spot checked. Any substitute should be detailed and quoted separately.

PAC, regular, and low vis. No technical grade polyanionic cellulose should be accepted. This can be spot checked. Vendor should state the degree of substitution for both PAC regular and low viscosity submitted.

Nonfermenting starch. State biocide and percentage present. If the vendor does not quote a bacterially stabilized starch, the percentage of the biocide to stabilize the starch should be quoted.

Guar gum viscosifier. Tenderer should quote a guar gum variant to provide viscosity. The type of variant should be stated (e.g., hydroxy propyl guar.)

Sulfonated asphalt. Quote a known trade name, e.g., Soltex. Otherwise, specifications should be provided.

Amine corrosion inhibitor. This should be a film-forming amine and if possible should have biostat properties. A brief description and specification should be provided.

Basic zinc carbonate. This must be a zinc carbonate and zinc oxide blend. The stoichiometric ratio should be provided.

Glycols provide useful inhibitive and lubricating properties. The use of glycols in drilling fluids has recently become popular. The required glycol is a polypropylene glycol, which has been manufactured specifically for use in drilling muds. Industrial waste products from a "glycol stream" should not be accepted. Vendors should state the suppliers trademark name of the glycol supplied. In the event that the vendor is not prepared to supply this information, the generic chemical name, the percentage, and the content of impurities must be supplied. Glycol can be spot-checked against a "finger print" of a reference sample. Glycol is a pale, straw-colored liquid with an SG of just over 1.0; a waste-stream glycol will probably be dirty brown or black and will probably be heavier.

2.5.14. References for Drilling Fluids Program

Formation damage

Allen, Thomas O., and Alan P. Roberts. "Production Operations," *OGCI.*

Archer, J.S., and C.G. Wall. *Petroleum Engineering, Principles and Practice*, Graham and Trotman.

Mahajan, Naresh C., and Bruce M. Barron. "Bridging Particle Size Distribution: A Key Factor in the Designing of Non-Damaging Completion Fluids," *SPE* 8792. Paper presented to the Fourth Symposium on Formation Damage Control for the SPE and AIME, Bakersfield, California, January 28-29, 1980.

Muds

Chenevert, M.E. "Shale Control with Balanced Activity Oil Continuous Muds," *SPE* 2559, October 1970.

Clark, Sheurman, et al. "Polyacrylamide/Potassium Chloride Mud for Drilling Water Sensitive Clays," *JPT*, June 1976.

Gray, George R., H.C.H. Darley, and Walter F. Rodgers. *Composition and Properties of Oil Well Drilling Fluids*, Gulf Publishing.

IDF mud company manuals, Baroid, Milpark.

Stuart Smith, drilling fluids consultant, contributed most of the material for section 2.5.

$$\left[\, 2.6 \,\right]$$

Casing Running Program

The casings to be run are defined by the well design. For the drilling program, the configuration of each string should be stated per the drilling program checklist given at the start of this section (2.1.1.). It is not necessary or helpful to include all the casing design calculations in the drilling program.

Equipment checklists for each casing job are useful to include as an appendix to the drilling program. This should ensure that any special equipment is not forgotten and that all the correct quantities of materials are sent out to the rig. The person responsible for logistics can note on the checklist who supplies each item and its current status.

Apart from the casing configurations resulting from the well design process, practical notes can be made in the program where special procedures or particular precautions are needed.

2.6.1. Normal Drilling Program Requirements for Running Casing

Once the casing point has been reached, the requirements for the mud change. Good gel strength (for suspending solids) and yield points (for cleaning the hole) are not needed and they are detrimental for the casing and cement job. The mud should be conditioned before pulling out of the hole for low gels and YP for lower surge pressures while running casing. Minimizing the MBT and low gravity solids will also be worthwhile considering the possibility of differential sticking. Optimizing rheology will improve mud displacement and minimize ECD during cementing (reducing the chance of inducing losses).

If the rig tank capacity permits, consider having a tank of clean, rheology-optimized mud standing by so that after circulating clean, the optimal mud can be displaced into the open hole prior to pulling out for logging/casing. This will save rig time compared to making several circulations while conditioning the mud. Only the open hole mud needs to be displaced.

If any fill is likely, then spotting 200-300 ft of viscous pill on bottom after getting the conditioned mud in the hole will help to keep the fill off bottom.

Normally on the last trip out before casing, a survey tool (Totco or single shot) will be dropped as a check and the drillpipe will be measured with a steel tape to confirm the tally.

Most operators throw in a "traditional" wiper trip before running casing, especially if logging is done. Wiper trips cost money and are often more detrimental than beneficial to the wellbore. Refer to the notes on wiper trips in Section 2.9.1, "Wellbore Stability." In practice, if the wellbore is stable, no mobile formations are squeezing in, and no problems are seen while tripping or logging, then a wiper trip is unlikely to be necessary.

Once on bottom, circulate 110% of casing contents while monitoring for losses to ensure that nothing in the casing can plug the float. The casing should be filled and circulated with conditional mud. If possible, use a cement head which allows all of the programmed plugs to be loaded in advance so that circulation can be maintained until the slurry is ready to displace and periods of no circulation are minimized.

The casing program needs to include precautions or procedures required for possible problem areas.

2.6.2. Addressing Potential Casing Problems in the Drilling Program

If the casing should become stuck while running, the two likeliest causes are differential or geometry related (high DLS or a ledge). Some preventative measures can be specified by the program; well-centralized casing run in a well-conditioned wellbore with properly conditioned mud is less likely to get differentially stuck. Careful reaming of problem sections will only be effective for geometry-related problems.

In high-angle or horizontal wells, the casing may push cavings or cuttings ahead and buildup a wall of debris, which may then get the casing packed off. A casing circulating packer that will allow circulating to start with a minimum delay should be included in the program. Consider washing every joint down from about 70° inclination, taking care not to initiate lost circulation with high AVs/ECDs. A casing circulating packer is cheap insurance on any casing job, but it makes no sense not to use one for a high-angle well (>45°), if there are any problems or if fill is anticipated.

An undergauge hole may be caused by mobile formations squeezing in (e.g., salt or some shales) or thick filter cake buildup. If logs are run before casing, a 4-arm caliper should be included in the logging program for drilling evaluation. This gives an accurate cement volume, and will indicate if there is an undergauge hole that needs to be reamed out prior to running casing. In this case, a wiper trip after logging before running casing is justified.

If filter cake buildup against a permeable sand is a problem, there should also be concerns regarding differential sticking. One thing that can be done is to ream down through these intervals to remove the filter cake and spot weighted, supersaturated salt pills across the sands. This will delay filter cake buildup by several hours as the salt slowly dissolves but meanwhile acts as a fluid loss control agent. Even if the filter cake has started to build up by the time casing is run in, it will not be as thick as it would be otherwise. This method can also be used as a nondamaging temporary fluid loss/filter cake control agent for taking RFT or MDT pressures and samples, and to minimize the risk of differentially sticking these kinds of tools.

If a build, drop, or directional motor assembly is planned, a relatively stiff BHA should be used to ream through these sections prior to running casing. When reaming a kickoff in softer formations, take precautions to ensure that a sidetrack is not drilled. Use low WOB and higher RPM and if necessary, make a special trip with a bit-sized hole opener and undergauge bullnose to ream out the hole.

Ledges can occur when there are large changes in formation hardness. Any resistance seen while tripping prior to casing should be reamed out straightaway using low WOB and high RPM.

$$\begin{bmatrix} 2.7 \end{bmatrix}$$

Cementing Program

Well cementing is a specialist area. While the main considerations for writing a drilling program are discussed, ultimately it is important to use the expertise of a cementing service company to design the optimum cement job. The objective of this section is to help formulate ideas for cement job design and to ensure that the most important factors are covered. Further reading is recommended; in particular, the book *Well Cementing*, written by Erik B. Nelson and published by Schlumberger Educational Services (order number SMP-7031), is probably the best practical reference work on the subject.

Normally the drilling program will contain only outlined details on the cement job. Many specific details must be sent out shortly before the job as a supplement or amendment to the drilling program. Information not available before spud, but which will have impact on the casing and cementing programs, includes:

- information from drilling (e.g., actual lithologies encountered, hole problems, and directional profile)
- logging (e.g., enlarged hole and formations requiring special considerations)

■   slurry recipes (designed using samples of cement, mixwater, and
     additives taken from the rig)

One of the most difficult and critical things to achieve is a good
primary cement job. Remedial cementing is difficult and expensive to
execute and in practice rarely achieves a satisfactory result. This means
that the primary cement jobs must be planned so that the objectives are
met the first time. A few dollars saved at the expense of job quality can
potentially cost millions in remedial work later.

Quality control procedures are vital—first, to ensure a good job
and second, to record all necessary data to evaluate the job afterwards.
Field quality control (QC) procedures are noted in Section 3.2.2,
"Quality Control." The program should specify the QC procedures to
follow and flag any particular recording requirements.

2.7.1. Slurry Properties

Density. An amount of cement powder will require a certain volume of water to hydrate it and make it pumpable. Excess water will be
left as "free water" after the slurry sets. The point where the correct
amount is used to completely hydrate the cement with no free water
is known as "neat" cement. For API class G cement, the water requirement is 5 U.S. gallons for each 94 lb sack and the resulting slurry
weight is 15.8 pounds per gallon.

To modify the slurry density, additives such as bentonite are used
to soak up extra water (extenders), allowing a lighter slurry, or
weighting agents are used (barite) to increase the density. Neat slurry will generally have a higher compressive strength than extended
slurry. There are other methods of modifying slurry density; for
instance, by using hollow glass microspheres in the slurry or foamed
cement for extremely light slurry (perhaps when cementing in zones
of total losses).

Usually a casing cement job will use two slurries: a light "lead"
slurry ahead and a denser (often neat) "tail" slurry around the shoe.
Using neat slurry for the whole job would increase hydrostatic and
circulating pressures in the wellbore and is less economical than
using an extended lead. If satisfactory shoe strength for drilling ahead

is the main goal, then this type of cement job should meet the objectives. Normally only one slurry is used for very small cement jobs, such as a liner.

The upper limit on hydrostatic pressure in the well will be dictated by any weak or loss circulation zones, allowing also for circulating pressure losses. The lower limit will need to maintain well control. Casing flotation and collapse should also be calculated.

When lab testing the slurry recipe, ensure that the slurry is still pumpable at 1 ppg more than the designed weight to allow for mixing inconsistencies on the rig. On the rig, a pressure balance should be available to measure the true density of the slurry while mixing.

Fluid loss. Fluid loss properties of the slurry should be measured and, if necessary, additives should be used to modify this. Fluid loss is important because when cement slurry is placed across a permeable formation, loss of filtrate into the formation will dehydrate the slurry. This will affect the setting time (even cause flash setting if extreme), set strength, and may lead to channeling. In addition, a high fluid loss will build a thick filter cake that will narrow the annulus, leading to increased annular pressure drop and possibly induced losses or fracturing.

A general recommendation for maximum API fluid loss is 100 cc/30 min for casings or 50 cc/30 min for liners to prevent channeling. For horizontal wells, use the lowest fluid loss that can be reasonably achieved (due to the large exposed permeable formation), <50 cc/30 min. For cementing against gas zones, fluid loss should be as low as possible; values down to 15 cc/30 min have been used.

Free water. Any water that is used in excess of that needed to completely hydrate the cement and additives is known as free water. An excessive free water property will increase settling of solids with water forming within and on top of the slurry.

In a deviated well this can cause a channel at the high side of the hole and in a horizontal well this will be even more critical. In general, the *maximum* free water should be 0.5%, less in a high-angle well (>45°), and zero in a horizontal well (>80°) or against a gas zone. Density readings taken at the top and bottom of a vertical 1 ft slurry sample should vary by less than 1/2 ppg once the sample has stood for long enough to gel.

Excess free water will promote gas migration and could form a channel on the high side of the hole if cement settles out with free water moving up. This problem gets worse as inclination increases; in a horizontal bore it will form a perfect communication channel on top of the cement. During cementing, samples are taken to show when the slurry has thickened. Often these form a pool of water on top, showing that the slurry was mixed with excessive free water. This is a sign of a less than optimum job.

When using recirculating mixers in the field, cement slurry densities can vary quite a lot. This can lead to parts of the slurry having significant free water. The slurry should not be mixed and pumped faster than is possible to closely control density. Check which type of equipment is available on the rig for mixing the slurry. If care is not taken on a manual recirculating mixer, cement density can vary by ±2 ppg!

The latest generation of field-mixing equipment has microprocessor-controlled mixing, leading to much more homogenous slurries than are possible with manual systems. If possible, batch mix critical slurries such as horizontal wells, tail slurries, and small-volume slurries.

Thickening time. The thickening time of slurry is tested as part of the slurry design. Procedures for testing are given by API specification 10, which measures the time it takes for the slurry to reach a consistency of 100 Bearden units (Bc) at downhole temperature and pressure. The Bc is a dimensionless value that cannot be directly converted to oilfield viscosity units such as poises. While the test measures the time to reach 100 Bc, it is generally accepted that the limit of pumpability is reached at 70 Bc. The test lab can be asked to provide both values.

The thickening time generally should be enough to displace the cement and circulate it back out if problems occur. Since it is measured at downhole temperature and pressure, the time taken for batch mixing on surface will not count minute for minute against thickening time. In critical jobs when the slurry has to be batch mixed but thickening time needs to be as low as possible, the test can be run by mixing the slurry, agitating it for approximately 30 minutes at surface temperature/atmospheric pressure, before applying downhole conditions.

Compressive strength. Compressive strength measurements do not give the actual point of failure of the placed cement. The measurement is used more for comparison purposes, but the tested compressive strength can be related to the application. API specification 10 defines the procedure.

Cement is placed in a 2 in cube mold and set. After removing it from the mold, the cube is subjected to a uniaxial force using a hydraulic press until it fails. The reason why it does not give actual failure pressure downhole is because it is unconfined at the sides. In the well, the cement is confined sideways by the formation and casing, and downhole conditions of temperature and pressure will alter the true compressive strength.

A more recent innovation is the use of ultrasound for measuring compressive strength. Since this is nondestructive and correlates well with physical measurements, compressive strength development can be continuously monitored during curing.

For supporting casings, a minimum measured compressive strength of 500 psi is recommended. For perforating, 2000 psi is recommended.

Temperature rating. The thickening time and compressive strength buildup are dependent on well temperature. The slurry must have sufficient pumpable time to complete the job, with a safety margin in case of problems. In addition, the thickening time should not be so long that rig operations are unnecessarily delayed while waiting on cement. The thickening time is determined in the laboratory using samples of cement and mixwater sent in from the rig.

Accelerators or retarders can be used to lessen or lengthen the pumpable time and will similarly affect the rate of compressive strength buildup.

Under static (nonpumping) conditions, the well will have a temperature gradient as the formations get hotter with depth. Circulating will decrease the local temperature around the wellbore. Thus at any particular depth, two working temperatures will be relevant to cementing operations: circulating and static. A temperature log run some hours after finishing circulating will give the bottom hole static temperature (BHST) at the bottom of the well. The bottom hole circulating temperature (BHCT) at depth can be calculated by reference to API specification 10, which contains temperature schedules. It is also possible to measure this temperature directly during circulating with small thermosensitive probes.

Of these two temperatures, BHST is relevant to investigating cement stability and compressive strength development with time. BHCT is used when calculating pumpable time. As a rule of thumb, the static temperature at the depth of the top of cement should not be less than BHCT used in slurry design. If it is significantly less, it may take an unacceptable length of time to cure; in this case, extra testing should be done at the actual TOC static temperature to see if the cement characteristics are still acceptable. Consider a multistage cement job if the slurry at required TOC will not cure as needed. Note that for deep, hot wells (BHST > 110°C [230°F]), the long-term stability of Portland cement requires the addition of silica flour.

Thixotrophic cement. A fluid is said to be thixotropic when it forms a structure exhibiting high gel strength when shear stress is removed but will thin again once shear stress is reapplied. The viscosity of the fluid will vary for a particular shear rate depending on whether the shear rate is increasing or decreasing; therefore, viscosity is also dependent on history as well as shear rate.

If the fluid is at rest and has gelled, then a high stress has to be applied as shown in Figure 2-8. However, if the fluid is moving and measurements are taken while reducing the shear rate, the fluid looks Bingham in nature.

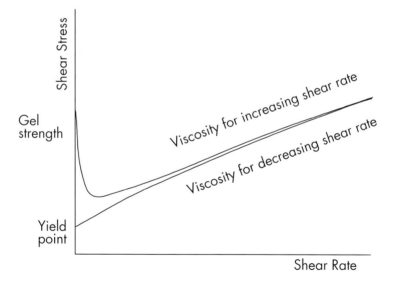

Fig. 2-8 Viscosity of a Fluid that Varies with Increasing or Decreasing Shear Rate

Thixotropic cements also tend to show increasing yield points and gel strengths after each rest-pump cycle. This makes it particularly important not to stop pumping unnecessarily during the job, otherwise high pressures can be imposed downhole when starting to pump.

Thixotropic cements are useful for ensuring that slurry does not move after placement (but not for cement plugs, as previously noted) or when curing lost circulation with smaller slurry volumes where the cement will gel up in the loss channels after displacing into the loss zone.

Expanding cement. Slurry can be made to expand after setting. Additives (most commonly, Ettringite) form crystals whose bulk volume exceeds the total volumes of the reagents. This will help form a good bond to casing and formation if a microannulus may be formed. Ettringite forms long needle crystals which impart thixotrophy to the slurry but if thixotrophy is undesirable, a dispersant will reduce the thixotropic behavior.

Low temperature cement (permafrost). Cement liberates heat during setting. If drilling in permafrost, cement heat will lead to thawing of the permafrost layer and unstable ground close to the surface. Slurries can be prepared that do not freeze, have a low heat of reaction, and develop sufficient compressive strength at low temperatures.

2.7.2. Chemical Washes and Spacers

Chemical washes (low viscosity pills with a density between mud and cement containing dispersants, surfactants, etc.) work by diluting and thinning the mud. This aids mud removal and the thinning action helps prevent mud flocculation and gelling. Chemicals in the wash also help to remove mud film from the casing, which will aid bonding, especially if oil-based mud is used.

Spacers could be high or low viscosity and have a higher solids content than washes. The abrasive action of solids may help clean surfaces to be cemented.

Correct design of washes and spacers is important and several factors must be considered.

1. The chemicals used should be compatible with the formation fluids, mud, and cement so they do not cause adverse reactions (formation of viscous fluids) or reservoir damage (precipitates or emulsions). The chemical wash is meant to disperse the mud and clean out residual solids. Spacers are used to act as a buffer between

mud and cement and to aid cleaning formation and casing surfaces.
2. Rheology must be correct. Turbulent flow is very beneficial in removing gelled mud and any solids left in the wellbore. Sometimes a turbulent spacer is used to disturb cuttings beds followed by a higher viscosity spacer to sweep up the solids.
3. Volumes of washes and spacers should ideally be enough to give at least 10 minutes of contact time at the planned displacement rate.
4. Alternating high and low viscosity spacers can be used in high-angle holes where solids will have to be removed.
5. One technique that is not so common nowadays is to place a scavenger slurry ahead of the main slurry. This is just a thin cement slurry, which may be too thin to actually set up in any reasonable time, but it will help prepare the wellbore for the cement behind.
6. If expensive mud is in the hole that could be stored and reused, it would be worthwhile to program enough spacers so that all the mud is displaced out.

Removal of mud and mud solids and water wetting the formation and casing surfaces are essential in creating a good bond with the slurry.

2.7.3. Factors for Ensuring a Good Cement Job

There are two main keys to getting a good cement job. The first is fully displacing mud from the intended interval to be cemented and the second is having a properly formulated, homogenous slurry placed correctly.

Mud removal. Many interrelated factors come into play when considering mud removal. Following are important considerations:

1. *Casing centralization.* Centralization must be assured if cementing outside a casing (as opposed to setting a cement plug). Fluids will flow preferentially around the largest space if the casing is cemented eccentrically, and fluids in the narrowest part of the annulus may not move at all. API recommends a minimum standoff of 67% (where 100% is perfectly centralized). Standoff percentage is obtained by dividing the shortest distance between the casing and the wall by the average radial clearance. (See Fig. 2-9.)

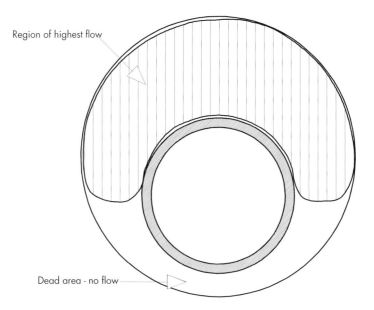

Region of highest flow

Dead area - no flow

Fig. 2-9 Fluid Flow in an Eccentric Annulus

2. *Pipe movement.* Rotating or reciprocating the pipe during displace-
    ment has a definite beneficial effect. This is probably because pipe
    movement causes side to side movement which effectively moves
    the area of preferential flow around, thus getting better movement
    of mud that might otherwise stay gelled up in a dead area.
    It is possible to use liner hangers, which allow pipe rotation after
    setting. With casings, reciprocation is easier than rotation, though
    casing rotation is possible by using a rotating plug container. Hole
    conditions need to be good to allow movement during displace-
    ment or the casing could get stuck off bottom.

3. *Gauge hole.* If there are significant hole enlargements, then the mud
    in the outer reaches will gel up and be very difficult to move. High
    flow rates giving turbulent flow, combined with casing movement
    and effective spacers and washes over a recommended 10 minute
    contact time, will provide a chance to move these mud gels. Also
    conditioning the mud well during the last bit run in the hole so
    that gels are minimized will help. However, with very large
    washouts, it may be impossible to displace mud from the outer
    reaches of the washout.

Accessories can be used on casings (turbolizers) that are designed to deflect mud flow into enlarged areas. A turbolizer is like a spring centralizer with fins added on.

4. *Turbulent flow.* Even if the mud and cement cannot be designed to become turbulent at the possible flow rates, the spacers could be thin enough to be turbulent. This would be the best flow regime for displacing mud.

5. *Chemical washes and spacers.* Chemical washes and spacers help to displace mud and mud solids and also prepare the formation and casing for the slurry. Wash and spacer design is covered later.

6. *Buoyancy.* The slurry density should be greater than the mud density and the spacers should come in between.

7. *Use of proper cement plugs and shoetrack.* The bottom plug, dropped ahead of the cement slurry, will scrape the film of mud off the inside of the casing that would otherwise adhere to the steel. If a bottom plug was not run, the top plug would wipe off the mud and this would lead to contamination of the cement slurry in the worst place—around the casing shoe. The shoetrack (usually two joints between the float shoe and float collar) contains a volume of slurry when the plug is bumped. This is to ensure that minor leakage of mud past the top plug, or minor wiping of mud by the plug, contaminates cement which ends up in the shoetrack, so as not to compromise the cement around the shoe.

8. *Continuous pumping.* After running to depth, casing is usually circulated for at least 110% of the casing internal volume. This checks that no foreign bodies in the casing will plug the floats before cement is pumped. While circulating for this extended time, gelled mud may be broken and moved. However, if pumping is stopped for any period before cement is placed, mud may gel up again in dead areas.

9. *Pump a good excess volume of lead slurry.* If the first part of the slurry is contaminated, it does not matter if there is enough uncontaminated slurry behind it.

Solids removal. In high-angle hole sections, especially if the hole is overgauged, solids may have settled on the low side. These cause problems running casing (solids pushed ahead by the casing can pack off and stick the casing) and cementing (leaving an uncemented channel which could allow communication behind the casing later). Sometimes even if the hole is clean at the last trip out, continued caving may lead to solid beds before the casing reaches bottom.

The possible solutions to this include:

■ Designing a mud system to minimize wellbore instability (refer to the notes discussed in "Drilling fluid" in Section 2.9.1).
■ Using a casing circulating packer and carefully washing down each joint once the shoe reaches any unstable zones. Be careful not to induce losses or fracture the formations since ECD can be high due to the narrow annulus.
■ Pumping turbulent spacers to try to disturb cutting beds.

2.7.4. Cementing Design for Casings and Liners

To correctly specify the cement properties, define what the job is to achieve. The most important objective is to have sufficient slurry (and spacer) density to control downhole pressures at all points of the displacement cycle. Further objectives may include minimum ECD (related to density and rheology) to ensure that a weak zone is not fractured, compatibility with formation fluids and formations, isolation of weak or permeable zones, protection of water sources, and low fluid loss to minimize filtrate invasion in sensitive zones, etc.

Having defined the objectives for each cement job, the cementing contractor should propose slurry designs to meet those objectives. Various slurry objectives and design properties that will be relevant to cementing casings and liners are discussed below.

The cost of cementing during the well will be a significant proportion of the total well cost. The slurry cost per barrel and the cost of each additive should be shown. If expensive additives are used, check to see if a cheaper additive is available that will still meet the objectives. One example is where gas-blocking additives are used only for fluid loss; cheaper fluid loss additives should be available if no gas-bearing permeable zone is present.

It is worth checking through the proposed cementing programs twice; the first time to check that all the parameters (volumes, depths, temperatures, etc.) are per program, and the second to compare the detailed proposals with the specific job objectives.

Cement job objectives for the conductor. Most conductors from nonfloating rigs piledrive the conductor in. For a cemented conductor pipe, the main criteria will be supporting the weight of the pipe and diverter (if used) and minimizing losses. The drill mud will usually be viscosified water/bentonite (spud mud) or seawater with viscous slugs. An inner string technique will be used and cement slurry may be mixed and pumped until cement is seen at surface. No spacers will be needed—just water ahead.

On a subsea well where slurry is to be pumped until the ROV detects returns, the slurry can sometimes be hard to see. A few handfuls of Mica LCM mixed in with the first part of the slurry or spacer can be more easily seen when it returns to the seabed and this is more effective than the dyes usually used. Both can be added if desired.

Cement slurry should not be too dense since losses may be easily induced. It is common to displace extended lead slurry and finish around the shoe with heavier tail slurry. This saves on cementing costs, puts strength where it is needed at the shoe, and gives ample support to the casing higher up as well as reducing bottom hole pressures. A compressive strength of 500 psi will be adequate to support the casing.

Cement job objectives for surface casing. This string is again usually cemented to surface. This cement job is not often complex, though a fairly large volume of slurry may be needed since annular capacities tend to be big.

Shoe strength will be important to maintain well control while drilling ahead. A competent shoe is probably the most important property of the final job. On land rigs the surface casing may penetrate water sources, which need to be protected from produced fluids.

Cement job objectives for intermediate casings. The deepest intermediate job tends to be the largest volume of slurry pumped in one attempt on the well. If the shoe is set deep, with cement high up (say into the previous casing), check that the well temperatures at TOC and shoe are close enough for one slurry design. A multistage cement job may be necessary if this cannot be done with one slurry. As a rule of thumb, the BHST at TOC should not be less than BHCT. Slurry designed for BHCT will then have a satisfactory setting time at the top.

A long section of heavy slurry may give enough pressure differential down low in the string (with mud inside after displacement) to collapse the casing and it is also possible for casing to float when the cement is displaced if there is enough difference between slurry and mud density. Check these when the slurry and spacer densities are known. Inducing losses or fracturing formation is also possible, especially since ECDs will be high during displacement.

As with the surface casing, obtaining a competent shoe to drill ahead may be the primary function of the set cement. The main reason for a high top of cement would probably be to isolate troublesome or permeable formations. Some regulatory bodies require abandoned wells to have cement across all exposed formations, while others only require cement against permeable zones.

Cement job objectives for intermediate casings: multistage cementing. If possible, it is best to avoid cementing up into the previous casing to allow more options for sidetracking later. Before deciding to perform a multistage cement job, consider whether it is necessary to have TOC so high that a multistage cement job is called for. One reason may be to isolate a higher permeable formation from the well annulus.

A multistage job could also be used to isolate a potential kick zone higher in the annulus while avoiding an excessively long cement column all the way from bottom. That is, the stage collar is set above the primary job TOC and below the zone of interest, and a separate job is performed. Therefore, there will be an uncemented interval between the two jobs.

There are two main disadvantages to multistage cementing: the stage collar will always be a weak point in the casing string (a potential leak point) and the equipment sometimes fails to function correctly during the cement job. It also adds to the cost of the well.

Cement job objectives for production casings. The main objectives for production casings (and liners) will be quite different from those for the previous jobs. First, isolating the producing formations without damaging them is vital for the whole production life of the well. This requires good cement-formation and casing-cement bonding and low fluid loss and filtrate-to-formation compatibility. Gas channeling through the gelled (unset) cement must be prevented.

The cement must be able to withstand perforation charges without shattering and be strong enough for any hydraulic fracturing, back-surging, or other mechanical treatments. Isolation within the zone may be important to prevent water or gas production along with oil. Compressive strength above 2000 psi is recommended for perforating.

In addition, maintaining well control during cementing is important as with the previous jobs.

Cement job objectives for liner cementing. One of the most difficult aspects of cementing liners is the small cement volume needing to be placed in a narrow annulus. However, several things can be done to improve matters:

1. Batch mixing the slurry will produce a more homogenous slurry and a better job.
2. Under-reaming below the previous shoe gives better annular clearance. (Ideally, hole diameter should be 3 in greater than casing/liner diameter.) This was done in the North Sea where a 5 in liner was normally set in a 6 in hole. Under-reaming the 6 in hole to $8^1/_2$ in avoided problems experienced while running liner on the earlier wells (none of the liners had reached bottom in the 6 in hole but they all did in the under-reamed hole) and cement logs were much better in the larger hole. A PDC under-reamer did the job in one fast run so a lot of time was not taken.
3. Liner rotation during cementing will improve mud removal.
4. Pump a good spacer volume ahead and as much cement excess as necessary.
5. Check that the mud or formation fluid does not contain chemicals, which may cause flash setting of cement—such as calcium chloride.

The cement job objectives for a production liner will be similar to those described for production casings.

Cementing against massive salts. When cementing in massive salts, the cement forms an essential part of the casing string integrity. Inadequate cement here will make shearing, distortion, or failure of the casing possible if the salt moves. The potential failure modes include:
1. Point loading of the casing due to uneven salt closure. The casing can collapse with much less force than would be the case for even loading.
2. Collapse due to overburden pressure being transmitted by mobile salt.
3. Shearing of the casing due to directional salt flow.
4. Corrosion of the casing, particularly if magnesium salts are present.
5. Long-term degradation of the cement sheath by ionic diffusion into the cement, if it is not salt saturated. If the cement sheath degrades, uneven loading may occur leading to eventual collapse.

The contributing factors to these failures include:

1. Salt creep causing hole closure. This occurs faster in bigger hole and is proportional to hole diameter; a 16 in hole will reduce in diameter twice as fast as an 8 in hole. Lowered hydrostatic pressures will increase the rate of creep.
2. Salt flow due to directional field stresses.
3. Leaching by mud and cement leading to overgauge hole and slurry chemistry alteration.
4. Ionic diffusion of salts into a nonsaturated slurry after setting. Magnesium is particularly detrimental.

The essential objectives are to cement throughout the whole salt body interval, to ensure that good cement completely fills the annulus, and to prevent long-term degradation due to ionic diffusion. Several things can be specified in the drilling program to maximize the chance of success:

1. Use a salt-saturated slurry. If the slurry is unsaturated at downhole temperature, substantial quantities of salt can be leached out by the slurry. This will give overgauge hole and significantly affect thickening time, rheology, and compressive strength. Supersaturating slurry may involve heating the mixwater to dissolve more salt. At these saturations, special additives (especially dispersants and fluid loss) are needed.

    Saturated KCl slurries give higher compressive strengths faster than saturated NaCl slurries. Setting time is important (see #2).

    Salt-saturated slurries can cause problems against other formations. If exposed long term to unsaturated formation water, osmotic forces will leach salt out of the cement slurry that can lead to cement failure. This may or may not be a problem, depending on what formations are exposed and where.
2. Use fast setting times. Once cement gels and hydrostatic pressure is lost, salt creep rate will increase substantially. With long setting times, the salt could creep in enough to touch the casing. Since salt does not creep uniformly, the resulting point loading on the casing will quickly collapse or deform it. Even the strongest casing cannot resist such point loadings.

3. Use suitable drilling fluids to minimize leaching out the salt. Large washouts will lead to the normal problems of mud removal and these will lead to an incomplete cement sheath. However, using oil or salt-saturated water muds can cause problems because the hole will close in while drilling. A bi-center bit to drill a slightly over-gauge hole may be needed.
4. Increased mud densities will reduce the rate of creep.
5. Low salt slurries have been used successfully in the Gulf of Mexico and other areas. These give a fast development of high compressive strengths. These slurries will avoid problems against other formations due to osmosis as mentioned in #2. However, washouts are still likely to occur and long-term ionic diffusion may be a problem later.

Clearly, cementing against massive salts is a complex problem if the well is to meet its long-term objectives. The success of this cement job starts when drilling through the salt (minimizing leached washouts). Good planning, expert involvement, and attention to every detail including slurry and spacer design, rig equipment, downhole casing configuration, and cement job supervision/quality control are vital.

Cementing against permeable, gas-bearing formations. Slurry design in this case will try to minimize the flow of gas into the setting cement. In order to understand the design requirements, it is first necessary to understand what happens as the cement slurry sets.

Cement after placement exhibits complex non-Newtonian fluid behavior. It has a yield point, plastic viscosity, and gel strength. It transmits full hydrostatic pressure from its own density and from the fluids and pressures above it.

With time, gel strength develops and the cement enters a transition state at the gel strength of around 21 lbs/100 ft^2. The cement solids start to form chemical and electrostatic bonds and gradually the hydrostatic gradient of the slurry decreases to that of the mix water. At this stage, the slurry is similar to a porous formation with mix water in the connected pores. Calcium silicate hydrates form as setting continues and this leads to a reduction in the bulk volume of the slurry because the products of the chemical reactions have less volume than the total volumes of the reactants. Portland cement shrinks up to about 5% depending on the exact constituents of the cement. The pore pressure within the cement matrix now drops rapidly due to the low compressibility of cement slurry. The transition state ends when the gel strength is enough to prevent large gas bubbles from percolating upwards due to buoyancy, at around 250 lbs/100 ft^2.

Gas can start to flow from the formation as soon as the pressure imposed on the formation drops below formation pore pressure. This may happen as the cement gradient drops to that of mixwater or, later, when the bulk cement volume reduces. The gas can enter the cement matrix and create channels through the cement, which cannot be closed by cement hydration.

There are several contributing factors to gas flow during or after cementing:

- If the hydrostatic pressure falls below pore pressure during displacement or before the cement is fully set.
- If there exists a channel of mud within the slurry (e.g., poor mud displacement or contaminated cement).
- If a microannulus exists between formation cement or cement casing.
- If gas movement into the slurry occurs, at best the cement will be porous and at worst there will be a channel allowing gas migration upwards to another zone or to surface. This could ultimately lead to a blowout.

Slurry design for gas zones therefore targets the causes of the above mechanisms.

1. *Effective mud removal is an absolute prerequisite for a good cement job.* This was previously covered in detail.
2. Low fluid loss is one of the key elements in cementing gas-bearing zones. If filtrate leaves the slurry for the formation, the resulting volume loss will cause pressure drop in the slurry, which is likely to allow gas flow. This is second in importance only to effective mud removal.
3. A good cement bond between formation/cement and formation/casing must be obtained and must not be broken. Pressure testing casing after cement setting, or displacing the casing to a lighter fluid after cement setting, may create a microannulus. Weak bonds may provide a path for gas to break through. Injection/frac pres-

sures or thermal stresses can also weaken cement bonding to casing during the well life, providing possibilities for gas migration.

4. Additives can be used that reduce and/or eliminate slurry shrinkage during setting.

5. Zero free water in the slurry is important, otherwise the slurry will not be homogenous as lighter, unbound water migrates upwards.

6. Shorter setting times (up to the end of the transitional state) are clearly beneficial. Longer setting times allow more opportunity for gas to flow.

Another technique is to set an ECP above the gas zone, on the basis that this mechanical barrier will form a limit to upward gas movement. If it seals against the formation, then this should indeed stop migration. However, the slurry beneath the ECP will lose pore pressure much faster than if the ECP was not used, which may lead to porous cement below the ECP.

It is worth noting that extended cement with a high water content can show permeabilities in the millidarcy range (up to 5 md) when set. Over extended time, gas can flow through the set cement matrix, which may lead to the slow development of annular pressure.

As explained above, effective cementing against gas zones requires definition of objectives for the life cycle of the well, gathering of all relevant information, expert input, careful planning, and proper execution.

2.7.5. Cementing Design for Cement Plugs and Squeezes

Programmed cement plugs are needed for jobs such as abandonment and kickoffs. Placement techniques are important and, with careful planning, the chances of first-time success are generally high. Plugs set in open hole require as much planning as would be put into cementing casing; this is covered in "Kickoff plugs" within this subsection.

The usual squeeze technique is a hesitation squeeze. Surface pressure below fracture pressure is applied and the surface pressure is monitored. As the pressure bleeds off, pump slowly into the well to restore surface pressure. When no more bleed off is seen, maintain pressure until surface samples are hard or earlier if decided to stop for any other reason.

Slurries will have particular design requirements depending on the objectives.

Zone abandonment. When a producing zone is exposed (either open hole or perforated casing), cement is usually placed across the exposed interval and a moderate squeeze is applied. Fluid loss is matched to the formation; a high permeability zone requires a lower fluid loss and a low permeability zone requires a high fluid loss. As squeeze pressure is applied, filtrate will move into the near wellbore zone. Cement wall cake will form on the face of the formation. When perforated casing is being abandoned, cement wall cake is desired to be built up inside the tunnels. However, if the fluid loss is too high against a very permeable zone, cement cake nodules can build up at the casing inside face rather than getting cake buildup inside the tunnels.

If a cement retainer is used, it can be set on wireline and the cement stinger can be stabbed in to pump the cement. For this to work it must be possible to easily squeeze off fluid into the zone or cement will not reach the formation face. Otherwise, a retainer can be run on drillpipe, cement can be pumped over the zone, pipe can be pulled back, and the retainer can be set above the zone. Squeeze pressure is applied and once no more slurry can be squeezed in, then the pipe can be released from the packer leaving squeeze pressure below the retainer. Any cement left in the string is dumped on top of the retainer.

Spot slurry across the zone a distance above (100-300 ft) if a cement retainer is not being used. Then pick up above TOC, circulate out excess slurry, close the BOP, and apply squeeze pressure.

During well abandonment, cement sometimes has to be spotted outside casings due to exposed permeable zones. The casing can be either perforated (4 shots at 90° phasing) or cut, recovered, and cemented, or can be placed across the cut off and over the exposed formation. In the first case, the techniques will be as described above. In the second case, see the next topic.

Well abandonment. Cement plugs may be set across liner laps, across cut off casings, or placed at an intermediate depth in a casing if a lower plug fails. The cement does not need any special properties, just retarded or accelerated to give a reasonable setting time.

Refer to the placement techniques described for kickoff plugs, which follows. The placement needs for a plug in casing are less severe than an open hole plug, but some of the techniques are applicable and easily used.

Kickoff plugs. Particular reference and acknowledgment is made to the article *Process Implementation Improves Cement Plug Success* in the Petroleum Engineer International cementing supplement.

When an open hole plug is set for directional kickoff (perhaps after abandoning a fish or plugging back to sidetrack to a new target), the objective is a correctly placed plug of high-compressive strength. If the cement is weaker than the formation, because the native strength is low or due to contamination, it will be hard or impossible to get away from the old hole. Other open hole plugs may not need high strength but require as much planning to assure first-time success.

Successful open hole plugs need to address the following design criteria:

1. Temperature at the setting depth
2. Mud removal
3. Slurry properties
4. Slurry volume
5. Slurry stability after placement

Setting depth temperature. If the temperature that the slurry will be subjected to during setting is not known, a temperature log should be run. Over-retarding the slurry will be detrimental, both for plug stability after placement and for wasting rig time waiting on cement.

Mud removal. In some ways, mud removal requires more thought than with a casing job. In washed out zones, the annular capacity around the cement stinger is much higher than if casing was in the hole.

If a traditional, muleshoe type cement stinger is used, cement exits the stinger jetting downwards where it will mix with mud before starting back up the hole. In addition, the only mechanism in this case for moving gelled mud in enlarged areas is pure speed/annular velocity. This also tends to contaminate a lot of the slurry since mud is slowly moved into the fluid stream as the job progresses.

The answer to these problems is a jetting tool. This is a joint of tubing or drillpipe, closed at the bottom and with holes or nozzles pointing outward. The best results are reported with ports placed at a tangent to the inside diameter and tilted slightly upwards, so that a flow pattern spiraling upwards is initiated. (See Fig. 2-10.)

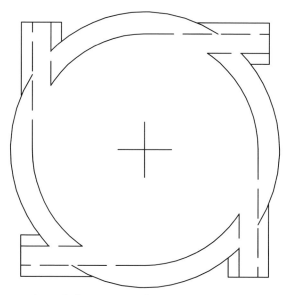

Fig. 2-10 Section through the Jetting Tool

Slurry properties. The considerations discussed previously apply to this slurry, with respect to rheology, compatibility with mud and formation, gas blocking if opposite a gas zone, free water, settling, and low fluid loss.

Settling and mixing is minimized with increased yield point and gel strength. However, high gel strength (or thixotropic slurries) may lead to the slurry not properly filling the void left when the stinger is pulled out.

Free water should be low and the slurry should not settle out significantly during setting so that the top of the cement plug is hard. Thickening time should not be excessive since long thickening times will give more time for the plug to mix with mud. Ideally, good compressive strength will buildup by the time the stinger has been pulled out and the directional assembly run in so that time waiting on cement is minimized. Running a temperature survey will help this.

Slurry density should be close to the mud density to prevent settling (covered in "Plug stability" within this subsection). Heavy slurries combined with dispersants and retarders should be avoided because they will settle downhole.

Fluid loss should be dictated by formation permeability. Refer to the "Zone abandonment" previously covered in this subsection.

Slurry volume. It is important to know the actual volume of slurry required. If a 4-arm caliper is run first, this will give accurate volumes and a temperature survey can be done at the same time.

If the job is planned carefully, it should not be necessary to pump large excess volumes that later have to be drilled out. Allowing 10 to 20% above open hole capacity will account for shrinkage and some mixing at the top interface as long as mud removal is effective and the plug is stable.

Plug stability. When slurry is placed in the open hole without optimizing mud, spacer, and slurry properties, the slurry will tend to slide down the low side of the hole (if more dense than the mud) or move upwards due to buoyancy (if less dense). High gel strength slurry will leave the slurry cored after pulling out the stinger. In vertical or deviated holes (not horizontal) a base for the plug should be used, a high yield/gel mud, and possibly a spacer that reacts at the mud/spacer interface to produce a viscous or gelled fluid.

The following factors are important for first time success:

1. Reduce the difference in density between mud and cement
2. Increase the yield point and/or gel strength of the mud below the cement
3. Use a reactive spacer above the mud as a base for the plug

2.7.6. Special Purpose Cementing

Curing total lost circulation with cement. Offset data sometimes indicates a high risk of total losses. The following general technique can be modified as needed and added to the drilling program. Thixotropic cement is not needed; the technique has worked well in practice with normal cements.

1. Drill ahead blind until it is anticipated that the loss zone has been completely penetrated. Do not pump pills, LCM, fibers, or anything similar that might impede the movement of slurry into the loss channels around the wellbore.

2. Run in with a smaller bit (e.g., 8^1/$_2$ in in a 12^1/$_4$ in hole) without nozzles on heavy wall drill pipe (HWDP) and drillpipe to about 30 ft above the loss zone. (Note: Do not position the bit below the loss zone. Any cement remaining in the wellbore below the loss zone is a waste; if slurry is left above the loss zone it can at least drop down and enter the zone.)

3. Mix and pump 100-200 bbls of extended, low fluid loss lead cement slurry.

4. Pump 100 bbls of extended, low fluid loss tail cement slurry with 0.5 ppb polypropylene fibers added, preferably batch mixed in advance of mixing and pumping the lead for the best quality slurry.

5. Displace with mud. If the annulus fluid level can be estimated, pump a quantity of mud that leaves a small quantity of cement in the string after U-tubing. It is vital to avoid mud entering the loss zone after displacing the cement.

6. Monitor for returns at surface while pumping and displacing cement. If returns are seen, close the BOP and displace with mud. Now that the annulus level is at surface, displace with the string capacity -5 bbls instead of the originally calculated displacement. Slow the pump if necessary so that excessive pressures are not imposed on the well.

7. Pull out of hole (POH). Pull back two stands without filling, then fill the annulus from the trip tank with just enough mud to replace the open-ended pipe displacement. Pull back to the shoe and wait on cement samples in the oven. It is better to add slightly too little mud rather than too much.

8. Run in hole (RIH). Carefully drill out the cement. If losses are seen right away, repeat the cement job.

9. If losses are later experienced in the same zone, repeat the process.

There are two key elements. First, place sufficient cement in the zone around the wellbore to flood the loss channels. Cement left in the wellbore after setting is of no benefit and may cause an inadvertent sidetrack. Second, ensure that the unset cement is not displaced further away from the wellbore by adding more mud than necessary to the well as pipe is tripped out.

The first slurry should enter the loss channels easily. It is a good sign if the fiber cement starts to plug the loss zone near the wellbore (indicated by returns while displacing) because the plugging will prevent migration of the lead slurry away from the wellbore.

If returns are seen at surface, then cement will have risen in the annulus around the drillpipe. If the loss zone is still taking fluid then this should drop when pumping ceases to the level supported by the formation fluid pressure and the cement that has moved up the annulus will probably drop down back to the loss zone.

Repairing damaged casing. The technique will be similar to spotting and hesitation-squeezing cement over perforations (see "Zone abandonment" in Section 2.7.5), usually without a cement retainer. Use slurry with low fluid loss and high strength so that cement penetrates through the split and forms a high-strength sheath in the annulus. Thixotropic slurry may be used so that it will not move away from the area of the split after spotting.

2.7.7. References for Cementing Program—Design

Nelson, Erik B. *Well Cementing*, published by Schlumberger Educational Services (order number SMP-7031).

"Process Implementation Improves Cement Plug Success," from Cementing the Future *Petroleum Engineer International*, Hart Publications, 1996. Supplement.

$$\begin{bmatrix} 2.8 \end{bmatrix}$$

Formation Evaluation

(The logging books referenced in this text are from Schlumberger, but other companies may offer similar books)

Formation evaluation includes several different technological approaches. These techniques are applicable at different stages of the well. The cost of each must be considered; most are fairly expensive. However, it is an investment that should repay more than it costs over the life of the well in improved production and reduced future drilling and workover costs.

2.8.1. Electric Logging and Sampling

Electric logs are specified by the exploration department to evaluate the formations that have been drilled through. Drilling also can gain a lot of useful information from logs; it seems in most operating companies that the drillers do not have much say in the logging program or in the interpretation work that is required.

New logging techniques are being developed all the time and of course what follows cannot be an exhaustive list. The reader is urged to become familiar with the logging company used, their products, what analyses can be done, and what it costs.

It is very important that the proposed logging program is reviewed by the drilling department and modified as necessary to provide the most accurate information for future well planning. The following brief summary shows what kind of information is needed and how it helps improve future plans. Note that for convenience the abbreviations are for Schlumberger wireline tools. Other logging contractors will provide equivalents for some or all of these.

Wellbore profile from a 4-arm caliper. Immediate use is to calculate cement volumes and to spot thick filter cake buildup, which may flag up potential differential sticking of drillstring or logs. Ledges and washouts will show where special care needs to be taken for logging or running casing. For later evaluation, the shape of the hole and the measurements of minimum and maximum sizes give information on how directional the field stresses are. Enlarged sections are worthy of definition and study so that the enlargement mechanisms can be identified and possibly mitigated on the next well. (Refer to Section 2.9.1.)

Sonic data. Sonic properties for shear and compressional waves can be evaluated to give rock mechanical properties and information on formation stresses. Stoneley waves from the dipole shear sonic indicator (DSI) can also be evaluated to give permeabilities. This will aid in bit selection, wellbore stability studies, fracture gradient, and pore pressure prediction. Further, it may indicate fractured formations. It is important to specify that the full waveform is recorded for rock mechanical property evaluation; sometimes the loggers do not record the full waveform. Long spaced sonic (LSS) is not sufficient, array sonic (AS) is better but at a slightly higher cost, and dipole shear (DSI) is the best. If the exploration program is for the LSS or AS, upgrade to DSI.

Sonic tools include the *ultrasonic imager tool (USIT)*. It gives a detailed picture of cement in the annulus (including microannulus and channeling). The USIT also evaluates casing thickness and therefore wear or other damage.

Temperature log. Determines geothermal gradients for cement, mud, and brine design; also detects lost circulation zones.

Resistivity, porosity, and density data. Gives indications of potential overpressure buildup. (Refer to Section 1.4.4.) May be useful for correlation if LWD required in future wells.

Microresistivity. Detects fractures and faults (wellbore stability and directional performance), bedding plane dips (directional), and may produce acceptable directional surveys.

Pressure measurements and formation fluid samples. Used to determine wellbore stability, well control, reducing reservoir damage, and/or cementing problems due to reaction of fluids with mud and cement.

Lithology and porosity determination. Determines bit selection, wellbore stability, mud and cement programs, drilling practices, and casing seat selection.

Sidewall core samples. Used for shale mineralogy studies; and is especially useful if mechanical rather than explosive sidewall cores have been taken. If an MSCT or CST is programmed, try to get one or two samples from each distinct shale lithology.

Gamma ray (GR). Used to determine bit selection, correlation, and wellbore stability. The gamma ray spectrometry tool (GRST) provides details of clay mineralogy and allows good well-to-well correlation.

Seismic tools. Detects the presence of faults. Walkaway survey can look ahead of the bit. Also used for overpressure prediction and bedding planes (directional performance).

This provides a general picture of types of logging tools and the general applications they have in well planning. It is important to study what tools are actually available and to recognize how they may apply to well planning. Wildcat exploration wells will be more thoroughly logged than development wells. In an exploration well, the objective is to gather as much information as possible to improve performance in the subsequent wells, primarily by optimizing bit selection and mud design. On development wells the logs run will have more to do with addressing known problem areas or addressing specific areas of interest.

Electric logs may be run on wireline in open hole or casing. Logging while drilling is self-explanatory; it allows the driller to have a real-time picture of the formations at the bottom before much reaction has taken place with the mud. Some tools offer the ability to look ahead of the bit (seismic while drilling) or to make real-time decisions on casing points or geosteering in horizontal wells, e.g., resistivity at the bit (RAB). Logging while drilling may be done in memory mode as well as surface readout; if memory mode only is used then of course real-time decision making is not possible.

Electric logs can also be run on drillpipe or coiled tubing if difficult hole conditions make wireline logging impossible. These techniques make logging much more expensive than on wireline.

2.8.2. Coring

Coring attempts to remove a sample of formation from the well-bore in an undamaged state. There are two basic classifications of coring operations: sidewall coring and full-hole coring.

Sidewall coring is done using a wireline tool to either blow a recoverable sample chamber sideways into the formation with an explosive charge (e.g., Schlumberger CST), or to use a small rotating corehead to cut a sample (e.g., Schlumberger MSCT). These were mentioned in Section 2.8.1, "Electric Logging and Sampling."

Full-hole coring uses special tools to cut and recover the core. There is a variety of tools for various conditions, which are described below. The single most important factor for a successful coring job is proper planning, which should involve the drilling engineer, drilling supervisor and/or toolpusher, geologist, mud engineer, coring company specialist, and core analysis specialist.

Cores can be taken in any formation, at any hole angle up to horizontal, oriented, in any circulating medium, using rotary or mud motor, and kept under downhole pressure. Coring is very expensive and selection of the best system to do the job is vital, as it ensures that the objectives of the job are fully defined and justified.

Cores have been taken up to 600 ft in length. If a long or very long coring run is planned, then the core barrel will require external stabilization and internal bearings to support the inner liner. Long core runs reduce rig time and therefore cost, but long core barrels should not be used if:

1. The formation contains gas of sufficient pressure to cause the well to kick as the gas expands when the core is pulled (except pressure coring)
2. The bit life will be short due to the hard or abrasive nature of the formation
3. The deviation profile of the well makes it difficult to run or pull the core barrel
4. The likelihood of differential sticking is high since the core barrel creates a large contact area with the formation

In general, it is best to cut and recover the largest diameter of core possible while still allowing the barrel to be fished. This reduces the area that the bit has to cut, improves ROP, and minimizes the likelihood of the core jamming.

Planning considerations. To give the best chance of a successful coring job, all phases have to be carefully planned. If this is not done then the required information may not be obtained and more cores may have to be taken on future wells at high cost. Planning actions include:

- Producing specific action plans and checklists
- Reviewing the plans with all concerned personnel
- Assigning responsibilities to specific personnel

Points to cover during the planning stages that may need answering before the program can be finalized include:

Objectives and justification

1. Which formations should be cored?
2. Has a list been created of the client departments requiring information from the core, as well as the information each client wants? (Do not forget to include drilling. It may be that a core includes nonreservoir material such as shales, and analyzing it could beneficial.)
3. Can the required information be obtained through cheaper alternatives rather than full-hole coring? (e.g., wireline logging, sidewall coring [rotary or explosive], cuttings analysis)
4. What are the criteria for coring? (e.g., on shows, after penetrating a particular marker)

Methods of coring

1. Is the core likely to be unconsolidated, fractured, or otherwise difficult to core?
2. What type of coring system will give the best chance of obtaining all the information needed?
3. If special techniques are called for (e.g., pressurized coring, oriented), is the equipment available in this country and at what cost?
4. Are there any special requirements for the bit, core catcher,

inner tubes material, and/or mud system? Define specific parameters of each.

5. What is the maximum practical core barrel length in the hole size, inclination, and formation?
6. How many coring runs are likely to be required?
7. Who is the person responsible for making operational decisions during coring?
8. What material is best if disposable inner sleeves are used?(e.g., aluminum, fiberglass, plastic)
9. Are there any offset coring runs that might aid in core bit selection?

Core recovery, preservation, and surface handling

1. Should the rate of pulling out of the hole be controlled to minimize core dilation due to hydrostatic pressure reduction? (Most likely to be a problem in the last 1000 ft; may consider slowing down for the last 10 stands when POH.)
2. How can damage to the core be minimized when surface handling? Are special containers called for? Are there special techniques for lying down?
3. Should the core be cut into particular lengths, and if so what?
4. Are there any special requirements for core cutting?
5. What core preservation procedures are required? (e.g., freezing and/or epoxy injection)
6. How are the cores to be marked for transport?

Wellsite analysis and reporting

1. What is required to be recorded and reported at the wellsite?
2. Who is responsible for recording and reporting at the wellsite?

Core transport and storage

1. Who is responsible for the core at this stage?
2. Will any special precautions be needed due to possible hazards from core material? (e.g., pressurized core, H_2S, other toxic fluids)
3. How heavy will the individual core sections be and are the containers, slings, etc. rated for this load?

4. Does the core have to be protected from freezing?
5. How sensitive will the core be to shock or vibration?
6. What documentation is required for transporting the core? (e.g., export documents, cargo manifests, governmental reports)

Laboratory analysis

1. Where does the core have to be transported?
2. Have arrangements been made so that the core can be examined quickly after arriving?
3. Are the laboratory analysis requirements fully defined?
4. Who is to receive copies of the lab reports?

Timing.

1. When should the coring start (earliest/latest dates)?

Safety and quality control

1. Who is in charge during the operation at the wellsite and in the office?
2. Who are the planning focal points from each department and service company involved?
3. Are there any inherent hazards in the operation (e.g., H_2S, pressurized coring)?
4. Can the identified hazards be eliminated or mitigated by changing the plan?
5. What precautions can be put in place to minimize the identified hazards and allow safe recovery?

Contingency planning

1. What events could take place that would prevent coring as planned?
2. Could core be taken as required in a smaller hole size? Are the smaller sizes of equipment available if drilling conditions require an extra casing string to be set higher up?
3. Is it possible that the coring requirements could change based on information obtained while drilling? What impact would that have on the plan/equipment/personnel?

Costs

1. What are the costs of coring equipment and services, rig time, transport and storage, lab analysis, mud treatments, etc.? How does this compare to the budgeted cost?
2. How accurate can the cost estimates be? (May be affected by knowledge of area, certainty of defined coring criteria, etc.)
3. What costs should be included in the drilling AFE (coring system on the rig) and other AFEs (lab analysis)?
4. Who is responsible for tracking all the various cost elements?

Coring system considerations—choice of inner barrel. The length of the core barrel and the type of formation will determine what kind of inner tube to use. Those available include fiberglass (GRP) and aluminum disposable tubes, chrome lined steel tubes and GRP, aluminum, sponge, and rubber sleeves that fit inside standard steel inner tubes.

A disposable tube is best for fractured or unconsolidated formations. The core and tube are recovered together and the tube is sewn up with the core still inside. This protects the core and keeps it as undisturbed as possible. GRP has a smooth ID, which reduces jamming from fractured cores and is resistant to corrosive fluids. Aluminum tubes are more robust than GRP tubes but are not as resistant to corrosive attack. Both types can be run in core barrel assemblies up to 270 ft in length.

Liners can be used inside conventional steel inner tubes. Liners are available in aluminum, plastic, and rubber. The main limitations to using a liner are that a smaller diameter core is recovered and, in plastic and rubber, temperature limitations apply.

Sponge coring (available from Diamant Boart Stratabit) uses an aluminum tube with a sponge lining. This catches fluids expelled from the core as it is pulled from the well. The core diameter will be less than if a disposable or unlined steel tube were used.

Standard steel liners may be used where the formation is consolidated and unfractured. The core is slid out of the liner on the catwalk, sawed up, and boxed for transporting.

Coring system considerations—general type description. Some typical coring systems and their applications are described below. The descriptions are based on systems available from Eastman-Christensen.

Conventional core barrel. Suitable for most formations. Can be run at high angle, though not recommended for horizontal cores. An oriented core can be cut. Usually available in 30 ft sections, which can be made up to 180 ft. Special core catchers can be used in the standard barrel to allow recovery of soft, unconsolidated cores.

Rubber sleeve coring. A rubber sleeve is stretch-wrapped around the core, supporting its weight and protecting it against fluid washing. This is able to recover cores from unconsolidated or badly fractured rock that will not support its own weight in a conventional barrel. Operation of the system is complicated and it is recommended that an experienced operator be available on the rig site to supervise the coring. The application is restricted to near-vertical wells and formations of soft to medium compressive strength.

Horizontal coring. Systems are available to cover medium (286-715 ft) radius and short (20-40 ft) radius wells. Outer barrel stabilization and roller bearings supporting the inner barrel are required. The system can be run on rotary, or can be driven by a positive displacement mud motor.

Applications include all formations except extremely hard and abrasive, high angle, or horizontal. Oriented cores can also be taken.

Slim hole coring. It is possible in slim holes to drill using a continuous coring method. The core bit has a drill plug that can be wireline removed and an inner barrel set on wireline. After coring the length of the inner barrel, the barrel and core are recovered on wireline. Another core can then be cut or the drill plug can be replaced so that full-hole drilling is resumed. This is only applicable in near vertical wells.

Low-invasion coring. This option is available where the core needs to be contaminated by the mud as little possible. The corehead has no inner gauge protection cutters. This leaves the initial filter cake that forms on the core OD undisturbed. The core moves up into a special inner barrel shoe positioned immediately above the last inner cutter, protecting it from washing by the mud coming through the bit.

The mud has to be designed to be low fluid loss with the cored formation, incorporating solid particles sized correctly for efficient bridging. It can be run in wells up to horizontal, oriented if required.

Pressure coring. This system allows the recovery of cores in nearly in-situ conditions. Advantages of this over conventional coring methods include better measurements of fluid saturation, permeability, gas content, and mechanical property data. The core barrel is suitable for all formations except extremely hard abrasive, in vertical to high-angle wells.

To use the pressure coring system a pressure coring service unit is positioned on the rig, which provides all required operations for the coring run and recovery. The unit is built into a 40 ft ocean-going container and is self-contained.

Oriented coring is available. A knife in the barrel makes a score mark along the core at a known orientation. This allows determination of dip and direction of formation and fracture planes. Allow time to order the oriented coring kit.

2.8.3. Mud Logging

Mud logging involves taking measurements and samples during drilling. This includes: taking samples of cuttings, mud, and formation fluid shows and analyzing them; supervising, collecting and boxing cores; logging and recording all important drilling parameters; detecting and warning of the presence of problems such as kicks, H_2S, and washouts; and producing analyses and reports. Recording quantities and descriptions of cavings will allow close monitoring of wellbore stability and mud effectiveness as drilling progresses.

The time-based record can be invaluable when a problem occurs and needs to be investigated later. Such problems that merit later investigation may include stuck pipe, kicks, losses, etc.

The usefulness of the service is strongly related to the quality of the personnel. They should help the drilling supervisor to make better decisions during drilling and will record information to improve planning for subsequent wells.

When selecting a mud logging service company, do not choose simply on the basis of cost. First examine equipment availability, safety records, equipment reliability records, and personnel qualifications and experience. Vendors who would be unable to provide a proper service should be eliminated, regardless of how cheap they may be.

Mud logging unit services specification example. The following example shows some of the services that might be specified for mud loggers.

A fully computerized unit with readout screens is used in the drilling supervisor and toolpusher offices. Calculate the D exponent, sigma log, and gas fracture gradient continuously and monitor the active tank levels and gas readings. Inform the driller (first) and drilling supervisor of any significant change.

The following readings should be taken. At the end of the well these should be available on a $3^1/_2$ in floppy disk as a spreadsheet file or a tab delimited ASCII file (depth-based table):

- Depth, in 5 ft intervals
- Drilling rate, in ft/hr
- Weight on bit, in klbs
- Rotary speed
- Rotary torque, in ft/lbs (Torque should not be reported in motor amps since future evaluation is impossible unless the motor and gearing are known to convert to torque)
- Pump output, GPM
- Pump pressure, psi
- Mud density in, psi/ft
- Mud density out, psi/ft
- Mud temperature in, °C
- Mud temperature out, °C
- Gas readings (total gas)

Printouts are required at the end of the well with a time-based record of bit depth, hook load, RPM, torque, flow rate, pump pressure, and active pit volume/trip tank volume (as applicable).

Place at least two ditch magnets in the possum belly tank. Clean each midnight and report Kg of metal recovery each day. Be careful to measure actual metal content without mud solids, etc. that will adhere to the metal. Describe the type of metal shavings seen (e.g., resemble lathe cuttings, flakes, fine powder, scale).

Cutting sampling. Ditch samples are to be taken at 10 ft intervals to the $9^5/_8$ in casing point and 5 ft intervals thereafter. Three sets of wet samples, three sets of washed and oven-dried samples, and one set of samples for drilling department shale analysis are required in all drilled sections. Show the lithological description for each 5 ft drilled from cuttings samples.

Cuttings testing. Visually examine the cuttings and cavings samples taken by the mud loggers. Look for evidence of avoidable problems and discuss mud treatment alternatives with the drilling supervisor before taking action. If a test kit is available, carry out extrusion tests and/or dispersion tests to check on mud inhibition performance.

Cavings sampling. Cavings give valuable information to optimize mud properties and drilling practices. Analyzed after the well, they allow better optimization of the drilling program for the next well.

While circulating with returns to surface, measure the amount of cavings in lbs/hr coming over the shakers, estimated from a representative sample gathered from one of the shakers over a 5-minute period. Preserve a part of the sample for later examination in the same manner as for the washed and dried cuttings.

Note a basic description for each sample of cavings: average length in inches, shape (e.g., splintery, blocky, evidence of rounding in the wellbore, angular, curved), and evidence of natural fractures (e.g., carbonate-filled veins, clear fracture faces visible). If two distinct types of cavings are present, estimate percentage of the total for each and give separate descriptions. Make note of whether the source formation can be identified by microscope examination and comparison with earlier cuttings samples. Inform the drilling supervisor of any significant change to the amount or appearance of cavings.

Preservation of shale formation samples for drilling department shale analysis. Oven drying samples has a great effect on the reactivity to drilling muds, which will affect analyses to be done in the lab. For shale samples (cuttings and cavings) required for shale analysis, rinse the samples in the base brine of the mud (mud liquid phase), then blot dry. Loosely fill an airtight jar, label with date and time of sampling, and bit depth. Store in a cool place until transporting it to the lab. Avoid rough handling.

$$\begin{bmatrix} 2.9 \end{bmatrix}$$

Drilling Problems—
Avoidance Planning

Many well plans seemed to be designed to get the rig into trouble. This section will look at borehole stability, losses, stuck pipe, etc., from the viewpoint of planning to avoid the problem or to mitigate the effect of it if it occurs.

2.9.1. Wellbore Stability

The following information comes from personal experience and from interesting lunches and E-mails with knowledgeable industry figures, especially Dr. Fersheed Mody of Baroid and Dr. Eric van Oort of Shell. Some of the following is based on a presentation by Dr. Mody at the AADE-Houston Chapter Drilling Fluids Conference and Exhibition.

Every drilling program should be written with wellbore stability in mind. An understanding of the mechanisms, which cause and accelerate instability leading to knowledge of what can be done to avoid it, is essential for all drilling engineers.

Shale comprises over 75% of drilled rock and causes over 90% of wellbore instability problems, ranging from washout to complete collapse of the hole, bit balling, sloughing, or creep. The problem is estimated to cost the industry at least $500 million per year.

Shale instability is closely connected with bulk properties of shales such as strength and deformation, which are a function of depositional environment, porosity, water content, clay content, composition, compaction history, etc. The mud bulk properties such as continuous phase chemical makeup and concentration, properties of any internal phase, additives associated with the continuous phase, and system maintenance are also important. Understanding the fundamental physics and chemistry of the mud/shale interaction is critical.

Other factors such as in-situ stresses, pore pressure, temperature, time in open hole, depth and length of open hole interval, and surrounding geological environment (salt dome, tectonics, etc.) directly impact drilling and completion operations.

This is an extremely complex subject in which our knowledge is not yet complete. For more in-depth information, refer to the many SPE/IADC technical papers and other authoritative sources of information.

The parameters influencing wellbore stability can be divided into five main groups: drilling fluid, rock properties, in-situ stresses, drilling practices, and drilling mechanics.

Drilling fluid. Shales are made up of layers of flat crystals. Between the crystals are tiny spaces (in the order of nanometers in size). Drilling through shales exposes these pore spaces to the wellbore.

With water-based muds, filtrate may enter into these pores and cause shale instability through two mechanisms:

- Causing a local increase in the pore pressure in the near-wellbore region (pore pressure penetration)
- Chemically hydrating the shales

Assuming for a moment that the pore pressure and the wellbore hydrostatic pressure are the same, then the formation stresses will act to push shale into the wellbore. If the shale is weak in tension (which it is), these stresses will be large enough to cause tensile failure and the shale will destabilize. If, however, the mud hydrostatic exceeds the pore pressure by a margin which equals or exceeds the stresses in the formation, then tensile failure will not take place and the shale will be physically stable.

If filtrate invasion is allowed to occur, then a tiny influx into the pore spaces will cause an increase in pore pressure. This pressure cannot dissipate far into the shale since permeability is extremely small, therefore, it will build up in the near wellbore region until it equals the total driving force pushing filtrate in. The extra pressure from the ECD will further increase the local pore pressure as drilling continues.

Note that the API fluid loss value has no effect on filtrate invasion of shales. The pore size is much smaller than the passages through filter cake (as previously noted) and in any case, filter cake will not build up on a shale surface because the amount of filtrate moving into the shale is too small.

The only ways to affect the level of filtrate invasion is either by making the filtrate more viscous and/or by blocking the pore throats with sufficiently small materials. See the material on inhibitive glycol and silicate water-based muds in Section 2.5.5, "Nondispersed or Polymer Water-Based Muds."

Oil-based muds are so inhibitive because capillary forces prevent oil filtrate from entering the pores with no local pore pressure change, and of course if there is no water, there is no hydration. If mud can both prevent hydration and give pore-pressure isolation, together with providing sufficient hydrostatic support against the in-situ rock stresses, then the wellbore will be stable. However, even with oil-based muds, problems can occur. Many shales are fractured and, in this case, oil mud will enter those fractures and cavings due to pressure fluctuations will result. If shale cavings are seen on bottoms up after a trip with correctly maintained oil mud, it is most likely due to fractured shales. With fractured shales, the trick is to minimize trips, minimize mud density (higher mud density will drive mud into the fractures), use good connection practices, and use sized additives to block off those fractures as they are exposed.

There are several chemical and physical forces tending to push filtrate into or away from the pores. The net resulting force will determine whether filtrate invasion may or may not occur; indeed it may be possible to tailor these forces to give a net dehydrating effect by moving pore fluids out of the near wellbore region.

The forces to consider are:

■ Hydraulic—overbalance
■ Chemical—if the mud is able to form a semi-permeable membrane

on the formation face (e.g., invert oil emulsion mud) then the salinity of the mud-water phase can be adjusted to change the net osmotic force. Briefly, if the mud-water phase salinity is different from the pore fluid salinity, then water will tend to move to the area of higher salinity. If the pore fluid is less saline, then a force to move water out of the pores is present. Refer to "Properties of the brine phase—osmosis" in Section 2.5.8.

■ Electrical potential—research indicates that this has some effect but practical recommendations have yet to be made. The forces involved are probably very small compared to the first two.

Rock properties. The strength of rock material is described by strength parameters such as the unconfined compressive strength and/or triaxial compressive strength. However, strength is not the only rock property controlling stability.

First, strong intact rock may behave weak when fractured. The rock strength is determined by the (weak) fractures rather than the (strong) intact material. They can be natural or drilling induced (e.g., drillstring vibrations). Second, borehole stability also depends on the ductility (the degree to which rock can plastically deform without losing load bearing capacity). Plastic deformation results in delay of failure due to transfer of excess load to rock located away from the borehole wall. A ductile rock remains intact under more severe loading than a brittle material of similar strength. The mode of borehole instability is related to the ductility of the formation. A brittle rock can lead to hole enlargement. The borehole wall material desegregates and detaches, i.e., sloughs into the hole as soon as its strength is exceeded. A ductile formation, on the other hand, may experience substantial plastic deformation leading to load distribution to the adjacent intact rock. Hole enlargement is observed in ductile formations as well due to erosion of weakened or damaged material. The strength and ductility of formation material depends on lithology.

The most important lithological parameters are mineral composition and porosity. Typically, high porosity material has a low strength and some ductility, whereas low porosity rock has high strength and low ductility. Given similar porosity, shales behave more ductile than sandstone. The strength and ductility of the shales is also influenced by the water content. In general, the lower the water content the higher the strength and vice versa.

When a good mud cake is formed, the mud pressure does not affect the pore pressure distribution. This is most advantageous from a borehole stability point of view. A good mud cake is usually formed if the permeability exceeds 1 md. If formation permeability is lower than 1 md, mud cakes do not form. They would be ineffective anyway since mud cake permeability is approximately 1 md. Field indications are that borehole instability is much less prolific in formations exhibiting reasonable permeability. Gauge holes have been drilled in weakly consolidated, highly permeable sandstones. Most drilling problems are observed in shales. One of the contributing factors is that shales have very low permeability, typically between 10^{-12} to 10^{-6} darcy and usually between 10^{-10} to 10^{-8} darcy. This makes mud cakes formation ineffective.

A further mechanism for the abundant drilling problems in shales is the water adsorption potential of its clay components. The intake of water causes shales to swell, weaken, and fracture leading to hole failure. The sensitivity to water is large for smectite (montmorillonite), medium for illite, and small for kaolonite and chlorite. This is associated with the clay surface area that controls the amount and effects of water adsorption.

Fractured limestones and coal seams can also cause serious problems. Any fractured rock can be further destabilized by the drilling practices employed.

In-situ stresses.

Collapse. The loads acting on the borehole region consist of the far field in-situ stresses, the wellbore mud pressure, and the formation pore pressure. The onset and severity of borehole collapse is determined by the magnitude of the in-situ effective stresses (i.e., total stress, pore pressure) and the mud overbalance relative to rock strength. Borehole collapse increases as the effective stress increases.

Fracturing. Borehole fracturing becomes less likely as the minimum in-situ stress increases. Borehole fracturing increases as the total minimum in-situ stress decreases and/or mud weight exceeds the breakdown pressures for intact rock, or mud weight exceeds the fracture extension pressures for fractured rocks.

Drilling practices.

Drilling practices can have a dramatic effect on wellbore stability. Consider the following points. Ensure that precautions are designed into the drilling program. The drill crews must be properly briefed on the effect of some drilling practices on wellbore stability. Connection and tripping practices in particular are within the control of the drill crew and a lack of knowledge and/or care here can make the problems much worse.

Surge and swab pressures—connection practices. Pumps should be started and stopped carefully. Kick in one pump at low speed, and as circulating pressure comes up, increase flow rate steadily. Similarly, do not switch off the pumps too quickly since this reduces ECD too fast for the formation near wellbore stresses to react to the removal of pressure.

Tripping practices. Tripping causes substantial mud pressure variations (swab and surge). This will promote hole failure—both collapse and fracturing; the more tripping, the more risk of hole instability. However, tripping serves other purposes such as bit changes and hole cleaning; these will usually override borehole stability considerations. The exception here should be wiper tripping. Routine wiper trips are often done without justification and can lead to wellbore instability with no improvement in hole condition. Do not wiper trip unless this can be justified. When tripping, do not try to run or pull pipe too quickly. Keeping rheology and gels low will minimize the effects of flow rate changes and pipe movement.

Open hole time.

Borehole collapse increases with open hole time. This is demonstrated routinely by comparing caliper runs taken at different intervals after drilling. Time-dependent rock deformation is especially noticeable in salt zones; creep inevitably results in hole closure. Other types of rocks show some degree of time dependency. Unconfined compressive strength may decrease by 50% as time progresses, and thus explains at least part of the observations when drilling with a WBM. Another important factor is the fluid invasion in low-permeability rocks (e.g., shale). In general, fluid invasion and the swelling associated continue with time, gradually enlarging the affected zone and reducing stability. Therefore, factors such as bit selection and parameters that influence the rate of progress can have an effect on wellbore stability.

Hole orientation

Collapse. The combination of in-situ stresses and hole orientation governs the stresses that act to destabilize the rock (collapse mode) surrounding the borehole. In a typical case when the in-situ stresses are S.Vertical > (S.Horiz$_{max}$ = S.Horiz$_{min}$), a vertical well has equal horizontal stresses acting along the cross section, compared to a horizontal well that has unequal stresses (S.Vertical and S.Horiz$_{max}$ or S.Horiz$_{min}$) acting along the cross section. Consequently, the stress concentration at the borehole wall is higher in case of the horizontal hole, making it more prone to collapse. Field data indicate more problems in deviated hole. Note that these are partly associated with increasing open-hole time, dogleg severity, and hole-cleaning problems. Changing the stress state around the wellbore certainly has an influence, but the magnitude can only be ascertained if good estimates of the in-situ stresses (magnitude and direction) are known. The optimum mud pressure in a tectonically relaxed environment (S.Horiz$_{max}$ = S.Horiz$_{min}$) tends to increase with hole angle. In general, field experience indicates an approximate increase of the mud pressure gradient by 2ppg (0.11 psi/ft) between vertical and horizontal. This relationship differs in a tectonically stressed environment.

A deviated well tends to be more stable when drilled in the direction of the principal horizontal stress, and it is least stable when drilled perpendicular to it. This may affect the surface location in severe cases.

Unequal formation horizontal stresses cause directionally preferential hole enlargement. This is why a 4-arm caliper should be routinely run; it allows evaluation of the severity of the directional stresses. In severe cases, the major axis may be off-scale and the minor axis in gauge.

Fracturing. Borehole fracture initiation pressure depends on the borehole orientation. Therefore, leak-off pressure and formation breakdown pressure are in part orientation dependent. The pressure required for sustained fracture propagation and lost circulation is controlled by the minimum stress and is therefore independent of hole deviation and azimuth.

Drilling mechanics. The bottom hole assembly and drill pipe scrape along the borehole wall possibly eroding (loose) damaged rock material. A smooth assembly, such as a barrel shaped stabilizer, reduces the scraping action.

During drilling, the BHA vibrates. In case of heavy lateral vibrations, parts such as stabilizers and long unsupported drill collars may hit the borehole wall, imposing substantial dynamic loads onto the rock material. This may lead to creation of cracks or cavings. The initiation of conductive cracks facilitates the invasion of mud and may jeopardize pressure isolation measures. Cracks also lead to a reduction in overall rock strength. Drillstring vibrations are more severe for hard rocks since the excitation at the drill bit is larger. It is feasible that severe hole enlargement observed in some hard rock sections may be related to drillstring vibrations. On the other hand, severe washouts enhance vibrations. Minimizing drillstring vibrations should reduce borehole instability.

A short story. Time to make a connection. The driller stops the pumps by winding the pump controls rapidly to off. ECD drops very fast (in a few seconds). The pore pressure in the near wellbore region, which has built up during the time since the last connection, now exceeds the wellbore hydrostatic. Since the permeability is very low, pressures have no time to equalize and so the shale has temporarily lost the hydrostatic overbalance that was keeping it stable. Cracks start to appear in the formation. After the connection, the driller winds up the pump control to the chalk mark he has made on the control panel. The pump kicks in rapidly. The mud in the annulus has a lot of inertia due to its weight and has gelled up a bit while static. These two factors resist the sudden initiation of flow therefore a rapid pressure peak is built up. Whole mud is driven into the tiny cracks and filtrate is forced into the pore spaces. The whole mud lubricates the crack faces and allows deeper filtrate penetration into the formation. The shale starts to *destabilize* and the problems have just begun; hole enlargement, cuttings beds buildup in the washed-out sections, packing off, stuck pipe, problems logging and running casing, and bad cement jobs.

Meanwhile, the drilling supervisor loves to wiper trip. It's the magic bullet, the answer to his prayers. By regularly wiping the hole he feels that those tricky shales will be caressed into submission and things will be better than they otherwise would be. So after every 24 hours drilling he calls the driller and tells him to circulate clean and do a 10 stand trip before continuing to drill.

Pipe movement up or down will always cause some pressure fluctuation in the wellbore. It is like death and taxes, impossible to totally avoid. The driller really likes making hole so he is keen to get this trip over with so he can get back to turning and burning. However, he is conscientious and so after circulating for around 90 minutes, he sees the shakers are clean; he slugs the pipe, racks the kelly and starts to pull out of the hole. Now instead of brief—if fairly severe—pressure peaks and troughs from the connections, there is a long, steady drop in pressure as each stand is pulled out. That whole mud that was forced into the cracks the driller just created in the shale now wants to get back to that lower pressure wellbore but this small lump of shale, by now cracked nearly all the way around, is in the way. It is not difficult to finish the job and push the lump out too, helped by the formation stresses behind it. As it exits, more small cracks are starting to appear in the freshly exposed formation face because the supporting hydrostatic no longer exceeds the heightened pore pressure.

Running back in the hole, the piston effect of the BHA is increasing wellbore pressure again. Those cracks now make a convenient escape route for tiny amounts of mud and filtrate. The crack faces are forced slightly apart, lubricating mud slides between the fracture faces.

The driller starts to drill again. On bottoms up he notices a lot of cavings on the shakers but probably does not bother to mention it to anyone. After all, that always happens after a wiper trip, doesn't it?

If the driller is well briefed on this mechanism, he will take a lot more care when making connections. The drilling supervisor must understand the destabilizing effect of wiper trips by listening to the hole. He should think carefully about the cost and benefits of those wiper trips. So, what are the answers they need?

- Use good connection practices. Start and stop the pumps slowly. Take a minute or two. Refer to the procedure in Section 3.3.7, "Making Connections to Minimize Wellbore Instability and Losses."
- Do not wiper trip unless it can be justified. Will the time and cost of the trip be repaid with better conditions?
- Listen to the hole. Cavings are a warning that something is not working. An increase in cavings after trips is saying that things are becoming unbalanced. If oil-based mud is in use, cavings indicate that the shales may be naturally fractured and possibly that the mud properties, plus the drilling and connection practices, are causing fractures.
- Ensure that drillers especially understand the mechanisms of shale instability. How they to their job has the greatest effect.
- Use mud systems that are tailored to address the mechanisms of

shale instability that may be present in the well.

■   Ensure that the mud system is correctly maintained.

■   Gather all available information. Analyze it carefully so that root causes are identified and can be addressed on the next well.

2.9.2. Stuck Pipe

Some years ago, I worked on the BP Stuck Pipe Task Force. We did a lot of research work on instances of stuck pipe, looking at causes. One of the conclusions from this work was that most cases of stuck pipe—well over 90%—are avoidable with good planning and listening to the hole. There are rare cases when this is not so. Sticking can occur when drilling in highly mobile salts or into a stressed fault, even while drilling ahead.

Two of the key factors in stuck pipe prevention are training and crew awareness. This explains how the well program should be used as a tool in stuck pipe prevention. Section 3.3.1, "Stuck Pipe," covers the practical rig site aspects of preventing and curing stuck pipe.

Sticking mechanisms summary. Causes of stuck pipe can be classified into three basic categories.

Geometry is related to dimensional problems. Circulation is usually possible—the problem will be seen with the string moving and only in one direction.

Solids are related to solid particles in the hole. Circulation may be restricted or impossible and hole cleaning may have been inadequate. Usually occurs when pulling out of the hole.

Differential sticking is related to differential pressure between formation and hole. Four conditions are identified that must all be present for differential sticking: the presence of a permeable zone covered with wall cake, a static overbalance on the formation, contact between the wall and drillstring, and a stationary string.

Stuck pipe: geometry related problems.

Undergauge hole. Ream tight spots when tripping in. Ream to bottom after tripping if dictated by hole conditions or if the bit laid out was undergauge.

Keyseating. Keyseating can occur if a dogleg section is followed by a long tangent section before running casing. For this reason, high doglegs should be avoided high up in the hole. Drillpipe fatigue and wear will also be accelerated by high dogleg severities.

Stiff assembly. Where a stiff BHA is run after a build or drop assembly, or after a directional motor, lay down enough drillpipe to ream the complete section drilled with the more flexible assembly. When reaming a kickoff in softer formations, take precautions to ensure that a sidetrack is not drilled. Use low WOB/higher RPM and if necessary make a special trip with a bit-sized hole opener and a $\frac{1}{2}$ in undergauge bullnose to ream out the hole before drilling ahead.

Ledges. Can occur when there are large changes in formation hardness. Note this in the program. Run slowly when tripping in at these points. Ream any resistance seen on trips.

Mobile formations. Sticking in salt may occur fast enough for the bit to get stuck as it drills. If this condition occurs, it is hard to avoid altogether but may be reduced by using a water-based mud that leaches the salt slightly as drilling progresses. Higher mud weights may (or may not) help to keep the salt under control. Bits are available that drill slightly off center and so cut an overgauge hole (bi-center PDC bits); these are helpful when drilling in problem flowing salts.

If the salt moves more slowly and causes problems, mainly when tripping back through it, try programming wiper trips through newly drilled hole after a short time. Sometimes a wiper trip is effective after 18 hours drilling—up to the previous wiper trip depth. Field experience will show the best way to handle flowing salts.

Stuck pipe: solids related problems. The deviation profile has an effect on hole drags/torques and on the potential for sticking. Computerized models may predict likely downhole drags and allow the wellpath to be designed for the lowest figures. Select drillpipe to give at least 100,000 lbs overpull over up drag at section TD (after applying a safety factor; 85% is common).

Unstable formations are better drilled at low inclinations, if possible, to reduce the likelihood of formation compressive failure due to overburden.

Cuttings beds. Ensure that mud rheology and AVs are sufficient to clean the hole. This will be more difficult as inclination increases. Circulate clean before pulling out, displacing a pill around if necessary. If cavings are causing drags on trips, a slight increase in mud weight may help to stop the caving (if this will not cause losses). Cavings may cause a dirty annulus during the trip, even if the hole was clean at the start of the trip. If there are indications during the trip of a dirty annulus, then stop and circulate clean before continuing out.

Top-hole collapse. This may occur since very shallow formations are unconsolidated and there is little overbalance to help stabilize the wall. If possible on an offshore well, drill with returns to the rig and use a mud with good wall cake-forming characteristics or pump slugs around to help plaster the hole.

Reactive formations. For hydratable shales, inhibit the mud with KCl, Polymer, and/or glycols; or use a silicate mud system to prevent or reduce the rate of hydration. (Refer to the notes on these mud systems in Section 2.5.5.) Keep the level of inhibition according to program or change it if necessary. Oil mud will prevent hydration. For brittle failure type formations, higher mud weight is the best stabilizing mechanism. Ensure that the hole is kept clean while drilling.

For shales with natural microfractures, use sized fluid loss additives to plug off the fractures since they are exposed and keep mud densities low. Minimize swab and surge pressures.

Wiper trips can initiate or accelerate instability in shales. See "Drilling practices" in Section 2.9.1.

Geopressured formation. In shales where the pore pressure is greater than mud hydrostatic, slivers of shale will be pushed into the wellbore. Use higher mud weight if possible to reduce or eliminate this and keep the hole clean using pills if necessary.

Where the formation is exposed for a long time, hole enlargement can occur which may lead to other problems with large chunks of formation falling in, big cuttings beds forming, fish becoming impossible to recover (if they fall over in the washout), and bad cement jobs. If the formation cannot be controlled with mud weight, if possible, ensure that casing can be run within a short time of drilling the formation.

Fractured and faulted. Some formations are already naturally fractured before being drilled into. This can again cause chunks of rock to enter the wellbore, causing mechanical sticking. As the damage is already done, avoid making it worse by minimizing swab/surge pressures and drillstring vibrations, drilling at low angle if possible, getting through it quickly, and casing it off.

Junk. Ensure all drillstring components are inspected in accordance with API RP7G and that good handling practices are used. Specify the correct size and grade of drillpipe to withstand drilling stresses. These will help avoid downhole failures leading to junk in the hole.

Cement blocks. Where a large pocket exists under a casing shoe, cement in the pocket after cementing the casing may fracture, as drilling progresses, and fall into the hole. Do not program in a pocket any larger than necessary. Consider adding fibers (such as Dowell D094) to the tail slurry.

Soft cement. It is possible to run into cement that is not completely set, then find it impossible to pull back out or to circulate. Monitor surface cement samples (preferably kept at bottom hole circulating temperature) to ensure they are hard before approaching bottom and run in the last couple of singles slowly and with the pumps on.

Stuck pipe: differential. The following four conditions are considered necessary for stuck pipe to occur.

Permeable zone covered with wall cake. Where such a zone is identified, the wall cake characteristics can be optimized with a good mud program. Additives can be used to make the cake thinner and less sticky or oil mud will form very little cake.

The desirability of low fluid loss is generally related to reducing contamination of formation fluids, as discussed at length under "Formation damage due to the liquid fraction" in Section 2.5.6, "Formation Damage with Water-Based Muds (and Cements.)" However, low solid content/high fluid loss muds tend to give increased rate of penetration in permeable formations. The fluid loss also affects wellbore clay hydration. Therefore, the desirable level of fluid loss is often related to whether or not damage to the permeable formations needs to be minimized.

Thick filter cakes are undesirable. Differential sticking becomes much more likely with a thick cake. Cement-formation bonding requires removal of the cake, which may be difficult. There are even cases of thick cake causing mechanical sticking by reducing hole diameter above the BHA. For reduced differential sticking, the API HPHT fluid loss in oil mud should be kept below 5 cc, and in water mud 5-8 cc is desirable. Generally, higher fluid loss increases sticking. Filtrate from oil muds should be 100% oil; the presence of water is a sign of reduced emulsion stability.

When drilling the final (payzone) hole section it may be possible to change to low solids, nondamaging drilling in fluid in a well-known development area if the formations allow. This will be especially useful if drilling in a depleted or low-pressure zone.

Static overbalance. Use the minimum safe mud weight to minimize the static overbalance on the formation. Condition the mud carefully before running casing to the minimum safe density.

Wall contact. Use a well-stabilized BHA. Spiral or square drill collars have less contact area than round ones. Run HWDP in compression to reduce the length of drill collars. Use a bit that requires less WOB and therefore less BHA (PDC, diamond) if the formation is suitable and the rig cost justifies it.

Stationary string. Minimize programmed wireline directional surveys. Top drive is an advantage, since fewer connections are required. Make the initial flow check on trips brief, pull out above the permeable zone, and carry out a full flow check. Keep the string moving slowly during flow checks. In extreme cases rotate out on connections and keep the string rotating until ready to latch or stab, ensure the crews are properly briefed to carry this out safely. Centralize the casing well over this interval.

Drilling jars and jar placement. Except in a horizontal hole, jars are normally run in the BHA. Some operate mechanically and some hydraulically. Jars may be able to jar both up and down—some only up. It may be possible to adjust how much overpull sets off the jar and internal mud pressure may have an effect (refer to the manufacturer's manual). They all work by allowing the driller to take a strain (or set down weight) on the drillpipe and then they suddenly open or close. This uses the stored strain energy in the string to accelerate those drill collars above the jar and deliver a hammer blow to the collars below. The strength of the blow is related to the weight of the drill collars above the jar and how fast they are moving at the end of the jar stroke.

Jar placement has been the subject of much research and is quite complex. As a general recommendation, the jar should be run with either one or two drill collars above it for maximum effectiveness, if no jar intensifier is used.

If the jar has adjustable settings for tripping, set it for a tension approximately equal to twice the BHA weight, unless field experience dictates otherwise. Be careful not to run the jars within 5000 lbs of the neutral point when drilling, otherwise the jar will wear rapidly and may not work when needed.

Ensure that a backup drilling jar is on site and that the jar in use is changed out at around the interval recommended by the manufacturer, or less if it is suspected that the jar is no longer functional.

Stabilizers or other full gauge tools should not be run above the jar, because if the string became mechanically stuck the jar could not be used. Keyseat wipers may be run above but should be sized halfway between DC OD and hole diameter.

Sometimes sufficient strain cannot be taken in the pipe. At shallow depths not enough DP stretch is available, in deep-deviated holes drag may restrict stretch available at the jar. In these circumstances, a drilling jar intensifier (accelerator) can be run in the string. In effect, it is a powerful spring that sits above the collars and provides the strain energy that the drillpipe, due to hole drags or string length, cannot. Therefore, for correct jar and intensifier placement, run Bit - BHA to Jar - Jar - 2 or 3 DCs - Intensifier - HWDP - DP. If the intensifier is run right above the jar, this eliminates the jarring force because the intensifier will absorb the jar stroke. The drilling supervisor and drillers must be familiar with the operation of the jar in use—its settings and limitations.

2.9.3. Lost Circulation

Mud losses are a major cost in drilling operations worldwide. Field experience will give good guidance as to what types and severities of losses may be expected. This will allow the well to be planned as far ahead as possible to avoid recurrence of losses and to be ready to react in the best way possible if losses do occur.

Some losses cannot be avoided. In this case, the choices are to drill ahead without returns, to use a lighter fluid column or an air assisted medium (such as foam) to case off the loss zone (that may need special cementing techniques), or to abandon the well.

Top hole, to surface or seabed outside conductor. Where this occurs, jack-up rigs may lose leg support. Offshore, this will not happen if returns are to seabed; however, where a riser is used the extra hydrostatic can fracture to seabed.

To avoid such losses do everything possible to reduce hydrostatic pressure on the formation:

1. Set the conductor as deep as is practical.
2. Drill with returns to seabed. Use subsea diverters if available. When drilling from a floating rig, drill riserless until surface casing is set and a BOP can be installed.

3. Maximize annular velocity. Use the largest liners at maximum SPM and consider an extra skid-mounted pump for top hole. Circulate viscous sweeps around each connection. If equipment problems restrict the flow rate, stop drilling and circulate until resolved. High flow rates in large, shallow hole will not add much ECD. It is more beneficial to clean the hole effectively than to restrict the flow rate. If restricting the flow rate does help, slow down the ROP as well to avoid loading up the hole.
4. Maximize AV with large drill collars and possibly large drillpipe.
5. Drill with controlled ROP to minimize annulus loading. In practice, as long as sufficient AV is maintained this should not be necessary.
6. Use bentonite or polymers to increase mud viscosity.

The easiest way to cure these losses is probably with large volumes of cement. Refer to the "Recommended Procedure for curing total losses with cement" in Section 3.5.2, "Slurry Mixing Options." Prevention is better than cure.

Shallow unconsolidated formations—severe or total losses. The principal cause of losses in these formations is very high permeability. The mud does not make an effective mud cake to seal the loss zone. Losses are likely to start as soon as the formation is penetrated. An annulus loaded with cuttings, excessive mud density, insufficient mud viscosity, high water loss (low solids content to plaster the wall), or excessive surge pressures will all contribute to the losses. The following precautions can be taken:

1. Use high rheologies by adding flocculating agents such as lime or cement. Thickening the mud will give greater resistance to flow into the formation, but in shallow large diameter holes will not make a significant difference to the ECD. This will also help keep the hole clean.
2. Drill with the minimum safe density to control formation pressures with a trip margin. This will have to be balanced against any formations that may destabilize with insufficient hydrostatic and cave into the hole. Some caving can probably be tolerated if this allows the losses to be avoided.
3. Use a high circulation rate to clean the well. If possible, increase riser booster pump output when in deep water with a floating rig.

4. Drill at controlled rates to reduce annulus loading.
5. Pump slugs around which contain solids to plaster the wall. Solids suitable for plugging off large pore spaces would be taken out by the solids control equipment, making it difficult to add to the whole system. With a slug this does not matter.
6. Modify the casing program to case off the problem zones as soon as possible.

Heavily fractured cavernous formations. Losses are likely to start as soon as the formation is penetrated. Attempting to cure the losses with LCM is likely to be expensive and ultimately futile and the best options will probably be to drill blind with water if the area is well known or to drill with foam if not. Set casing as soon as possible.

Cement can work well if done properly. Refer to the "Recommended procedure for curing total losses with cement" in Section 3.3.2, "Lost Circulation."

Normally pressured, deeper formations. These formations may be unconsolidated, naturally fractured, become fractured by the drilling operation, or consolidated but highly permeable with pore sizes too large for the mud solids to plaster. The loss zone can be anywhere in the open hole—not necessarily at the formation just drilled in to. Several factors will contribute to the mud loss, such as annulus loaded with cuttings, high ECD, excessive mud density, high water loss (low solids content to plaster the wall), excessive surge pressures, breaking the formation during a formation integrity test (FIT), or closing in the well after a kick.

The following techniques will reduce the incidence of losses:

1. Use low rheologies to minimize ECD.
2. Drill with the minimum safe density to control formation pressures with a trip margin. This will have to be balanced against any formations that may destabilize with insufficient hydrostatic and cave into the hole. Some caving can probably be tolerated if this allows the losses to be avoided.
3. Use the minimum AV that will effectively clean the well. If high ROP is expected then it may be necessary to control the ROP to avoid excessive annulus loading. Displace viscous (low inclination) or turbulent (over 45°) pills to clean the well if annulus loading is suspected.

4. Ensure that good drilling and tripping practices are used to minimize swab and surge pressures.
5. Keep fluid loss low.
6. Use the recommended procedure for a formation integrity test, not the continuous pumping method. Refer to the "Formation integrity test recommended procedure" in Appendix 2.

[Section 3:]

Practical Wellsite Operations

This section deals with practical matters at the well-site. Much of the material in Section 2 is also relevant to the wellsite; cross-references are made within the text where appropriate.

[3.1]

Well Control

Some of the practical aspects of well control are discussed in this section. A certain amount of knowledge is assumed here—equivalent to at least a driller level well control certificate; this is not meant as a "start from scratch" course in well control.

Refer to the following topics covered earlier in this book: prediction of pore pressures and fracture gradients and other casing design topics in Section 1.4, "Casing Design," well control from a drilling program writing perspective in Section 2.2, "Well Control," kick tolerance calculations in Appendix 1, "Calculating Tick Tolerances," and formation integrity test procedure in Appendix 2, "Formation Integrity Test-Recommended Procedure."

3.1.1. Kick Prevention

A kick is an uncontrolled entry of formation fluids into the well-bore due to formation pore pressure exceeding the hydrostatic head of the fluids in the wellbore. If this situation is left to develop, a blowout will eventually result, where formation fluids enter the wellbore and blow out unconstrained into the atmosphere.

Not all formation fluid entry is a problem. For example, as permeable formation is drilled, the fluids contained within the cuttings can enter the drilling fluid, causing an increase in background gas. In very low permeability formations such as shales, gas contained in the tiny pore spaces may be at a higher pressure than mud hydrostatic, but the flow is very slow.

Sometimes gas-bearing cuttings can liberate enough gas to cause serious gas cutting of the mud. In shallow tophole sections, this may cause a shallow gas blowout due to the reduction of a few psi in hydrostatic—it could be so finely balanced. Deeper down, even serious gas cutting is unlikely to cause a kick as the actual bottom hole pressure (BHP) reduction is proportionally very small.

Kicks are generally prevented by ensuring that the mud hydrostatic exceeds formation pore pressures when the formation permeabilities can allow a significant volume to flow into the well. At first view it might seem that the answer is simply to drill with a very heavy mud. However, this is not practical due to the increased possibility of losses, stuck pipe, instability in fractured shales, reduced ROP, and reduced MAASP should be a kick result. Therefore, it is necessary to predict what pore pressures are likely to be and to track this while drilling in order to maintain a safe but small overbalance on permeable formations.

The amount of overbalance we maintain is related to what is needed to eliminate influxes due to swab pressures when tripping out of the well. It is therefore generally termed the "trip margin." Trip margin can be calculated by relating mud rheology to the hydraulic diameter of the hole/drillpipe annulus by the formula:

$$TM = \frac{YP}{225(Dh - Dp)}$$

where TM is trip margin in psi/ft, YP is Yield Point, Dh is hole diameter, and Dp is pipe outside diameter.

The trip margin is added to the mud gradient required to balance the formation pore pressure.

We can monitor various indicators while drilling to see whether the amount of overbalance is reducing (pore pressure gradient increasing with depth). These warning signs indicate an impending problem and preventative measures should be taken, such as increasing the mud density to restore the overbalance. Indicators that may warn of decreasing overbalance include:

- Increasing rate of penetration in homogenous formations
- Decreasing D exponent in homogenous, compressible shale formations
- Increasing background gas level
- Connection gas and/or trip gas
- Increase in mud chlorides, if the source cannot be accounted for from surface sources
- Appearance of tensile failure mode shale cavings
- If LWD tools are used, trends in resistivity or sonic velocities may warn of pore pressure gradient changes

These trends need to be continuously monitored by the drillers and mud loggers. Any changes should be cause for heightened monitoring. Refer also to the topics on predicting pore pressures in Section 1.4.4, "Pore Pressures and Fracture Gradients."

3.1.2. Kick Detection and Response

Once flow from the formation has actually started, it must be detected and action taken right away. The time between the flow starting and the well being shut in will determine how much influx is taken. Detection and response depends on maintaining the kick detection systems and continuous training of the drill crews.

Detection systems. These vary from low to high tech. It is easy to get carried away with the high tech stuff, but do not forget the value of the simpler systems.

When I was a trainee driller in Holland, the driller used to send me and a fellow trainee down to the mud pits while we drilled. It was winter and very cold; we had to hug the desilter feed pipes to keep warm. Every so often someone would come along and adjust the nut hanging on a rope attached to the handrail so that the nut just kissed the surface of the mud in the tank. We did not know why this was done—no one explained these things to mere trainees—and every so often if he did not appear, we would do it for him. Of course what he was doing was monitoring the level of mud in the active tank; an increase could be accurately seen as long as the flow rate was constant (and no one messed with the nut).

Tank level monitoring systems using sound waves feed into the driller's totalizer system and also to the mud loggers computers. Alarms can be preset at loss/gain levels to assist monitoring. These systems are generally accurate and reliable. One in each corner with the signals averaged may be needed on a floating rig to account for rig (and hence mud surface) movement.

Floats that ride on pipes mounted in the tank also can feed computerized detection systems. Float systems can stick on the pipe or may leak; these are therefore less reliable than sonic (or nut on a string!) systems.

A flo-sho is installed in the return flowline coming up from the riser and bell nipple. When circulating, the return flow hits a paddle that is pushed up. The position of the paddle is related to the flow rate and mud density. If the well kicks, the first primary indication is an increase in return flow rate in the flow line and a change in the position of the flo-sho paddle, which will show on the driller's panel. The paddle can stick due to gumbo shale or other solids in the flow line, which is often seen as an indicated flow even when flow has stopped. Therefore, the flo-sho can give false indications of flow or may not show a change even if the flow out changes.

All of these ways of monitoring the mud level suffer from one major drawback; they only work when the system dynamics are not changing. The flow rates have to be constant (could be zero) and have to be stable (must have been constant for some time). If the pumps are sped up, first the active tank level drops and then the flow at the flowline increases some time later. This increased flow feeds into the shaker tanks and sand trap, eventually reaching the active tank. It can take several minutes on a deeper well for the flow into the active tank to equal the flow out of it. A good driller will have a feel for how much is gained by the system when the pumps are stopped for a connection (could be over 20 bbls) and will monitor the totalizers to ensure that the loss/gain indicator comes back to zero after a connection.

Computerized data acquisition systems have been developed for slimhole drilling where very small influxes of around a barrel can be detected. These monitor different parameters such as pump rates, flow-line levels, and tank levels over time so that at any stage, even when changing the pump rate, the computer can compare actual with pre-dicted levels and flows.

The Mk 1 Eyeball should also be used whenever the driller has any reason to think that the well might be kicking. If a secondary indication such as a drilling break (sudden increase in penetration rate) or a primary indication (increase in flow) is seen, a flowcheck should be done.

Flowchecks are often difficult. With the pumps off, colder mud that has just arrived downhole from the surface system will warm up downhole and a small flow may result. On a floating rig, heave can make it very tricky to decide whether or not the well is flowing. The flowcheck must be kept going for long enough to be sure. A tight for-mation may kick with a very slow influx rate. Further, the ballooning effect (described below) can give a realistic "false kick," which can develop fairly high surface pressures if shut in.

MWD tools now play a significant role in kick prevention. Sperry Sun offers a pressure detection service; a dedicated pressure engineer stays in the logging cabin and monitors resistivity from the MWD tool. A decrease in the resistivity trend with depth may indicate increasing pore pressure gradients. The engineer will also monitor the levels of gas dissolved in the mud, pressure cavings, D exponent, and other indi-cators to buildup a picture of the pore pressure trends. If any increase in pore pressure is indicated, it can be reported to the driller and drilling supervisor for action to be taken.

Anadrill has an MWD tool that measures resistivity right at the bit (RAB tool). The tool measures five resistivity values: at the bit, using a ring electrode (for compensating certain formation effects), and at three button electrodes (when configured as a stabilizer). The button electrodes mea-sure formation resistivity Rt at depths of investigation of approximately 1 in, 3 in, and 5 in in an $8^{1}/_{2}$ in hole size. Gamma ray, axial, and transverse shock load measurements are also available from the tool.

Resistivity at the bit is obtained by inducing a voltage above the bit and measuring the axial current flowing through the bit and into the formation.

Response training and drills. All the best detection equipment is useless if the driller does not respond when a kick is indicated. Regular drills should be initiated to train the drill crew and test their response. The two most common drills are a pit drill and a trip drill.

For a pit drill, the toolpusher or drilling supervisor will do something to cause the pit volume totalizer system to show an increase in pit volume. When float type detectors are used, these can be pulled up. Mud can be pumped into the active system from a reserve tank or the trip tank. The driller should respond by stopping the rotary and pumps, picking up off bottom, and doing a flowcheck. The driller will not normally close in the well on a pit drill; the flowcheck will show that all is well and the drill can be stopped once the flowcheck is initiated.

A trip drill is normally initiated by the driller to train the crew. It is good practice to stop for 10 minutes in the shoe when tripping out, which can be combined with a trip drill every time. The objective here is not speed but correct action; it is much more important to do it in the correct sequence than it is getting it fast. Speed is the ultimate objective, but if they practice the sequence on every trip, when it is done for real it should be both accurate and fast.

This is one suggested way of holding a regular trip drill. The driller pulls into the shoe, breaks off the stand. The block is stopped halfway down and the driller shouts "trip drill." The crew picks up the full opening kelly cock and stabs it into the drillpipe. The kelly cock is made up hand tight and closed.

(At this point in a real kick, if strong backflow is coming up the drillpipe, the crew has to get that kelly cock on and closed. They may have to do this by feel in an extreme case. Once the kelly cock is closed then the flow will stop—apart from maybe some minor flow from the hand-tight connection, which will stop once the tongs are on and the connection tightened. For the drill, it is not necessary to torque up the connection.)

The driller now lowers the block and latches the elevator, picks up the string and the slips pulled. The casing shoe flowcheck can now be done with the kelly cock in place. If a flow is detected then the driller only has to close in at the BOP as the kelly cock is already on. After confirming that the well is static, the string can be set in the slips, the kelly cock removed, and the trip out resumed.

Precautionary flowchecks should also be done when tripping back into a horizontal wellbore. A swabbed influx can stay in the horizontal section to be displaced out when tripping back in. Since the hole may have been static for several hours, the drill crew may not expect a problem. Now that the influx is out of the horizontal section it moves up and, if gas, can cause the well to start flowing as the gas expands and hydrostatic reduces.

False kicks; the ballooning effect (Formation overpressures due to ECD). A permeable formation may become locally pressured up around the wellbore by the extra pressure exerted when circulating. When pumping stops, the formation returns mud to the annulus. This is known as the "ballooning effect." If this is not recognized and the well is killed, it will be found that it flows again after circulation ceases.

This happens when a tight permeable zone takes slow losses while circulating with a high ECD. The flow will take some time to develop and may be hard to recognize at first because nothing may happen for 15 or 20 minutes. Stopping pumping for connections may not cause a flow, so the formation becomes locally charged as high as the ECD while drilling and then comes in during tripping. In deeper, smaller holes the closed-in surface pressure may reach several hundred PSI. Warning signals may include:

- Slow losses while drilling
- Cuttings or LWD logs show sandstones of low permeability
- Flow takes a long time to develop
- Pressure buildup takes a long time when the well has been closed in
- There is little difference between pressure on the drill pipe (Pdp) and pressure on the annulus (Pan), unless the returning mud brings with it some formation gas
- There is a reduction in Pdp and Pan after opening the well and allowing some mud out. For example, release a measured amount through the choke (5 bbls), let pressure stabilize, and repeat until a definite trend either way is seen. Increasing annulus pressure indicates a kick.

Once the condition is confirmed, the well can be depressured in controlled stages, monitoring pressures, and volumes. Do not increase mud density; this will make the problem worse.

3.1.3. Drilling Below Normal Kick Tolerance Levels

Refer to Appendix 1 for calculating kick tolerances. The level of kick tolerance required for drilling a hole section depends on several factors:

■ Company or government regulations may specify a minimum kick tolerance, usually by hole size
■ Kick detection equipment on the rig (larger influx tolerance is needed for less accurate equipment)
■ Level of crew training
■ Mud system in use; oil mud may dissolve a gas influx that comes out of solution as it moves close to the surface. A smaller kick tolerance for the initial influx size may be needed and this may in turn require improved equipment and crew training.

There will be times when you may need to drill ahead to find a competent formation to set your next casing shoe in with a kick tolerance that is below the normal minimum. In this case, extra precautions can be taken to minimize the chance of a kick and to catch it early if it does occur. An alternative may be to set cement on bottom and run casing prior to drilling ahead.

Precautions may include all or some of the following:

■ Restricting the rate of penetration
■ Flowchecking every drillpipe connection or even more often
■ Circulating bottoms up after drilling a specified distance
■ Drilling for a specified distance, making a two- or three-stand trip, and circulating bottoms up for trip gas
■ Making a short trip and circulating for trip gas before pulling out of the hole

3.1.4. Well Killing in a High-Angle Well

Refer also to the notes on planning considerations for high-angle wells in Section 2.2.4, "Well Control in High-Angle and Horizontal Wells."

Operational considerations Once a kick is taken, plan for a driller's method kill in two circulations if any of the following conditions apply:

Bit-to-shoe volume is less than surface-to-bit volume. In this case, the heavy mud will give no reduction in open hole pressures while circulating out the influx.

A swabbed kick was taken, or difference in TVD between the casing shoe and the horizontal wellbore is small (well kicked off to horizontal just below shoe or through a casing window).

If a balanced kill is to be used, the Phase 1 calculations require the static and dynamic pressures to be calculated separately. The increased dynamic pressure loss through the drillstring is calculated as before, but the reduction in surface pressure during Phase 1 due to heavy mud being pumped will be different.

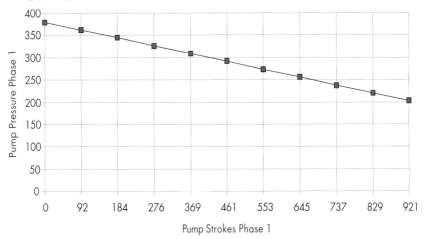

Fig. 3-1 Horizontal Phase 1 Circulating Pressures Using Conventional Kill Formulae

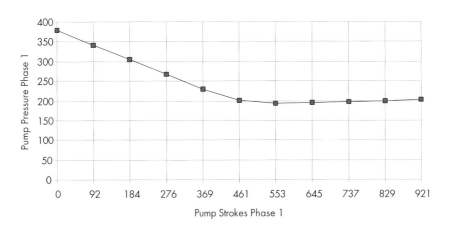

Fig. 3-2 Horizontal Phase 1 Circulating Pressures Calculated for Maintaining Constant BHP

321

The differences in the kill graphs are illustrated in Figures 3-1 and 3-2 for a 2000 m MD, 1000 m TVD well, mud gradient = 0.495, Pdp = 200 psi, and the well is kicked off to horizontal at around 850 MD.

It can clearly be seen that if the conventional kill graph is followed, the BHP will rise by 90 psi at 50% of Phase 1 strokes before falling back to the balanced BHP at the end of Phase 1.

Calculations Note: These calculations are incorporated in the spreadsheet *wellcalc.xls* available from the web site at *http://www.drillers.com*.

To calculate Phase 1, balanced method graph for a horizontal or high-angle well, first establish Pc1 and Pc2 in the conventional manner. Divide the measured bit depth into 10 approximately equal segments and calculate the increase in circulating pressure drop at each depth. The depths do not have to be at exactly equal spacing but may be at convenient depths where both MD and TVD are already known, such as survey stations. The intermediate circulating pressure (Pci) at each depth will be:

$$Pci = Pc1 + \frac{MDstation \times (Pc2 - Pc1)}{MDbit}$$

Now for the measured depth at the end of each segment, take the survey depth or calculate the TVD from the directional plot and work out how much extra pressure will be required to balance formation pressure (Ps) over the mixed mud hydrostatic. The extra surface pressure Ps will be

$$Ps = Pf - ((TVDwell - TVDstation) \times (\rho2 - \rho1))$$

For each station, add together Pci and Ps to get the total Phase 1 balanced BHP circulating pressure. If circulation is stopped at any point, the closed-in surface pressure will equal Ps for that depth (see Table 3-1).

The easiest way to make these calculations is with a spreadsheet. The one included as a sample (*wellcalc.xls*) assumes 10 equal spaced MD and TVD stations, but allows manual overwriting of MD and TVD entries and plots the pressures and Phase 1 graph automatically.

In many cases, a driller's method kill may be preferred. If a balanced kill confers no clear advantages of reduced open-hole pressures

Table 3-1 Calculation of Surface Circulating Pressure During Phase 1

MD	TVD	Pressure Drop, Pci	Hydrostatic	BHP	Ps	Pump Pressure	Pump Strokes
0	0	180	1624	1824	200	380	0
200	200	182	1664	1824	160	342	92
400	400	184	1704	1824	120	304	184
600	600	187	1744	1824	80	267	276
800	800	189	1784	1824	40	229	369
1000	950	191	1814	1824	10	201	461
1200	1000	193	1824	1824	0	193	553
1400	1000	196	1824	1824	0	196	645
1600	1000	198	1824	1824	0	198	737
1800	1000	200	1824	1824	0	200	829
2000	1000	202	1824	1824	0	202	921

during Phase 1, the driller's method will maintain a steady BHP whatever the well profile and is easier to control during the kill. Since this is easily determined in advance (see the three conditions stated above), the well plan should include a recommendation to this effect.

3.1.5. General Considerations for BOP Equipment

Most operators and drilling contractors follow the recommendations specified in the current *IADC Drilling Manual*. Reference should be made to that publication for specifications. What follows summarizes some of the main points from the IADC recommendations and normal practices.

The BOP stack should be capable of closing in the well with or without pipe in the hole by remote control.

The equipment should be rated to at least the maximum anticipated surface pressure. If the equipment is over 5 years old, the rated pressure shall exceed the maximum anticipated by at least 10% (i.e., its rating shall be downgraded by 9%, so a 5000 psi system will not be used over 4550 psi). No screwed connections are permitted on systems rated to over 2000 psi—only hubs, flanges, or welded connections.

For sour service the complete high-pressure BOP system is to be of metals resistant to sulfide stress cracking. Refer to the current issue of the *National Association of Corrosion Engineers*, NACE Standard MR-01-75.

Dedicated kill lines are to be a minimum of 2 in nominal ID and are to have two valves and a non-return valve (NRV) between the line and the BOP stack. Choke lines or dual-purpose choke/kill lines are to be at least 3 in nominal ID and should have two valves from the BOP, the outer of which should be hydraulically operated. For drilling and workover operations, blind/shear rams are required that can shear the drillpipe/tubing under no-load conditions and subsequently seal the well.

Any measuring instrument needs to be regularly checked and calibrated to ensure accuracy. On many rigs, pressure gauges stay in place for years without getting checked; they go through rig moves, weather, exposure to hot drilling fluids, physical knocks, vibration, etc. Yet when it comes to a well kill situation the safety of the people on the rig and the rig itself can depend on those gauges reading accurately. Check that they have valid test or calibration certificates—preferably not more than six months old and never more than a year old. If necessary, demand that they be replaced with recently certified gauges.

Diverters: general considerations. A diverter is installed to allow time to evacuate the wellsite in the event of shallow gas. Current systems are not designed to handle a sustained erosive flow from the well.

A diverter must be installed on a well when both of these conditions apply:

1. There is a possibility of losing primary control
2. The conductor or casing shoe is not strong enough to close in the well with a BOP, which would cause fracturing to surface if the well were closed in

The diverter lines must be designed so that produced fluids and solids can be taken clear of the rig with the minimum back pressure and without eroding or plugging off. The lines should be of minimum 12 in ID, and they should be as straight as possible without flow restrictions or bends that would induce turbulence and increase erosion. Minimum pressure rating for the diverter is 500 psi, but a higher rating may be necessary if the anticipated surface pressure is greater.

Full opening valves or bursting disks should be used. Modifications to the rig may be needed to ensure straight running lines.

Generally, two lines out of the diverter spool should be available so that the downwind line can be selected if the well kicks. One may be used in cases where there is a reliable prevailing wind, or on DP floating rigs that can "weathervane" with the wind. Lines also need to be as short as possible—just long enough to clear the rig structure (offshore) or keep away from obstructions (land).

Checklist for selecting a diverter system.

1. The equipment shall be selected to withstand the maximum anticipated surface pressures.
2. All connections must be welded or welded flanges must be used. Lines must be straight, properly anchored, and sloped down to help avoid cuttings settling.
3. Installation requirements for wellheads and BOPs also apply.
4. A diverter system may be a BOP stack with diverter spool, lines, valves, and control system—or a specific diverter setup. Closing speed is very important. Either way, the diverter and mud return lines should be separate to avoid gas entering the rig mud system.
5. Diverter valves must be full opening with a pneumatic or hydraulic actuator. (Note: bursting disks have also been successfully used.) Valve bore shall be the same as the lines.
6. A kill line with NRV must be incorporated for pressure testing and to allow water to be pumped into the diverter.
7. The control system should preferably be self-contained or it may be an integral part of the BOP accumulator and controls. It shall be located in a safe area away from the drillfloor and have control functions clearly identified.
8. If a surface diverter and subsea BOP are used, two separate control systems are required. The diverter should be operable with the minimum of functions, preferably a one-button or one-lever action required to divert.
9. A $1^1/_2$ in hydraulic operating line should be used for diverters with a $1^1/_2$ in NPT closing chamber port size. The opening line may be 1 in.
10. All spare lines not in use must be plugged off.
11. The control system must be able to close the diverter or annular

within 30 seconds (under 20 in), or within 45 seconds for larger units. Diverter valves must open fully before the diverter or annular closes.

12. It should be possible to control pumping operations at the pumps as well as on the drillfloor.

13. Telescopic joints on floaters should have double seals to improve sealing when gas is coming out of the marine riser. In this situation, the TJ is the weakest link, especially if there is significant heave.

14. All fans (including in accommodation) should stop automatically in the event of a gas alarm.

15. At least one windsock, visible to the driller, should be installed prior to spudding the well.

3.1.6. Surface BOP Stack Configurations

2000 psi WP classification.

1. One annular preventer, or a double hydraulic operated ram type preventer (one with pipe rams and one with blind/shear).

2. One full opening drilling spool with two $3^1/_{16}$ in bore side outlets. However, if a dual ram unit is used instead of an annular and is equipped with proper sized side outlets, the drilling spool may be omitted and the kill and choke lines connected to the lower preventer outlets.

3. Two dual purpose kill and choke lines, both connected from the drilling spool to the kill line and choke manifold. If dual-purpose lines are not used, the stack will have one dedicated kill and one dedicated choke line.

3000 and 5000 psi WP classification.

1. One annular preventer.

2. One double or two single ram preventers, one with blind/shear and the other with fixed or variable bore pipe rams.

3. One full-opening drilling spool with two $3^1/_{16}$ in bore side outlets. However, if the bottom ram unit is equipped with proper sized side outlets, the drilling spool may be omitted and the kill and choke lines connected to the lower preventer outlets.

4. Two dual purpose kill and choke lines, both connected from the drilling spool to the kill line and choke manifold. If dual-purpose lines are not used, the stack will have one dedicated kill and one dedicated choke line.

10,000 psi WP classification.

1. One annular preventer with a working pressure of 10,000 psi. However, a 5000 psi annular preventer on a 10,000 psi WP BOP stack is acceptable on existing stacks.
2. Three single, or one double and one single ram preventers. One must have blind/shear rams, one fixed pipe rams, and the third either fixed or variable rams. If a double ram unit is used, it should be on top so there is room to land a tool joint on the bottom rams.
3. One full opening drilling spool with two $3^1/_{16}$ in bore side outlets. However, if the middle and lower ram units are equipped with proper sized side outlets, the drilling spool may be omitted and the kill and choke lines connected to the lower preventer outlets.
4. Two dual purpose kill and choke lines, both connected from the drilling spool to the kill line and choke manifold. If dual-purpose lines are not used, the stack will have two dedicated kill and two dedicated choke lines. In this case, each line will have two full-bore valves, the outer of each choke line hydraulically operated. The lower kill and choke lines connected below the bottom ram shall act as spares.

15,000 psi WP classification. This may be a 3-ram unit of identical configuration to the 10,000 psi stack, except the annular preventer is rated to 10,000 or 15,000 psi and the BOP stack to 15,000 psi.

A 4-ram setup is preferable (though the 3-ram stack is acceptable, as was outlined in the previous subsection, "10,000 psi WP classification"). In this case the fourth ram unit may have fixed or variable pipe rams installed; however, at least one ram unit must have fixed pipe rams.

3.1.7. Surface Stack Control System Specifications

An independent automatic accumulator unit of 3000 psi WP with a control manifold showing "open" and "closed" positions for preventers and hydraulic valves should be equipped with regulator valves similar to the Koomey TR-5, which will not "fail open" and cause complete loss of operating pressure. With the recharging pumps disabled, the unit must have capacity to close and open all preventers, plus once more close one annular and one ram preventer. The unit should then have at least 200 psi pressure over precharge remaining and at this pressure should be capable of holding the BOP units closed against the rated WP.

The accumulator shall be located in a safe area away from the drillfloor. It must have a low-pressure alarm and hydraulic fluid level indicator or low-fluid level alarm.

On a surface stack, the control equipment must be able to close a ram preventer, or an annular preventer smaller than 20 in in under 30 seconds. Annulars of 20 in or larger must close within 45 seconds. For subsea stacks, the rams must close within 45 seconds and annulars within 60 seconds.

All BOP stack installations shall have two graphic remote control panels, each showing open and closed positions for each preventer and pressure operated choke valves. Each of these panels should have a master shutoff valve and controls for regulator and bypass valves. One panel must be located near the driller's position and the other near the location exit, or near the rig supervisor's office.

All four-way valves should be either in the full-open or full-closed position and be free to move to either position. For example, the Shear ram operating handles should not be locked, and four-way valves should not be left blocked or in the center position.

Control hoses should preferably be high pressure and fire resistant with a 3000 psi working pressure. Steel swivels are acceptable. All spare operating lines and connections that are not used in the system should be properly blocked off with blind plugs at the hydraulic operating unit. Test hoses to 3000 psi on BOP tests (except annular, or ram units with variable rams inside to prevent damage) by using the bypass valve, allowing full accumulator pressure in manifold.

3.1.8. Surface BOP Stack and Accumulator Testing

Testing of any part of the BOP equipment must be done under the direct supervision of the drilling supervisor. Pressure recording charts are required of any pressure tests for later reference.

The complete BOP system should be tested at the following intervals:

■ After installation of the wellhead and BOP prior to drilling
■ After 7 days, which may be extended to 14 days depending on the operations in progress and pending
■ Prior to drilling into expected hydrocarbon reservoirs or overpressured formations
■ Prior to a production test
■ At any time if any doubt arises over the stack integrity, such as after repairs

The accumulator should be tested at the same time as the BOP stack. They can be done together by switching off the accumulator charge pumps and using the stored energy to operate the BOPs while carrying out the pressure test. In addition, the accumulator should be tested after repairs have been made to the system, such as bottles, bladders, pumps, etc.

The accumulator bottles precharge pressure shall be checked prior to drilling out cement in the casing shoe. Unless otherwise specified the precharge pressure for a 3000 psi system shall be 1000 psi.

After nippling up, the wellhead and BOPs should be subjected to a 500 psi pressure test followed by a high-pressure test to the lower of rated BOP or wellhead/casing pressure. Subsequent pressure tests are to the maximum anticipated wellhead pressure, with a maximum on the annulars of 70% of working pressure. Kill and choke lines and manifolds are tested to the lesser of ram or manifold working pressure, but no tests shall be made against closed chokes. Kelly/TDS and kelly cocks are tested to their rated pressure. Test pressures in each case are to be held for at least 10 minutes; this implies that if the pressure drops initially then settles down (air in system), then it can be repressured and held. However, a steady drop indicates a leak that must be investigated and fixed.

The suggested BOP test procedure below shows a suggested method of testing a 3-ram surface stack, where the test pressures are the same as the casing test pressure. Kill and choke manifolds, kelly and stab-in valves may be tested during tripping or other operations to save time, so long as the normal shut-in procedure can be used in the event of a trip drill or kick.

During testing, compare readings on the various pressure gauges to ensure that they are all working and reading the same. Sensators need to be kept pumped up, and it is a good policy to have them knocked off and to, at some point, check them before the stack test.

If the water in the stack cannot be incorporated into the mud system (due to density or OBM in use, for example), the water can be dumped as it returns when pipe is run in the hole. See Table 3-2 and Figure 3-3 for a suggested BOP test procedure.

Table 3-2 BOP Test Procedure—Typical 13⁵/₈ in Surface Stack

Check accumulator fluid pressure, manifold pressure, regulated pressure, and reservoir level. Switch off the charge pumps. Drain the stack and fill with water. Pull the wear bushing. Operate the BOPs from the accumulator, noting times to close BOPs and resulting pressure drops on accumulator manifold gauge.	
Run a cup-type tester on open pipe or HWDP to no deeper than the third joint of casing. A stand of HWDP below the CTT may be needed to enter the casing. Open the previous casing annulus side outlet. Check that the drillpipe can take the force exerted on the CTT at casing test pressure. Check that the stack is full. Close annular, flush through the kill, and choke lines with water. Pressure test casing spool outer valves through the kill line against outer choke manifold valves to the casing test pressure when plug was bumped. Check side outlets for leaks.	1a Ann
Close the casing spool inner valves, open the outer valves. Bleed off after the test period at the choke manifold, open annular, and pull the CTT. Close the previous annulus.	1b
Set the test plug in the casing spool on a test sub and drillpipe with a stand of HWDP below. The test sub should be made up on the test plug hand tight only (if the test plug is not solid, torque up a closed kelly cock underneath). Make up a side outlet sub on the top of the drillpipe with the stab in kelly cock then the kelly made up on top. Make up chiksans from the cement manifold to the side outlet. Lift the kill line NRV off its seat and remove the kill line. Open the casing spool side outlet in case of a leak past the test plug. Test the TPR, stab in kelly cock, kill line outer manual, and choke line manual by pumping down DP with the stand pipe manifold bleed off or kill valve open. (If there are spare kill and choke outlets below BPR, test in parallel with the main outlets.)	2a TPR

Close the kill line inner valve and open the outer valve. Bleed off at the cement pump when the test is complete.	2b
Reseat the NRV, open kill line inner manual, open choke line manual, and close HCR. Open stab in kelly cock and close lower kelly cock. Repressure to full test pressure (no LP test needed against valves) and test NRV, HCR, and lower kelly cock. Bleed off at cement pump.	2c
Open TPR, close BPR, open lower kelly cock, and close upper kelly cock. Test BPR and upper kelly cock then bleed off at the cement pump.	3a BPR
Close the outer standpipe valves, open upper kelly cock. Test kelly hose and stand-pipe outer valves. When test is good, close standpipe valve and bleed off standpipe pressure to test the standpipe valve to the kelly hose. Bleed off at the cement pump and open the rams.	3b
Back the drillpipe and test sub out of the testplug, reconnect the kill line, close in at the choke manifold outer valves. Test the blind rams, kill, and choke lines via the kill line and choke manifold outer valves. Close next row of choke valves when HP test OK, open previous in turn until all choke manifold valves are tested. Bleed off through the choke manifold.	4 BR
Open rams, close the side outlets, run in the drillpipe, and screw into the test plug. Shut the BPR and the annular recording pressures on the accumulator; the final pressure should be at least 200 psi above precharge. Start up the pumps and measure how long it takes to recharge the unit to full working pressure. Compare the performance results with the manufacturer's specifications. The accumulator should be fully recharged within six minutes.	
Open BOPs, close HCR, line up all valves for drilling. Run the wear bushing back in.	
With the air pumps off, bleed down the manifold pressure and ensure that the electric pumps kick in when the pressure drops below 90% of working pressure.	

Record the results on the blowout preventer test form.
The remote panels may be function tested as follows: switch off the recharge pumps, close the valves from the accumulator bottles. Depressure the manifold with the bleed valve and leave open until finished. Operate each function on the remote panel and check whether the four-way valve moves correctly. When finished, line up all four-way valves correctly, open up bottles, and switch on recharge pumps. This method is fast since it prevents actual operation of the BOP units and preserves accumulator pressure.
Check the active tank volumes against the driller's gauges indicated volumes. Test inside BOP with test sub while RIH.

Figure 3-3 BOP Test Procedure Diagram

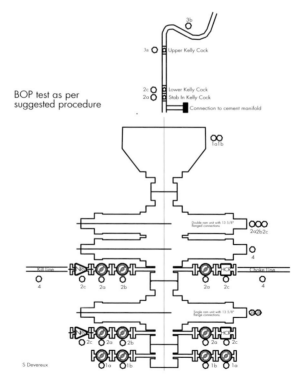

BOP test as per
suggested procedure

3.1.9. Well Control: Other Equipment Requirements

During operations where the hole is open to formation and the BOPs are nippled up, all the well-control equipment is to be kept working correctly and ready to use. The choke manifold should have the valves painted one color for normally closed and another color for normally left open when lined up for drilling in accordance with the current shut-in method. Spare needles and beans and an extra adjustable choke should be stored near the manifold, protected from the weather.

A full-opening kelly cock must always be below the kelly. This must be tested as part of the normal BOP test, be easy to operate, be functioned regularly, and have the operating spanner kept handy.

An upper kelly cock must always be above the kelly or be made up on a top drive system. This must also be tested as part of the normal BOP test, be easy to operate, be functioned regularly, and have the operating spanner kept handy.

A full-opening Kelly cock stab-in valve should be ready to stab into the drillstring, with either a lifting bar, a lift cap and winch already attached, or a line and balance weight. The spanner must be somewhere close and the valve easy to operate. If it has to be done for real, it could be with a lot of mud flowing back through the pipe and it has to be as easy as possible to stab, screw-in, and close under difficult conditions. If a X/over is required for stabbing into drill collars, etc., then this must be readily available on the drillfloor where the kelly cock can be made up into it before stabbing.

An internal BOP (grey valve or equivalent) must also be kept standing by on the drillfloor ready to stab and with the valve held open by the handling tool prong. All active tanks must have pit level indicators visible to the driller. These should be checked each BOP test by physically measuring the amount in each and comparing it to the indicator reading. The gain/loss gauge should be constantly checked while drilling to ensure it is working; natural movements when the pumps are adjusted or when deliberate changes are made to the active volume can all be monitored by the driller.

The flo-sho becomes affected if cuttings start to settle out in the flowline. If the flo-sho still shows an apparent flow on connections, check to see if the flowline has settled cuttings, and clean if necessary.

Stroke counters must be accurate and (if digital) display correctly. Check also the SPM and total stroke counters on the remote choke manifold for accuracy and correct display.

D exponent calculations should be made on all exploration wells where shown in the drilling program.

3.1.10. Suggested Rig Takeover Checklist

It is good practice for the drilling supervisor in charge of an operation to use a short checklist of things to look over. It is not possible to check everything on the rig, but those items essential to safety and efficiency can be checked out with a short walk around. A checklist is a guide for your eyes; for instance, while looking at the breakout lines you should also look at the guide rollers and the hoses to the cathead.

Figure 3-4 provides some ideas for a regular routine. The list can be photocopied if desired.

3.1.11. Minimum Mud Chemical Stock Levels Held on Rig

Below are some suggested minimum levels. Company policy or government regulations may dictate different amounts.

Barite. The amount of barite available on site must not fall below that needed to weight up the complete active system to the equivalent mud weight (EMW) of the last casing shoe test. If stock does fall below this while working with open hole, operations are to be suspended until the stock is at the necessary level.

The minimum level of barite should be 50 tons during appraisal/development drilling, or 100 tons during exploration drilling.

Cement. A minimum of 50 tons should be kept available on site. Inform base immediately if stock falls below this level.

LCM. Sufficient LCM materials of various grades/types should be available as appropriate for the drilling operation in hand, but a minimum of 500 sx is to be kept on site. A suggested mix of LCM types is two pallets of Kwikseal, one medium and one coarse pallet of fine mica, and one pallet of DF-Visc (to make the pill viscous before adding LCM).

Figure 3-4 Rig Takeover—General Checklist

Date: _____
Notes/Items to Rectify

_____

_____

_____

_____

_____

_____

Drillfloor:	Kelly cock correct size, ready to stab, left open, spanner close by, easy to close? Choke manifold correctly lined up for type of shut in?
	Is the drillfloor kept as clean as is practicable?
	Tong lines all in good shape? Sensator pumped up? Dies clean and sharp?
	Slips and safety clamps all OK?
	Check winch wires for general state, kinks, corrosion, or other damage.
	MAASP and ram sizes/positions displayed in driller's doghouse? Slow pump rates being taken each shift?
	Geolograph pens all working, charts attached correctly, unit wound up?
	Remote BOP panel looks OK?
	Driller's instruments intact and in apparent working order?
	Driller fully briefed on the current and planned operations?
	Pit levels, mud wt, and viscosity being monitored?
	Solids control equipment all working OK?
BOP area:	Positions and sizes of rams and annular preventers correct?
	Wellhead valves; which are closed or open? If gauges are attached are they open to the annulus and is there pressure on any annulus?
	Accumulator pressures and tank volume correct, lined up as required?

Figure 3-4 Rig Takeover—General Checklist (cont'd)

Chemical storage:	Is area (especially hazardous chemicals) clean, tidy, appropriate notices displayed? Wash station ready to use? Crews using correct safety clothing? Are damaged chemical containers correctly disposed of?
Rig housekeeping:	Is rubbish stored correctly and disposed of regularly? Is the area around the compactor/burn pit/rubbish skip also clean?
Office:	When is next BOP test due? Kick sheets prepared? Current and planned operations, logistics. Pipe tally correct? Strap pipe next trip if any doubt.

$$\begin{bmatrix} 3.2 \end{bmatrix}$$

Drilling Fluid

Section 2.5 covers drilling fluid types and properties in detail. This should be referred to as required. Maintenance of that mud on the rig involves maintaining the programmed mud properties by various chemical treatments and by using the solids control equipment to remove detrimental drilled solids from the system.

Chemical mud maintenance and treatment is a large and specialized area, outside the scope of this book.

3.2.1. Solids Control

For rig site drilling engineers, a thorough understanding of the mechanical processes and equipment used to remove undesirable solids is necessary. The equipment is simple in principle, but in practice problems can sometimes be hard to diagnose. Drilling crews often do not have a good understanding, and so it is essential for the rig supervisors to check on the equipment and educate the drill crews so that problems are spotted and solved early.

Effective solids control requires a chain of processes that remove progressively finer solids. Failure of any particular piece of equipment will adversely affect the equipment downstream by overloading it and therefore reducing its efficiency. If the shale shaker screens become torn and go undetected, the desanders, desilters, mud cleaners, and centrifuge can become overloaded and can discharge heavy amounts of mud—or plug up completely and pass solids back into the active system.

Ineffective solids control adversely affects mud rheology, plugs lines, damages pumps, and fills pits. Downhole the problems become large filter cake buildups (increasing the risk of stuck pipe), large surge and swab pressures leading to wellbore instability, and higher annular circulating losses (increasing the risk of lost circulation). Mud treatment becomes more difficult and expensive. Clearly, the solids control equipment is a key component of an efficient drilling operation and it must be kept operating at top efficiency. Each part of the system is discussed in detail.

Shale shakers. The first and most important item of solids control equipment is the shale shakers. Correctly set up, they remove the bulk of the solids, from the coarsest cuttings and cavings down to fine particles. If they are working correctly, they reduce the load on the hydrocyclones and centrifuges, allowing them to work more efficiently and with less wear.

The ideal shale shaker setup is a "cascade" system whereby the flowline routes to a header tank through a set of coarse screen shakers, then onto another header tank and the main bank of fine screen shakers. This allows finer final screens to be used without blinding off, which may occur if only one set of shakers were used, therefore optimizing the performance of the rest of the equipment.

Each set of screens should be as fine as possible without blinding off, avoiding the loss of large quantities of mud over the end. The top set of screens should be coarser than the bottom set (if the shaker has double screen banks), so that the work of removing the solids is shared between the screens, leading to greater efficiency.

Choice of screens will of course be limited to what is available on the rig, which in turn may be determined by the contract with the rig contractor. The following factors affect screen sizes:

- Higher flow rates require coarser screens
- Higher rate of cuttings generation (related to bit size and ROP) requires coarser screens
- Higher mud weight and viscosity both require coarser screens

It is not possible to give definite sizes to use for each hole section due to the number of variables and the different types of shakers in use; however, some general recommendations may be given as a starting point. These assume only one bank of shakers with double screen banks.

For a top hole of 20 in and above, high flow rates and large amounts of cuttings require very coarse screens: 40 top and 80 bottom. For hole sections below 20 in down to 12 in, 80 top and 100 bottom. From 12 in down, 100 top and 120 to 150 bottom.

Shale shaker screens may blind due to solids of a size close to the mesh size of the screen plugging up the mesh. In this case, finer screens are called for. They may also blind due to very small cuttings coating the wire, which effectively reduces the mesh opening. If this happens then coarser screens should control the blinding.

Screen cloths are available with a square mesh design or a rectangular design. It is generally recognized that the rectangular mesh is superior, giving greater fluid throughput capacity while removing the designed range of solids. "Pyramid" screens are a recent innovation which give a greater mesh area for the same size of screen; experience so far has been favorable.

The shale shaker screens must be held in the frame at the correct tension as specified by the manufacturer. If the screens wave up and down (as opposed to vibrating), then they are either insufficiently tensioned or have been stretched or torn. Remedial action must be taken quickly to avoid high levels of solids bypassing the shakers and affecting the downstream equipment.

Torn screens must be recognized and replaced immediately. Where high flow is passing over the shakers a torn screen can sometimes pass unnoticed for some time, so the drilling supervisor and toolpusher must both be on the ball when checking the screens. If this requires bypassing the shaker while the repairs are made it is important that the sand trap is not bypassed as well; this will hold the larger solids for a short time while bypassing the shakers and therefore protect the other solids removal equipment. In general, it is preferable to stop drilling, continue to circulate at half rate over the other shaker(s), and fix the split screen as soon as possible.

Splits can often be detected even when under a depth of mud; look at the surface for unusual flow patterns. Sometimes it looks as if mud is welling up from below; sometimes a pattern of mud splashing up is in a ring around the hole. Investigate any unusual signs.

Sand trap. After the shale shakers, the mud should pass through into the sand trap and possibly from there into a settling tank. These tanks should not be agitated nor used as suction feed tanks for hydrocyclones or centrifuges. Flow from them should pass over a weir to the next tank so that they are kept full to maximize settling.

The sand trap should have a discharge butterfly valve located at the bottom of the tank that can be quickly opened and closed again, allowing settled solids to come out while minimizing the loss of expensive whole mud. It is detrimental to clean out the sand trap except when necessary for changing mud systems. The solids will settle on bottom at an angle to the discharge gate, and if these bottom solids are cleaned out more whole mud will be lost whenever the sand trap is dumped, until this layer has built up again.

The sand trap is an important part of the system, protecting other equipment from shale shaker inefficiencies. If the shakers always worked perfectly and screens never tore then the sand trap would lose much of its purpose. However, this is unlikely to be the case for a considerable time, therefore, the sand trap must be designed and used properly.

Hydrocyclones: principle of operation. Once the mud passes through the shakers and sand trap, the coarse particles should have been removed. Finer solids will remain in the mud, which are still capable of damaging equipment and adversely affecting mud properties: abrasive sands, fine silts, and other low gravity solids. Correctly designed and adjusted hydrocyclone banks can remove most of these undesirable solids at the full flow rate used while drilling.

Fluid is fed into the feed inlet at the side of the cylindrical section on top of the cone. This flow enters the feed chamber directed along the inside wall. The curve of the feed chamber forces the mud to swirl around the inside. The overflow opening at the top extends down into the feed chamber so the inside of the chamber is an annulus, not an open cylinder. This pipe is called the vortex finder. As the mud moves around and reapproaches the feed inlet, it is forced downwards (as the top is closed) by more mud entering the chamber. The mud then follows a spiral path down along the cone, with its circular velocity increasing due to the decreasing diameter of the cone. This imparts high centrifugal forces on the mud, forcing solid particles to the outside of the mud stream and against the cone wall.

The volume of mud flowing down through the cone cannot exit at the relatively small hole at the bottom. As mud is forced towards the bottom of the cone, the pressure in the mud stream increases, forcing the mud to turn back on itself, spiraling upwards inside the downward stream (forced against the wall by centrifugal force). This mud exits at the overflow pipe at the top.

In the balanced design the fluid changes direction just above the opening. However, the solids, being heavier, cannot change direction so readily and are forced to exit from the cone at the bottom. The cone is adjusted by changing the size of the bottom opening.

Dissolved gas will reduce the efficiency of the hydrocyclones. The degasser should be lined up to suck from the tank immediately downstream of the sand trap and may discharge into the desander suction tank.

Unlike shale shakers, the operational part of a hydrocyclone is hidden from view. When problems occur, they may pass unnoticed or even be deliberately ignored by the crew assigned to the tanks and solids control equipment during drilling. It is imperative that the drilling supervisor is familiar with the operation of cyclones and knows how to recognize and correct problems. Ideally, the rig contractor should take steps to train drillers, ADs, and derrickmen to properly maintain this equipment, since malfunctions can be as costly and unacceptable as shale shaker problems. Figure 3-5 illustrates the process.

Hydrocyclones: types available. There are two types of hydrocyclone design: balanced and flood bottom. The flood bottom design has a constant discharge at the bottom, the same size as the adjustable bottom opening. Adjustment is a compromise between insufficient solids removal and excessive mud loss. No modern rig should have flood bottom hydrocyclones as primary solids removal equipment. The only place they should be found is upstream of a decanting centrifuge, where they are used for increasing the solids content of the mud sent

Hydrocyclone solids-removal process

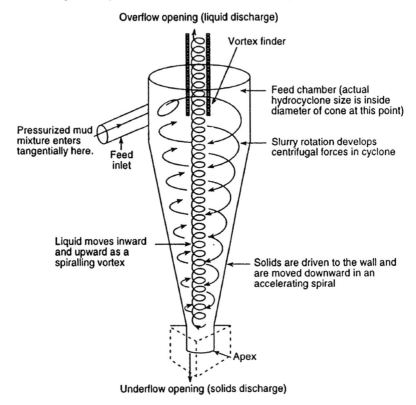

Overflow opening (liquid discharge)

Vortex finder

Feed chamber (actual hydrocyclone size is inside diameter of cone at this point)

Pressurized mud mixture enters tangentially here.

Feed inlet

Slurry rotation develops centrifugal forces in cyclone

Liquid moves inward and upward as a spiralling vortex

Solids are driven to the wall and are moved downward in an accelerating spiral

Apex

Underflow opening (solids discharge)

Fig. 3-5 Hydrocyclone Solids-Removal Process (courtesy of Baroid)

to the centrifuge (i.e., the centrifuge works on the underflow from the flood bottom hydrocyclones).

A balanced hydrocyclone, properly adjusted, will separate out solids that are discharged at the bottom, taking with them only fluid that is bound to the surface area of the solid particles. The size of a hydrocyclone is stated as the inside diameter of the top of the conical section. Smaller cones of the same design will have lower flow capacity than larger cones but will remove a greater percentage of the solids in the feed mud, including smaller particles.

Balanced hydrocyclone adjustment. When the hydrocyclone is adjusted with too small an underflow opening, the point at which the mud changes direction will be too far above the opening and very fine solids will then be deposited on the cone wall instead of being discharged. The end result will be a plugged cone and no solids discharge. The opposite can also occur, resulting in a discharge of a hollow spiraling cylindrical shape of mud and discharged solids. This second condition gives fairly efficient solids removal and does not lead to high mud losses unless it is excessive. As with spray discharge, suction can be felt at the bottom, which may be used to differentiate the condition with rope discharge.

The ideal adjustment of opening, feed rate, and feed pressure with correct chamber design (and an unworn lining) will give a spray discharge of damp solid particles and suction of air into the underflow opening, which can be felt with a finger. This leads to the most efficient, lowest cost operation with low cone wear. If this suction can be felt then the cone is working well.

Use of hydrocyclones and mud cleaners with oil-based mud. Desilters are not usually used with OBM because the discharge will contain a relatively large percentage of oil. If a cone malfunctions then large quantities of expensive mud can be lost. Environmentally this is unacceptable as well as expensive.

If the discharge from the desilters can be routed over a vibrating screen (such as a mud cleaner), then the desilters can play a valuable part in removing undesirable solids.

Mud cleaners are simply banks of desilter size cones (4 in to 6 in) over a fine mesh vibrating screen.

Hydrocyclone cone capacities

4 in	50 GPM/cone	12-60 micron cut size
6 in	100 GPM/cone	20-60 micron cut size
8 in	155 GPM/cone	30-60 micron cut size
10 in	500 GPM/cone	30-60 micron cut size
12 in	600 GPM/cone	30-60 micron cut size

Problems with hydrocyclones. Rope discharge from a hydrocyclone—a characteristic solid stream of spiraling mud with no suction—

may be caused by overloading the cyclone. The cyclone is removing more solids than can exit at the underflow discharge. The cure is to reduce the solids content of the mud (by using other mud cleaning equipment upstream or changing to finer shaker screens if possible), or reduce the feed volume. If the desilters are showing rope discharge and the desanders are not in use, starting up the desanders will be beneficial and may solve the problem. The desanders allow the desilters to work more efficiently and should normally be used, if available.

If the cone lining is worn it will affect the operation of the cyclone and also could lead to rope discharge. Ongoing wear will be rapid and eventually the lining will wear through if it is not changed.

Sometimes solid particles accumulate around the bottom of the cone inside and affect the action of the cone to produce rope discharge. Remove and replace the outlet pipe, flush with water inside the cone, pipe, and manifold inlet.

If the feed inlet becomes partially or completely blocked, there will be no spiral action and mud will simply fall out of the bottom. This may be mistaken for rope discharge, but the mud will not have any spiral motion. This wasted mud will contain mud that has already been cleaned by the other cyclones in the bank (since mud will be flowing back through the overflow discharge pipe from the discharge manifold). Therefore, not only is mud being wasted, but the work of the remaining cyclones is being compromised. The blockage must be fixed or the cyclone removed from service and the inlet and outlet capped as soon as possible.

Between wells it is worthwhile to remove the end cap from the feed pipe and clear out the inside. It is amazing how often you come across wood, rope, bits of cement plug, etc. in these feed pipes. This debris can move and block the feed pipe.

If the underflow discharge should become blocked (or deliberately plugged with a bolt by someone who cannot be bothered to fix a cone with excessive discharge), then the result will be high wear on the cone lining and no discharge of solids. Uncleaned mud will be discharged at the top to mix with the mud cleaned by the other cyclones in the bank. The problem must be fixed, not hidden.

Decanting centrifuges. While there have been several types of centrifuges used in mud solids removal applications, the only one likely to be used on the rig is the decanting centrifuge. A detailed description of the workings is not necessary here.

The size of the centrifuge is expressed as the bowl diameter and length, for instance 18 in x 28 in. A larger centrifuge can process more mud to the same size of separation, or the same amount of mud to a finer separation, than a smaller centrifuge.

All the other equipment described above processes the full mud flow rate; the centrifuge only processes part of the mud stream. When drilling is slow and relatively frequent trips are made (during which centrifuging continues), then one centrifuge is usually sufficient to keep up with the small solids entering the mud. It may be necessary to use some mud dilution when drilling fast with few trips in order to maintain the programmed properties, unless a second centrifuge can be obtained or the deterioration in properties tolerated.

Adjustment and maintenance of the centrifuge is best left to a technician who is familiar with the particular model of centrifuge. You can choose to have the feed rate adjusted to process relatively high volumes to a coarser particle size, or reduce the feed rate to remove finer particles—down to 2 micron (barite size) should be possible in water-based mud. These adjustments are not obtained simply by changing the rate of mud supply to the centrifuge, but changes are made in the centrifuge to the overflow port settings as well.

3.2.2. Quality Control

Are you getting what you pay for? If not you may get things you did not bargain for, which can cost you a lot both in mud treatment costs and in unnecessary drilling problems. Section 2.5.13, " Tendering for Mud Services," contains some information to incorporate at the tender stage to ensure the correct quality chemicals are supplied.

Many problems are caused by contaminated chemicals. This may occur accidentally with bulks (e.g., using a hose previously used for bulking cement, when loading barite, has happened!) or may occur deliberately when the mud contractor has supplied chemicals "cut" with something cheap and bulky, to increase profit at your expense.

Logistical tricks of the trade include sending out drums that are not full and loading pallets so there is an empty space in the center where it cannot be seen until the pallet is used. The pallet trick is a good one especially where the whole pallet is covered in a plastic sheet for weather protection; it cannot be checked when first received and you may not want to remove the plastic earlier than necessary. These may be done in conjunction with a ticket that states that a certain volume of mud was mixed but in reality a lesser amount was made up. If the mud engineer is smart, the apparent mud mixed will equal what would be mixed to include the nonexistent chemicals, so an audit later will apparently balance. One mud engineer tried this trick on me, but when he presented his job ticket for me to sign, it said he had mixed 600 bbls of reserve mud when the only tank he could have used was 400 bbls maximum capacity!

The biggest deterrent to mud chemical fraud is to take and test samples regularly. Sample each batch of chemical that arrives. Take a 1 L sample of the mud each day which can be tested later, if necessary, and return it to the shore base or a storage facility. Have someone independent of the project witness covered pallets being opened or see it for yourself. Knowledge that you are checking these things is a real incentive to more honest reporting!

[3.3]

Drilling Problems

Following are some of the most common drilling problems that you are likely to encounter while drilling. *Many of these problems are avoidable.* A good driller will look for all the signs coming from the well: pressures, weights, torque, mud properties, shaker material, etc. This is called "listening to the hole." It takes experience to be able to listen properly to the hole, but an experienced hand does develop almost a sixth sense of what is happening in the well. There is nothing magical about this, it just takes a lot of practical experience and theoretical study to do it. In fact, this ability is one of the major benefits gained from working on the brake as a driller. There is no doubt that this experience later on gives a drilling supervisor or superintendent a greater ability to make correct and timely decisions which will save the operator money.

3.3.1. Stuck Pipe

Refer to the notes on stuck pipe in Section 2.9.2 for information on the various categories of stuck pipe.

Tight hole is potential stuck pipe. Do not try to save time by ignoring the signs; this approach will usually result in getting stuck.

Where hole conditions exist that make stuck pipe likely, the drilling supervisor must be on the drillfloor while tripping through the problem sections. In any case, the supervisor should be present for the initial flowcheck and the trip out as far as the previous trip depth.

BP put together a task force some years ago to examine the stuck pipe problem and came up with recommendations to drastically reduce the incidence of stuck pipe. One of the more interesting findings we made was that a disproportionate number of cases occurred within the two hours after a shift change. This led to the advice "brief your relief." Clearly the handover between drillers was somehow deficient, which led to the new driller on shift being unaware of vital information needed to recognize incipient stuck pipe in time.

Most cases of stuck pipe are avoidable with proper supervision, and once stuck pipe is recognized, fast action is necessary. The best chance of getting free is just after getting stuck. Take the time to fully brief the drillers on the procedures you expect to be followed, including when you should be called if there are problems.

A good handover between drillers is essential since a disproportionate number of stuck pipe incidents occur within two hours of a shift change. The toolpusher or night pusher should be on the floor at handover time to allow the drillers time for a good handover, and it is a good idea for the drilling supervisor to be around at this time as well.

The drillers, toolpusher, and drilling supervisor should always know the allowable pull on the pipe (with and without the kelly on). There will not be time to work it out when you get stuck.

Undergauge hole. Ream any tight spots on trips. Ream back to bottom after changing the bit and/or BHA, the last 10 to 30 m depending on hole conditions. If deviation or other considerations permit, ream once each single before making a connection, which will also clean around the BHA before shutting the pumps off (which is good practice anyway).

If the hole should become undergauge above the BHA due to wall cake formation or mobile formations, it will probably be necessary to circulate and backream out. If the slips are used to backream with a kelly, take all proper safety precautions; tie the slip handles together when set, and *never* pick up out of the slips with torque left on the table. Keep personnel off the drillfloor as much as possible. Avoid picking up so much that the string torques up and stalls out while reaming.

Undergauge hole due to squeezing shales. In shales that "relax" some time after being drilled through, high overpulls may be seen when tripping out. If a top drive is not being used, it is a time consuming and risky operation to backream out and it may be possible to work out carefully without having to make up the kelly. Start by overpulling no more than half the BHA weight then slack off until free—even if this means slacking off all of the string weight. If you can come free easily going down, repeat with slightly more overpull (about 5-10 k more), but ensuring each time that you can come free by slacking off. Note the distance the string has to be lowered to come free; this will normally be at least the amount to lower the string from the up-drag weight to the down-drag weight. *Do not* pull up enough to trip the jar while working the string; this is likely to pull the BHA into the undergauge hole with enough force to completely stick the string.

Continue to work out of the hole. When the connection to break for racking back is above the table, keep working up until the connection is so far above the table that if the hole becomes tight after breaking off the stand, the string can be lowered enough to come free. This will be the distance noted above plus the normal tool joint height above the table when breaking the connection.

With this type of problem, note the depths on the daily report and take special care when passing those depths on the trip in. It often happens that once you have worked out through the tight spots, the problem will not be seen again.

If salt is squeezing in above, try programming wiper trips back through the salt after a short time, say 15 to 20 hours after the previous trip or wiper trip. This may give some overpulls on the first wiper trip through, but often these do not subsequently recur.

If the string becomes stuck in undergauge hole, work and jar the string in the opposite direction to that when the string got stuck. Do not jar the string both ways, this is a waste of time. If the string became stuck pulling up, set down/jar down and, if possible, apply torque to the string if setting down alone does not free the string. This can only be done if the kelly bushing is already in the rotary table or if using a top drive; do not attempt to use the slips, since there will be insufficient weight available to set down on them.

Often when working through these troublesome shales, it is better not to circulate but to work out carefully without rotating, taking care not to get stuck. Circulating sometimes seems to make it harder to work out. If you have problems picking up to make a connection in these shales, sometimes you will find that after a period of time the shales cave slightly into the wellbore and the problem cures itself while you are drilling. Field experience will dictate whether this is the case.

You can experiment (in a development area) if previous wells have had problems picking up to make connections by only working/back-reaming enough to make the connection. Drill ahead and when you come to POH, trip carefully through the troublesome intervals and note how much you have to work to get through. This may save time on future wells.

Keyseating. Pipe stuck in a keyseat will only occur when pulling out of the hole. The top drill collar or stabilizer will be the most likely item to hang up in the keyseat.

If the deviation profile and formation type make a keyseat likely, consider running a keyseat wiper on top of the drill collars or prefer-ably, if one is available, run a string reamer in the drillpipe (refer to "Keyseating" in Section 3.7.2).

If stuck in a keyseat, the jar will not work if the stuck point is at the top of the drill collars (above the jar). In this case, work the string in compression, letting it drop and catching it sharply with the brake to send a shock wave down the string. If one is available, a surface jar does the same thing only more efficiently. However, if the stuck point is deeper than about 700 m (or less if the well is crooked or significantly deviated above this), then the shock wave will probably not be great enough at the stuck point to have much effect.

If the downhole jar is working, the prognosis is much better. Limit the overpull at surface to just enough to cock the jar (if it jars down-wards) and work/jar down. Do not allow the jar to fire upwards, since this will force the BHA tighter into the stuck point.

Be prepared to backoff above the drill collars and run a fishing assembly consisting of overshot - fishing bumper sub - 2 drill collars - fishing jar - 3 drill collars - accelerator. A sub with the correct pin to make up into the fish plus a safety joint could also be run instead of an overshot. Bump/jar down to free.

Stiff assembly. Where a stiff BHA is to follow a more flexible one, or where a rotary assembly is run after pulling a motor, lay out enough drillpipe to ream the complete section drilled with the more flexible assembly.

When reaming a kickoff in softer formations, take precautions to ensure that a sidetrack is not inadvertently drilled. Use low WOB/higher RPM and, if necessary, make a special trip with a bit-sized hole opener and a bullnose to ream out the hole before drilling ahead.

If stuck in this situation, jar and work the string upwards. The jar should be free.

Ledges. Ledges may form where soft and hard formations alternate. Run slowly when tripping in at these points. Note and ream any resistance seen on trips.

To free the string if stuck by a ledge, jar up if stuck while running in or down if stuck while pulling out.

Mobile formations. Sticking in salt may occur fast enough for the bit to get stuck as it drills. This condition, if it occurs, is hard to avoid altogether but may be reduced by using a water-based mud that leaches the salt slightly as drilling progresses. Higher mud weights may (or may not) help to keep the salt under control. Bits are available that drill slightly off center and therefore cut an overgauge hole; they may be helpful when drilling in problem-flowing salts. If the string is sticking while drilling or has high drags on connections, try drilling one or two meters at a time, then pick up and ream down before continuing.

If the salt moves more slowly and causes problems mainly when tripping back through it, try programming wiper trips through a newly drilled hole after a short time (refer to "Undergauge hole" in Section 3.3.1, "Stuck Pipe").

When stuck in salt, start jarring immediately and if not free within a few jar blows, prepare a freshwater pill while continuing to jar. If it does not jar free quickly, it is unlikely that jarring alone will free the string. Analysis of cases of salt sticking indicate that once stuck in salt, jarring alone only frees the pipe within the first 10 minutes.

If the well is likely to have other mobile formations present, wiper trip and ream as dictated by hole conditions. If stuck, start jarring in the opposite direction right away.

Cuttings beds. Ensure mud rheology and annular velocities are sufficient to keep the hole clean while drilling. This will be more difficult as inclination increases. The most difficult hole to clean is between 45° and 55° inclination. Circulate clean at the highest practical no-loss rate before pulling out, not just bottoms up, displacing a pill around if necessary. If cavings are causing a problem, a slight increase in mud weight may help to stop the caving, depending on the cause of the caving, (if this will not cause losses), and cavings may cause a dirty annulus during the trip, giving problems even if the hole was clean at the start of the trip.

If the well cannot be cleaned by normal circulation for holes up to 45° inclination, circulate around a high viscous pill while moving the string to disturb the cuttings beds. In higher inclinations, turbulent flow is more effective, therefore, a low vis pill at high rate while moving the string may work better. In very high-angle holes using OBM, pumping a base oil pill for turbulence can strip off the wall cake and cause the hole to collapse as the hydrostatic will not then fully support the bore wall.

On tripping out, steadily increasing drags will indicate dirty hole. This could be caused by not circulating clean, or possibly by higher formations caving into the hole even after circulating clean. In a deviated well, cuttings beds may accumulate on the low side, which can be hard to shift; this problem gets worse with increasing hole angle and washouts.

The immediate actions are to stop pulling out and to clean up the wellbore. If circulating normally has not cured the problem, consider additional measures. This could include pumping at higher rates, rotating and reciprocating the pipe to disturb cuttings beds, pumping high, or low viscosity pills as noted above.

Getting stuck by packing off on cuttings beds is a difficult situation. Circulation will probably not be possible and the string may not move at all. The sticking force will increase with time as more solids settle around the string. Keep pressure on the standpipe (not so much as to induce losses) and try to work the string down, using torque if possible. Once circulation is re-established, circulate as fast as possible and continue to work the string down. Once free, clean up the well before continuing with other operations.

It is possible that the string cannot be freed by these methods. The alternatives are then to backoff and either wash over the string, sidetrack, or abandon the well.

If a decision is made to wash over the string, it is best to washover and recover in one run without stopping circulation. Otherwise, while the washover string is being pulled and the fishing string run, solids may again settle around the fish. Backoff at a convenient depth just above the stuck point and then run the following fishing assembly:

Washover shoe and pipe, long enough to reach below the bottom stabilizer or just above the bit sub to stab into the top fish connection - circulating sub with the sleeve already open - drill collars - fishing jar - drill collars - accelerator – HWDP, etc.

Washover the fish, screw into the fish without interrupting circulation, continue to circulate, and pull/jar free. By having the open circulating sub inside the top of the washover pipe, circulation is maintained all the time, even if the fish is plugged inside with solids (as is likely).

Reactive formation. There are two types of reactive formation: those that hydrate and slough in (Gumbo) and those that collapse due to brittle failure. Both types are likely to be time sensitive, therefore, the casing program should allow them to be cased off within the time that they stay stable.

In the first case, ensure that the level of inhibition is carefully maintained as per program and increase if necessary. Monitor the condition of the cuttings for signs of hydration. Minimize swab and surge pressures. Plan operations to minimize the time spent with open hole before running casing, but do not take risky short cuts.

For brittle failure type formations, higher mud weight is the best stabilizing mechanism. Ensure that the hole is kept clean while drilling. Avoid drilling at critical rotary speeds to minimize string vibrations. Minimize swab and surge pressures.

For both types of reactive formation, consider wiper trips carefully. Wiper trips should only be done when the wiper trip is likely to help improve the overall situation. Refer to Section 3.3.8, "Preplanned Wire tripping," for general information on wiper trips. Trip carefully and start up and stop the pumps slowly to minimize swabs and surges, all of which can help destabilize the formation.

In the case of pipe sticking due to reactive formation collapse or sloughing, refer to the method discussed in the Section 3.3.1, "Cuttings beds." Also see the wellbore stability topics in Section 2.9.1.

Tophole collapse. This may occur since very shallow formations are unconsolidated and there is little overbalance to help stabilize the wall. If possible, drill with returns to the rig (for greater hydrostatic) on an offshore well, and use a mud with good wall cake-forming characteristics or pump high solids slugs around to help plaster the hole.

To get free, try to establish circulation while working the string down and applying torque. If off bottom, a surface jar may help to get movement and circulation. Once circulation is established, pump as fast as possible without inducing losses and try to clean up the well. Keep the string moving. On a floating rig, use the riser boost pump to help lift the debris to surface.

Geopressured formation. In shales where the formation pore pressure is greater than the mud hydrostatic, slivers of formation will tend to be pushed into the wellbore. Use higher mud weight, if possible, to reduce or eliminate this and keep the hole clean, using pills if necessary.

Where the formation is left exposed for a long time, very large washouts can occur that may lead to other problems with large chunks of formation falling in, big cuttings beds forming, fish becoming impossible to recover (if they fall over in the washout,) and bad cement jobs. If the formation cannot be controlled with mud weight, then ensure that the casing can be run within a short time of drilling the formation.

It may also be possible to displace cement into the washouts and drill through it, leaving the cement to stabilize the formations. Adding fibers to the cement may be prudent to reduce the possibility of cement blocks. Also, fiberglass pipe can be run as if it were casing, placing cement in the washout and allowing a "pilot hole" through the cement for drilling out.

In the case of pipe sticking due to geopressured formation collapse, the method of recovery is discussed in Section 3.3.1, "Cuttings beds."

Fractured and faulted formations. Some formations are already naturally fractured before being drilled into. This can again cause large chunks of rock to enter the wellbore causing mechanical sticking. Since the damage is already done, avoid making it worse by minimizing swab/surge pressures and drillstring vibrations, drilling at low angle if possible, getting through it quickly, and casing it off.

Circulation may well be possible if mechanically stuck with large pieces of formation above the BHA. Maintain circulation and work/bump the pipe down until free. Start with light bumps because shock waves from the string may cause further bits of formation to come loose. When the string can be moved down, keep the string moving and circulate clean. It may be necessary to backream up to pass the obstruction. Ream carefully when running back in.

If the formation is particularly bad, it may be better to attempt to drill through it, then case it off as soon as possible, even if this means having to run an extra casing string. It may also be possible to set cement across the formation and drill out through it to stabilize the formation.

Junk. Ensure all drillstring components are inspected before the well in accordance with API Recommended Practice RP7G and that good handling and running practices are used. Use the correct size and grade of drillpipe. Keep a good check on the condition of slips, slip dies, rotary bushings, tong line pull sensators, and elevators. These will help avoid downhole failures leading to junk in the hole. Ensure that the crew follows good procedures to avoid junk falling through the table and that rams are closed when out of the hole.

If stuck on junk it is important to try to determine what the junk is, where it may be positioned, and how it is sticking the string. How to get free will depend on the situation.

Cement blocks. Where a large pocket exists under a casing shoe, cement in the pocket after cementing the casing may fracture as drilling progresses and fall into the hole. Do not program in a pocket under the casing to be any larger than necessary. It may help to add polypropylene fibers to the tail slurry when cementing (e.g., Dowell product D094). In the case of pipe sticking due to cement blocks, the method of recovery is detailed in Section 3.3.1, "Fractured and faulted formations."

Soft cement. It is possible to run into cement that is not completely set, then find it impossible to pull back out or to circulate. Monitor surface cement samples (preferably kept at bottom hole circulating temperature) to ensure they are hard before approaching bottom and run in the last couple of singles slowly and with the pumps on.

If stuck in soft cement try to jar free while maintaining pressure on the standpipe. If unable to wash/jar free, a backoff will be called for, followed by a washover fishing string, sidetrack, or abandonment.

Differential sticking—avoidance. As was noted in Section 2, there are four conditions that must all be present for differential sticking to occur. Avoidance and cure focuses on reducing or eliminating these four conditions.

A permeable zone covered with wall cake is one of the four conditions. Where such a zone is identified, the wall cake characteristics can be optimized with a good mud program. Of particular concern is development drilling through a depleted reservoir because the overbalance can be very high. Additives can be used to make the cake thinner and less sticky or oil mud will form very little cake.

Keep all the solids control equipment running at full efficiency to minimize low gravity solids buildup. Use a centrifuge if possible. Walk around and check the solids control equipment several times a day. Dump the sand trap and dilute, if necessary, to achieve LG solids control. MBT level should be kept as low as possible, preferably below 8 ppb equivalent.

Static overbalance is the second of the four conditions. Use the minimum safe mud weight to minimize the static overbalance on the formation. Condition the mud carefully before running casing to the minimum safe density.

Keep the hole clean to minimize loading up the annulus with cuttings. Watch rheology and pump rate to achieve this.

Wall contact is the third of the four conditions. Use a well stabilized BHA. Spiral or square drill collars have less contact area than round ones. Run HWDP in compression to reduce the length of drill collars. Use a bit that requires less WOB and, therefore, less BHA (PDC, diamond) if the formation is suitable and the rig cost justifies it. Wall contact will be greater as inclination increases. Centralize casing well where it is run past the problem zone.

A stationary string is the last of the four conditions. Minimize programmed directional surveys, especially if run on wireline. Top drive is an advantage since fewer connections are required. When POH make the initial flowcheck brief, pull out above the potential sticking zone, then carry out a full flowcheck.

If differential pressure is causing the pipe to start getting stuck when in the slips for connections or pulling out of the hole, try the following procedure. Ensure the crews are properly briefed to carry this out safely first.

Land in the slips, break the connection keeping about 30,000-40,000 lbs pull over block weight. Leave on the breakout tong, rotate out slowly, and leave the table rotating slowly. When ready to stab the next connection or latch the elevator, stop the rotary, ensure no torque is left on the table, make up connection, or latch elevator, and carefully pick out of the slips. Once the slips are out, move the string down rather than up; if differential sticking has already started, moving down will be more likely to get you free than pulling up, which may increase the side force onto the sticking formation.

It is good practice on trips out and connections to set the slips on a "down" drag. Pick up an extra 7-8 ft, come back down, and set the slips. This does two things. It reduces the tension and stretch in the drillstring, which reduces string side forces and sticking and also makes it possible to cock the jar as soon as the slips are taken out. It also confirms that you are not going to pull out of the slips straight into a stuck point. If you see signs of the string getting tight when picking up the extra height, go a bit further and work out the tight spot first before setting the slips while you have plenty of height to work with.

Centralize casing well over this interval. Plan the casing job and brief the crews well so as to avoid excessive time with the casing stationary.

In the event of equipment failure, try to keep the casing moving slowly while the problem is sorted out. If the casing cannot be picked up due to weight plus drags, or to a drawworks problem, have the driller lower the full joint very slowly, if possible, which could take several minutes of constant slow movement.

If the weight of the casing plus drags is likely to be more than can be picked up, consider floating in the casing (that is, running it partially empty from a certain depth to reduce the hook loads). In this case, check the collapse rating of the casing against the collapse pressure that will be exerted.

Differential sticking—cure. If a likely differential sticking zone exists in the well, it is especially important to avoid getting stuck by another mechanism first. This requires good planning and close supervision.

If differentially stuck, then the procedure for getting free involves reducing or removing some of the necessary conditions. Fast reaction is needed as the differential sticking force increases with time, so the best chance to free the pipe occurs when initially stuck. Circulation will be possible unless another mechanism also exists; for instance the string may be mechanically stuck, then become differentially stuck due to the pipe being stationary for a long time.

Start to work the pipe immediately after you become stuck. First slack off almost all the string weight; this may cause the pipe downhole to slump off the wall. Differentially stuck pipe comes free going down more often than by overpulling. If torque can also be applied, this is even better. Pull up to about 30,000-40,000 lbs above the up-drag weight and, if not free, then repeat several times. If unsuccessful start to jar up and bump down. This is one of the few situations where working the pipe both ways will not make the sticking problem worse.

If the stuck point depth is unknown, use stretch data to determine the approximate depth.

If not free within 10 minutes of working the pipe, continue to work the pipe while preparing a Pipelax or similar pill. These pills work by shrinking the wall cake, thus reducing the contact area between the pipe and hole. Make enough to cover the BHA (if stuck at the BHA) and to leave about 10 barrels inside the string.

Displace out the pill sufficient to cover the BHA, leaving the excess inside the bit. Pump about one-quarter barrel every 10 or 15 minutes while working the pipe.

In some cases the well has been displaced to oil-based mud to free the pipe. This would have to be fully discussed with the drilling office and would probably take a few days to organize logistically.

The sticking force will be roughly proportional to the overbalance at the stuck point. If the mud density were reduced then the sticking force would also reduce. This can be done as follows:

> Displace the complete well to lower density mud. If the annulus is loaded with cuttings at the time the pipe stuck, then circulating clean would reduce overbalance. Otherwise the mud can be conditioned to a lower weight, possibly over several circulations to ensure that well control is not compromised.

Some company policies prohibit the following method but it is worth considering as long as a float is not in the string and well control can be maintained.

Displace some of the drillstring to water, diesel, base oil, or some suitable lightweight fluid. Allow to U-tube out so that the level in the annulus drops and the bottomhole pressure decreases. Have the kelly or top drive on (bleed off on the standpipe manifold) and pull/jar to maximum. Full hydrostatic is restored by pumping in enough mud to replace the backflowed fluid; the remaining light fluid can be circulated out either conventionally or reverse as appropriate to the situation.

3.3.2. Lost Circulation

In some areas there may be offset information or "Field operational notes (Sec. 1.12)," which give specific guidance on how best to handle losses in particular formations based on previous experience. While this does not guarantee that past successful techniques will work again even in close offset wells, it gives a good starting point which is better than general guidance notes such as are given in the following topics. Therefore, these should be the first point of reference, if they exist for the problem formation.

Many instances of lost circulation are self-inflicted and avoidable. Keep the mud in good shape with rheology and gels low to minimize surge pressures and ECD. Trip in at moderate speed. Use good connection practices as detailed in Section 3.3.7, "Making Connections to Minimize Wellbore Instability and Losses."

There are many causes of lost circulation and various techniques to deal with them. Following are some of these techniques in detail, under the situations leading to the type of losses.

Where severe or total losses occur in relatively unknown areas, special attention must be paid to the possibility of taking a kick, either due to loss of hydrostatic or while drilling ahead with an actual or cured loss zone higher in the well. If an internal blowout is suspected, then the losses have to be cured before the well can be killed. In this situation, the drilling office will be closely involved in planning the strategy to bring the well under control.

Losses to surface or seabed outside conductor. Where this occurs the stability of bottom supported units can be seriously compromised. Offshore, if returns are to seabed, this is very unlikely to happen, but where a riser is in use the extra hydrostatic imposed can be enough to cause this.

Drill ahead until the losses become severe (unless drilling with a diverter—see last paragraph of this subsection). POH and run open-ended drillpipe to 30 m below the shoe. Spot a weighted viscous pill into the conductor and pull back to 5 m below the shoe.

Pump a neat cement slurry from 5 m below the shoe to 20-50 m inside. POH and make up the drilling assembly. When the surface samples are hard, run in and drill out the cement. The level should have dropped and cement displaced into the loss channels around the conductor.

It is better to wait until losses are fairly severe before doing this to ensure an easy flow path for the cement. This should not take long once losses have started.

Losses when drilling in divert mode must be cured; it is not safe to drill blind. If the cement does not cure the losses, consider running casing or respudding the well.

Severe or total losses in shallow unconsolidated formations. The principal cause of losses in these formations is extremely high permeability. The mud does not make an effective mud cake to seal the loss zone. Losses are likely to start as soon as the formation is penetrated. Several factors will contribute to the mud loss, such as annulus loaded with cuttings, excessive mud density, insufficient mud viscosity, high water loss (low solids content to plaster the wall), or excessive surge pressures. One or more of the following actions may be appropriate:

1. Add solids to the mud to increase rheology and plastering characteristics, such as bentonite, lime, cement, or polymer (CMC HV). Do not add high gravity solids such as barite, which increase the mud density.
2. Reduce mud density if possible by dilution and/or maximizing the use of solids control equipment.
3. Increase the circulation rate. The increase in ECD will be small compared to the reduction in annular density due to cuttings loading in large shallow holes. If in deep water with a floating rig, increase riser booster pump output if possible.
4. Drill at controlled rates to reduce annulus loading.
5. Add coarse LCM to the mud system. High strength is not necessary, so use the cheapest bulk material available, such as sawdust or ground nut shells.
6. Drill ahead if familiar with the area, confident that there are no well control problems ahead, and can replace the volume lost to the hole. Several precautions must be taken to drill blind: a float should be in the string and you must circulate fast enough to lift the cuttings to the loss zone (minimum AV around 50 fpm outside the DP). Monitor carefully the torque and drags because the string may get packed off. Drill at controlled ROP. A good supply of water must be available, which is a problem in the desert. Do not drill blind with a diverter in use.

7. If drilling ahead blind is not an option, consider using a foam or air-assisted circulating system.
8. Use of barite plugs, diesel oil bentonite plugs, or cement plugs may be considered as methods of last resort before attempting to drill blind or switch to foam. The cost of the rigtime lost will have to be considered against how likely success is, and whether the losses will simply reoccur after drilling ahead a few meters.

There are some problems that will have to be considered. Well control has already been mentioned as has packing off, but large washouts are likely and may lead to blocks of formation falling in.

Casing should be set as soon as a competent formation is drilled far enough to have a strong shoe. Apart from the well control considerations, loss of hydrostatic may destabilize other formations in the open hole. Cementing may be a problem if serious losses persist, in which case special cementing techniques may have to be considered such as adding LCM or using extended or foamed cement.

Losses in heavily fractured cavernous formations. Losses are likely to start as soon as the formation is penetrated, unless fractures are created while drilling. One option may be to drill blind with water if the area is well known or to drill with foam if not. Use the precautions as in the preceding topic and set casing as soon as possible.

Another option is the careful spotting of large volumes of cement. This has been successful in the field but requires careful planning and supervision. See "Recommended procedure for curing total losses with cement" later within this section.

Losses in normally pressured, deeper formations. These formations may be unconsolidated, naturally fractured, become fractured by the drilling operation, or consolidated but highly permeable with pore sizes too large for the mud solids to plaster. The loss zone can be anywhere in the open hole, not necessarily the formation just drilled into. Several factors can contribute to the mud loss, such as annulus loaded with cuttings, high ECD, excessive mud density, insufficient mud viscosity, high water loss (low solids content to plaster the wall), excessive surge pressures, breaking the formation during an FIT, or closing in the well after a kick.

It will be useful to identify the type of loss zone and the mechanism causing the losses to start. Knowing the depth and type of loss zone will help formulate a strategy to cure the losses.

It is possible (though not always necessary) to run logs to identify which zone is responsible. A temperature log may identify the formation, or radioactive tracers can be pumped and a log run to detect the radioactivity caused in the loss zone. The time and expense taken to do this must be justified by the severity of the situation and the improvement in decision making that should result.

In general, several techniques should be useful in most loss situations deeper in the well in smaller holes. In order of attempting, these are:

1. Decrease circulation rate for lower ECD and drill with controlled parameters to minimize annulus loading. The losses may well decrease over a period of time (a few minutes to a few hours). The severity of the loss may dictate whether this is acceptable, for instance with losses of over 60 bbl/hr using expensive mud the cost may be too high. Restricted replacement mud supply may also preclude this.
2. Reduce mud density if possible by dilution and/or maximizing the use of solids control equipment.
3. Pump a 100 bbl LCM pill around with around 45 ppb mixed fine/medium/coarse LCM. Spot across and above the loss zone and observe the well. When the well becomes static, break circulation cautiously, monitor active volume, and resume drilling.
4. Add solids (LCM) to the mud to increase plastering characteristics.

Further action may include setting barite plugs, diesel oil bentonite plugs or cement plugs, drilling with partial or total losses, setting an extra casing string, or plugging back and sidetracking if the severity of the losses warrants it.

Recommended procedure for curing total losses with cement. You should adjust this procedure to the particular circumstances. The general procedure is:

1. Drill ahead blind until it is anticipated that the loss zone has been completely penetrated. Do not pump Zonelock pills, LCM, fibers, or anything similar that might impede the movement of slurry into the loss channels around the wellbore.
2. Run in with a smaller bit (say $8\frac{1}{2}$ in in a $12\frac{1}{4}$in hole) without nozzles on HWDP and drillpipe to about 10 m above the loss zone.

(Note: Do not position the bit below the loss zone. Any cement remaining in the wellbore below the loss zone is a waste; if slurry is left above the loss zone it can at least drop down and enter the zone.)

3.   Mix and pump 100-200 bbls of extended "lead" cement slurry.
4.   Pump 100 bbls of extended "tail" cement slurry with 0.5 ppb polypropylene fibers added (Dowell product D094), preferably batch mixed in advance of the job for the best quality slurry.
5.   Displace with mud. If the annulus fluid level can be estimated, pump a quantity of mud that leaves a small quantity of cement in the string after U- tubing. It is vital to avoid mud entering the loss zone after displacing the cement.
6.   Monitor for returns at surface while pumping and displacing cement. If returns are seen, close the BOP and displace with mud. Now that the annulus level is at surface, displace with the string capacity–5 bbls instead of the originally calculated displacement. Slow the pump if necessary so that excessive pressures are not imposed on the well.
7.   POH. Pull back two stands without filling, then fill the annulus from the trip tank with mud to just replace the open-ended pipe displacement. Pull back to the shoe and wait on cement samples in the oven. It is better to add slightly too little mud than too much.
8.   RIH, carefully drill out the cement. If losses are seen right away, repeat the cement job.
9.   If losses are later experienced in the same zone, repeat the process.

There are two key elements. First, placement of sufficient cement in the zone around the wellbore so as to flood the loss channels. Cement left in the wellbore after setting is of no benefit and may cause an inadvertent sidetrack. Second, to ensure that the unset cement is not displaced away from the wellbore by adding more mud than necessary to the well as pipe is tripped out.

The first slurry should enter the loss channels easily. If the fiber cement starts to plug the loss zone near the wellbore (indicated by returns while displacing), this is a good sign since the plugging will prevent migration of the lead slurry away from the wellbore.

If returns are seen at surface, the cement will have risen in the annulus around the drillpipe. If the loss zone is still taking fluid this should drop when pumping ceases to the level supported by the formation fluid pressure, and the cement that has moved up the annulus will probably drop down back to the loss zone.

3.3.3. Washout Detection Procedure

A washout occurs when a hole appears in the drillstring. This may come from erosion, a damaged tool joint face, or a crack in the string. If left uncontrolled a washout can quickly become a drillstring failure because most drilling fluids are fairly abrasive and will soon cut metal away under high pressure. An otherwise avoidable fishing job will result.

Washouts can be prevented by regular inspection of the drillstring components, ensuring correct handling and torquing up of the drillstring and avoiding excess fatigue stresses (see Section 1.5.2, "Dogleg Severity Limits—Combined Buildup and Turn").

A washout may be suspected when a steady drop in circulating pressure is seen with steady pump rates, no change in mud properties, and no other apparent symptoms. This can usually be seen by looking at the geolograph chart recorder. Determining that a washout has occurred is a process of elimination.

1. Eliminate a formation fluid influx. Reducing surface pressure could be caused by a kick. Flowcheck before doing anything else. If that is OK, then:

2. Eliminate a leak in the high-pressure rig system or a change in mud properties. Close the lower kelly cock and pressure test the system from the pumps to the kelly cock. With pressure on, compare the gauge readings on the driller's console, standpipe manifold, and choke panel to ensure that a gauge problem has not caused the alarm. While that is happening, check that no fresh mud, water, or chemicals were added to the system giving a change in mud rheology. If that is OK, then:

3. Eliminate a drop in pump efficiency. Take pressures with each pump at the same slow circulating rate that was last taken (usually at the beginning of the tour or when the bit got to the bottom). If one pump shows lower pressure than before, check out the fluid

end for leaks. If both pumps show the same pressures as on the slow circulating rates (SCRs) taken earlier, resume drilling cautiously, watching for further signs of problems. If both pumps show lower pressure, something in the circulating system must have changed.

At this point, the surface equipment and mud have been isolated and so the problem must be downhole. Do not slug the pipe, POH wet to allow the washout to be seen more easily. Some old hands would throw small rope strands down the drillstring before tripping out. The string might be seen hanging out of the washout.

3.3.4. Backing Off

In some circumstances, when the pipe is stuck, it becomes necessary to backoff the pipe at one of the downhole connections, either to fish the remaining pipe or to sidetrack or plug and abandon.

The most common method used is to apply left-hand torque to the string and set off a small explosive charge at the desired backoff depth. The vibrations from the explosion cause the joint to unscrew with the left-hand torque stored in the elastic drillstring.

Sometimes an explosive backoff cannot be done. If the string is plugged higher up or if explosive is not available, it is possible to backoff at or close to the desired depth. The method is less reliable than using explosives.

Planning and precautions. A pre-job safety meeting must be held in advance since this is a high-risk operation that the crew may be unfamiliar with. Subjects to be addressed should include:

1. The pipe should be worked under torque and high tension with the tongs. If the tong becomes detached or the tong line breaks, the tong can fly across the drillfloor causing serious injury. The tongs should be tied between the jaws when on the pipe.
2. Torque should be held in the string by the slips unless using a top drive. If the slips were picked up with the table locked and torque in the string, the slips can fly across the drillfloor causing serious injury. The slips should be tied around the pipe when set in the table. No

one is to approach the table with the slips set until the driller has released the rotary brake and picked up the string slightly.

3. Whenever working the pipe in tension and torsion, no one should be on the floor other than the driller on the brake. Only when setting or pulling the slips or tongs will anyone be needed on the floor.

4. All entries to the drillfloor, except past the driller, should be barriered off with breakable tape. Rope or other solid barriers are not to be used because they will impede escape off the floor. Place notices and barriers at the bottom of any stairs that directly access the drillfloor (V-door, from mud pits, etc.). The deck crew must be informed that access to the drillfloor is restricted.

5. If working with explosive primer cord, keep unnecessary personnel clear and follow the instructions of the driller and logging engineer.

Procedure. Ideally the backoff is done in such a way that damage to the connection is minimized and it is possible to run in with a fishing assembly and make up to the backed off top of the fish.

Backing off can be done mechanically or with a downhole explosive charge. Where a charge is not used, the backoff cannot be controlled well enough to break at a specific connection. Even when a charge is used, a different connection will sometimes backoff instead, or several connections may become loosened.

First determine the stuck point depth, either by using stretch data or by running a wireline tool to measure stretch in the string (such as the Schlumberger Free Point Indicator Tool [FPIT]).

The logging engineer will determine how much explosive primer cord to run, given the depth, mud weight in use, and connection size. Run this in the string through a kelly cock with a cut-out protector to cover the thread to the planned breakout connection; a slimhole casing collar locator (CCL) is used to correlate depth. Meanwhile put the tong line sensator on the breakout tong where the breakout and backup lines attach to the tong. This allows torque to be monitored both while working the pipe up and down and also when putting in left-hand (LH) torque with the breakout cathead. If the kelly or top drive is still on the string, the backoff charge may be run through a removable plug on the gooseneck top. Do not forget to protect the thread on this from cutting out by fabricating a small protector to cover the thread.

Pick up with the elevators to a weight equal to the block + weight in air of the string above the stuck point. Mark the string level with the rotary table. Pick up another 20-40 k lbs and make another mark on the pipe. Set the slips with the minimum of downward string movement and tie the handles together to prevent the slips from being thrown across the floor if they should come out. Lock the block. Dress the backup tong to fit around the pipe body and check that the dies are clean and sharp.

With the rotary table apply left-hand torque to the string—about half the DP connection make-up torque. Count the number of turns. Lock the table with the rotary brake, then put the breakout tong on the pipe, a couple of feet above the table. Make the tong bite, take up some tension on the breakout line with the cathead, then carefully release the rotary lock. Allow the breakout cathead to come back so that the pipe tension is held by the backup line without shock loading it.

Keep everyone clear of the drillfloor. With the table fully unlocked, carefully pick up the pipe to release the slips. This ensures that the slips are not holding any torque when the crew pull them. Now send two crew members to untie and pull the slips, avoiding standing behind the tong. Stand the slips out of the way of the tong.

Work the torque down the string by moving the pipe up as far as the tong line, or pipe tension allows, and back to the lower line marked on the pipe earlier. *Do not lower the string beyond this line until ready to backoff.* Monitor the pull on the tong-line gauge; the tension should drop as torque is worked down the string. When the line tension does not decrease further, set and tie the slips with the pipe lowered to the lower line. Take a pull with the breakout cathead and then lock the table so that the tong can be unlatched. Remove the tong.

Apply more torque and work the string as before, increasing the torque each time until after three or four times you have worked in about 50% of drillpipe make-up torque but left-hand into the string. For instance, if the drillpipe tool joint make-up torque is 20,000 ft/lbs, aim to work in about 10,000 ft/lbs of left-hand torque. Keep a count of the total number of left-hand turns in the string. If the rotary table stalls out or cannot be controlled finely enough, take a bite with the tong, pull the tong, lock the table, move the tong back for another bite, etc., until enough torque is in the string.

Once enough left-hand torque is worked in, lower the string until the upper of the two lines is at the rotary table.

If a backup charge is to be used, it should now be set on depth and fired while holding torque with the tongs (i.e., without the rotary table locked). Watch out for the string lashing back as the torque comes off and watch the weight indicator for indications that the string is free.

Often the drillstring will loosen off in several places other than the actual depth that backed off. Be very steady pulling out of the hole, set the slips gently and do not rotate the string. Always check that all the connections are fully torqued up when running back in. If any connections are visibly backed out when tripping out, set the slips gently underneath and make up the connection right away so that a very loose connection is not racked back to be picked up later with string weight underneath it.

If attempting a purely mechanical backup, work more torque down the string a bit at a time, lowering the string this time to the lower of the two lines until something backs off downhole. Do not exceed the combined torque/pull limit of the pipe, or stop earlier if damage to the pipe body by the tong is causing concern.

If backed off in the wrong place. It is sometimes possible to screw back into the top of the fish and attempt a backoff at a lower depth. Work in RH torque in the same manner as described above, to ensure all connections are screwed in before making your next attempt.

The chances of success are lower than for the first attempt. Hold more tension when working the pipe to increase the chance of getting a backoff lower than before.

3.3.5. Fishing Operations

What follows is general advice on fishing jobs. It is not meant as an exhaustive guide to fishing! Where the fishing job may be complicated and/or take a long time it may be worth getting out a specialist fishing hand and tools.

When anticipating a fishing job, record as much information about the fish as you can. In particular you need to know:

1. Condition of the top of the fish: cleanly broken or severed or backed off, cut with explosives, drilled on, bent over, covered with cable or cavings, etc.

2. Dimensions of the top of the fish with a view to latching on with an overshot, spear, die collar, etc.
3. Weight and total length of the fish
4. Depth of the top of the fish
5. Whether the fish is stuck in the hole or if it should come free easily if latched on

If the condition of the top of the fish is unknown, a lead impression block (LIB) can be used to get an impression of the top of the fish. The fish should only be tagged once with the LIB, otherwise conflicting impressions are likely to be made. When the TOF is tagged, set down 10,000-20,000 lbs and POH.

Fishing for small pieces of junk. There are several tools that will catch bit cones and similar sized junk. A reverse circulating junk basket directs hydraulic flow onto the bottom and blows junk up into the basket. Spring-loaded fingers allow the junk to enter but lock closed to stop them from dropping out again.

The apple junk retriever is like a hollow cylinder run over the junk. By dropping a ball and pressuring up against a seat, a set of fingers is closed below the junk with sufficient force to cut a bit of a core in softer formations. Excellent for recovering this kind of junk.

A mill tooth bit and junksub can sometimes drill up junk or push it into the side of the hole. Quite effective in larger hole sizes/softer formations. Drill on bottom with low WOB/RPM and high flow rate. Shut off the pumps, pick up 10 m and go back to bottom. Repeat several times. Small metal parts may fall into the junksub.

A flat-bottomed mill and junksub may be better than a mill tooth bit in harder formations where the junk should stay put on bottom while milling. Work the junksub as before. This may take some time if the junk is able to roll around on bottom under the mill; pick up often then go rotating back down. Setting some cement on bottom may hold the junk in place long enough for you to mill it up.

Fishing for drillstring components and other tubular items (catching outside). The best tool is the overshot. It is simple to run, can exert tremendous pull on the fish, be jarred with and yet released easily. It can also apply LH torque to backoff the fish lower down. Refer to the Bowen section of the *Composite Catalog* or to the Bowen manuals for comprehensive information on the tool.

The overshot is dressed with the correctly sized basket or spiral grapple. The spiral grapple is used when the OD of the fish is at the largest sizes of the overshot catch range. With a basket grapple a mill control can be run, which gives some beveled milling action to the top of the fish. Unless there is a good reason for not doing so, run a mill control with a basket grapple.

At the bottom part of the overshot is the guide shoe. This guides the overshot over the top of the fish. It may open out to almost full hole size and have a spiral cut bottom end to aid in rotating over the top of a fish leaning onto the low side of the hole.

It is a good idea to paint the inside of the grapple and guide shoe. If the fish is not caught, you will be able to see if the grapple went over the fish but had perhaps slightly too great an ID to latch on.

Where the overshot has to catch on an OD some distance below the top of the fish, extensions can be run above the grapple to allow this. Be careful not to add too long an extension because if the grapple slips over the fishing neck and onto a section of smaller OD (such as at the top of a zip type drill collar), it may catch the fish but be impossible to release from the fish. The grapples are not designed to open up over the catch diameter going upwards because the back face is not beveled at all. If this is likely, then run a safety joint to allow the string to be pulled out if the overshot cannot be released and the fish stays stuck.

If the fish is likely to be plugged and circulation through it impossible, it is advisable to run a circulating sub at some convenient point above the overshot and/or leave out the packoff seal.

When above the fish, break circulation at a slow rate. Note the circulation pressures, drags, rotating weight, and torque. Run in slowly and tag the top of the fish, marking the string and noting the exact depth. Pull back a couple of meters, start circulating, and rotating slowly. Come down over the fish watching pressure, weight, and torque carefully for indications of going over the fish. The action of rotating to the right opens up the grapple as it contacts the fish. (If the top of the fish is clean, it will usually be possible to latch on simply by lowering onto the fish and applying weight without rotating.) If the fish cannot be circulated through, the pump should be stopped when pressure starts to build; bleed off the pressure if necessary to prevent pistoning off the fish.

When the grapple is far enough over the fish, stop rotating. Pick up slowly while watching the weight indicator. If the fish is caught in the grapple, the harder it is pulled the more it grips. Pull and jar as required to free the fish. Pull out of the hole without rotating. Set and pull the slips gently.

The die collar is another tool that catches on the fish OD. This must be run with a safety joint since it cannot be released downhole. It is not as strong as the equivalent overshot. Die collars and taper taps are only run as a last resort, due to their limitations, and also because they distort the top of fish, making it more difficult to subsequently run other fishing tools.

Releasing an overshot from the fish. To come off the fish either downhole or with the fish at surface, first set down some weight. This releases some of the grip. Start to rotate to the right (with chain tongs if at surface) and then pick up slowly. As stated previously, turning to the right opens up the grapple.

Work up slowly until the fish comes out of the grapple. If the fish can be set in the slips, carefully rotate to the left with the table while backing up the overshot with the make-up tong and picking up very slowly with the block. Tie the slips and use a safety clamp, but do not do this if the fish is long or if it may backoff below the rotary table. Close the blind rams below the fish first!

Fishing for casings and other tubular items (catching inside). If for some reason an overshot cannot be used, a releasing spear may be able to latch the inside diameter. It works very much like an overshot in reverse and may be released downhole. The releasing spear is most often used to catch casing, especially during abandoning if the casing is cut to retrieve it and a slip and seal hanger were used.

The weak point of a releasing spear is the inner mandrel. In large spears (say for $9^5/_8$ in casing) this is not an issue, but if you are catching a small ID (say a drill collar) then the mandrel could be quite weak.

Another tool that can catch on an inside diameter is a taper tap. This is stronger than a releasing spear of the same catch size but cannot be released downhole, therefore, it must be run with a safety joint. It looks like a long cone with threads on the outside and it is screwed into the fish ID.

Fishing after backing off. Where the fish can be screwed back into, a short fishing assembly can be run to try to screw back into the fish and jar it free.

Sometimes part of the BHA can be deliberately backed off in the drill collar section after getting stuck and packed off. This is a difficult fishing job, but it is possible to successfully recover by washing over.

If a decision is made to wash over the string, it is best to washover and recover in one run without stopping circulation. Otherwise, while the washover string is being pulled and the fishing string run, solids will again settle around the fish. Backoff at a convenient depth just above the stuck point and then run the following fishing assembly:

> Washover shoe and pipe, long enough to reach below the bottom stabilizer or just above the bit sub to stab into the top fish connection - circulating sub with the sleeve already open - drill collar - safety joint - two drill collars - fishing jar - three drill collars - accelerator - HWDP, etc.

Washover the fish, screw into the fish without interrupting circulation, continue to circulate, and pull/jar free. By having the open circulating sub inside the top of the washover pipe, circulation is maintained all the time, even if the fish is plugged inside with solids (as is likely).

The washover shoe may have to mill over several stabilizers, which may take more than one run to achieve.

Fishing BHAs. A fishing assembly will almost always include a jar and an accelerator. The Bowen "Z" fishing jar is the most commonly used jar with the matching accelerator. These will jar for many hours without losing effectiveness and can be torqued up in both directions. The Z jar jars up and only bumps down.

A fishing bumper sub can be run to allow bumping up or down, or to apply a specific downhole weight by running the required number of collars below and working with the bumper sub partially closed.

A drilling assembly will include a number of drill collars to give weight on bit. Fishing assemblies normally do not need to run much weight. This would be detrimental because it gives less overpull available for jarring and more weight below the jar would reduce the jarring force on the stuck point.

A typical fishing assembly would consist of whichever latching tool is to be run followed by fishing bumper sub - jar - two to four drill collars - accelerator - X/Over - HWDP (optional) and DP. A safety joint could be run above the latching tool as required. Sometimes a circulating sub is run as a precaution in the drill collars; check that the opening ball will drop through the accelerator and drift the string while running in.

For milling, more weight will be needed, though not nearly as much as would be run on a drill bit. The weight on mill may be recommended by the supplier or, as a guideline, run 1000-2000 lbs per inch of mill diameter. When running a mill shoe on a packer picker or washover pipe, less area is milled and so less weight is used. Refer to the recommendations of the supplier for these because the weight will also depend on what is used to dress the mill.

For fishing wireline tools, the logging company should provide the correct fishing equipment to run and crossovers to drill collar and drillpipe connections, if nonstandard connections are used.

Fishing jars, accelerators, bumper subs, and fishing tools to catch every sized tubular to be used should also be on location, along with the consumables required such as grapples. Normally they will be provided by the drilling contractor.

Bowen has a computer program to optimize jar placement, but it is only applicable to Bowen jars. Other vendors have similar programs.

3.3.6. Using Cement to Stabilize the Wellbore

Where the wellbore becomes unstable and leads to significant hole enlargement, problems tripping, running casing, and logging will increase well cost. Cement can be used to isolate the unstable formation while drilling ahead, however, this needs to be planned carefully to maximize the chances of success.

Do not wait until hole size has become extremely large. It is better to spot cement shortly after drilling through the enlarged zone. This will also reduce the chances of inadvertently sidetracking while drilling through the cement and will allow the cement plug to be set on bottom.

Run a 4-arm caliper log to evaluate the failure extent (depth of top and bottom) and to accurately determine cement requirements.

Follow the procedure for setting a kickoff plug as described in Section 2.7.5.

In extreme conditions (deviated well, long open hole times ahead, etc.), then polypropylene fibers (Dowell D094 or equivalent) can be added.

Run in with the next bit and wait above the top of cement. Check the surface samples. Once the cement has developed enough strength to drill through, continue drilling ahead.

In severe cases, an Expandable Slotted Liner™ from Petroline could be considered. This is set in place across the enlarged zone, an expander plug is pulled through it to expand it, and it is cemented in place.

3.3.7. Making Connections to Minimize Wellbore Instability and Losses

Surge and swab pressures can be seriously detrimental to your wellbore. A surge pressure occurs when a temporary increase in pressure is exerted on the formation; a swab pressure occurs when a temporary decrease occurs.

These pressure fluctuations are especially damaging in naturally fractured formations but can also destabilize nonfractured shales. It is often not recognized that how you stop the pumps is almost as important as how you start them up.

Starting the pumps. Kick in one pump at a few strokes a minute (say about 10-15% of your normal drilling flow rate). Watch for standpipe pressure and returns—you should see some pressure first then flow starting. Once you have pressure and returns, kick in the second pump to the same speed. Let the pressure stabilize. Smoothly increase pump speed over about 20-30 seconds to the desired flow rate (more slowly in critical hole sections).

Stopping the pumps. *Do not* wind both pump controls right away to zero unless you are stopping the pumps for a potential kick. Reduce flow rate first to around 75%. Wait until you see a decrease in flow on the flo-sho (note approximately how long this takes), then come down to 50%. Wait as long as you had to wait the first time, then reduce flow rate down to 25%. Wait the same time again then shut off completely.

Significant surge pressures can be created by running in too fast or starting up the pumps too quickly. The amount of pressure increase on a connection will depend on how fast the pumps are started up, depth (i.e., inertia of the weight of mud in the annulus), mud density, mud gel strength, mud rheology, and hydraulic diameter (which is related to hole size, BHA configuration, and any packing off around the BHA drillpipe OD). Significant swab pressures can be created by pulling out too fast or stopping the pumps too quickly. The amount of pressure decrease on a connection will depend on how fast the pumps are stopped, depth, mud density, mud rheology, and hydraulic diameter.

3.3.8. Preplanned Wiper Tripping

In certain circumstances, wiper trips are programmed into the drilling program:

1. Where offset experience shows that certain formations will deform plastically into the wellbore and a wiper trip, after a certain time period, justifies the costs involved because it reduces time later spent tripping and running casing. Massive salts are one example (such as the North Sea Zechstein, where wiper trips back to the previous wiper trip depth, after 18 hours drilling, prevent further tripping overpulls and stuck pipe).
2. In deviated wells, wiper tripping may disturb cuttings beds before they become great enough to cause a problem.

Wiper trips cost time (money) and help to destabilize the wellbore (increased open-hole time; surge and swab pressures), thereby initiating or worsening many common hole problems.

The decision to wiper trip can be summarized as: "Will the time taken to wiper trip lead to a greater time savings when tripping, logging, and running casing?" Experience has shown that most wiper trips are at best only time consuming and at worst may cause significant problems, without giving any benefits in return. This decision is best left to the discretion of the drilling supervisor, however, this must be justified with a properly engineered analysis of the costs and expected benefits.

If you do program in "routine" wiper trips, at least analyze whether any benefit has been obtained. If on your wiper trip you saw no drags or holdups then it is reasonable to say that the wiper trip probably had no beneficial effect. Therefore, on the subsequent well, you could extend the period between such wiper trips if the hole angle and mud properties are similar. Even if some drags or resistance were seen, it is not necessarily true that the wiper trip was necessary, except in mobile "plastic" salts and shales, as previously noted. Apply wiper trips intelligently to minimize cost and problems associated with wiper trips.

3.3.9. Barite Plugs

Barite plugs work by allowing barite to settle out rapidly after spotting the pill, forming a solid mass that is capable of holding some pressure. They can be pumped through a drill bit if necessary. Mixed at 22 ppg, the hydrostatic head of the plug will inhibit flow from a lower kicking zone. Barite plugs can be used to seal off a high pressure/low permeability kicking zone while casing is set, for some lost circulation problems and also for isolating a loss zone from a kicking zone during an underground blowout.

The plug slurry is usually mixed with barite, fresh water, sodium acid pyrophosphate (SAPP) and caustic soda. The SAPP lowers the slurry rheology, increasing the settling rate and the caustic soda raises the pH (to 10) to increase thinner effectiveness. Desco can also be used instead of SAPP, which does not need caustic soda to work.

A one liter pilot test should be done first since some supplies of barite do not settle well. This may be due to trace impurities. Mix in the correct proportions, stir well, and pour into a clear glass or plastic container. Half of the barite should settle out within 15 minutes (i.e., the bottom quarter of the container will have settled barite).

For one barrel of a 22 ppg slurry, mix either $1/2$ lb SAPP and $1/4$ lb caustic soda, or $1/2$ lb Desco in 21 gallons of fresh water. Add 750 lbs barite using a high pressure jet mixer. Mix the slurry using the cement unit. The volume may vary from a minimum of 40 up to 450 barrels, depending on the hole capacity and problem. If the treatment has to be repeated, use a larger volume than the previous.

Displace as soon as possible after mixing, at 5 bbl/min or more to prevent settling out in the drillstring. U-tubing will tend to occur; pump fast enough to maintain pressure on the drillstring until displacement is complete. Have cement and rig pumps lined up during displacing; if one pump fails switch immediately to the other or else you may end up with a plugged drillstring.

Underground blowout situation. In this situation, it is more likely that you will leave the string in place after setting the plug. Moving the pipe up out of the plug may mean stripping out and is likely to disturb the plug (which should be set around the outside of the drillstring). If the pipe becomes stuck higher up, subsequent barite plugs will be spotted higher and will be less effective.

The drillstring should be positioned below the loss zone if possible. However, if this would mean making up an inside BOP and stripping in (i.e., no float and pressure on the string), then the string will have to be left where it is since the inside BOP will prevent running a temperature log. If the bit is above the loss zone, a larger plug will have to be used (to give the desired length in open hole instead of around the drillstring), and hopefully the plug will move down and bridge off below the loss zone or inside it.

1. Mix, pump, and displace the barite plug slurry right out of the drillstring. Overdisplace by 5-10 barrels.
2. Wait about 6 hours for temperatures to stabilize. Meanwhile pump a few strokes every 15 minutes through the drillstring to keep it clear, unless there is still pressure on the string from the kicking formation.
3. Run a temperature survey through the drillstring (pressure equipment will be needed on the string). The loss zone should appear hotter than normal.
4. Wait another 4-6 hours and run another temperature survey. If the underground blowout has stopped, the temperature in the loss zone will have decreased. If the blowout is still going, set another (larger) plug.
5. If the plug has worked, bullhead cement below it. If the pumping or casing pressure fluctuates significantly, it may indicate that the barite plug has not held. Under-displace the cement to plug the drillstring to the theoretical top of the barite plug.
6. Wait on cement; pressure test inside the drillstring. Bleed off and check for backflow.
7. Perforate the drillstring, using the pressure lubricator in case flow starts.
8. Attempt to circulate. If possible to circulate, displace out any formation fluids with a mud density that will stop the losses. Then

when the well is static, backoff or blow off the drillstring at or above the perforation depth.

If not possible to circulate, set a wireline plug above the perfs, pressure test, and reperforate higher using steps 7 and 8.

Ongoing operations would probably be to P&A the well or to set casing to below the loss zone prior to sidetracking. If the barite plug fails, also consider using a diesel oil bentonite plug. If the kick is water, a DOB plug will have a higher chance of success.

High pressure, low permeability kicking formation. In this situation, the intention is to isolate the kicking formation without weighting up the entire mud system for a normal kill and leaving the drillstring clear and free after the operation. Situations making this necessary may include, for instance, problems with barite supply or rig equipment preventing a normal kill operation.

Preparation. You need to pull above the plug quickly after displacing. A hose or chiksans on the standpipe can be rigged up so that the first stand can be pulled and racked immediately after pumping ceases.

1. Mix, pump, and displace the slurry right out of the drillstring; keep working the pipe slowly (stripping through the annular if necessary). If a float is in the string, overdisplace by 5 barrels, otherwise leave about 3 barrels inside the string to prevent backflow.
2. Backoff the hose and circulating head from the drillpipe as soon as pumping is stopped. Pick up in stands to about a stand above the planned top of plug, stripping if necessary. Make up the hose again.
3. Start to circulate slowly over the chokes; continue to work the pipe slowly through the annular. If the plug works, influx/gas levels in the returns should decrease to near zero. If this does not happen after 2-3 circulations, set another plug.
4. When the well is dead, run back in to the theoretical top of the plug. Set a balanced cement plug on top of the barite plug.

Ongoing operations would probably be to P&A the well or to set casing to the top of the cement plug prior to drilling ahead. If the barite plug fails, also consider using a diesel oil bentonite plug. If the kick is water, a DOB plug has a greater chance of success.

3.3.10. Diesel Oil Bentonite Plugs (Gunk Plug)

Diesel Oil Bentonite plugs are also known as "gunk plugs." They work by holding bentonite in suspension in diesel until the plug is placed and then arranging for water to hydrate the bentonite. The bentonite yields rapidly, becoming extremely viscous.

They can be very successful in shutting off flow in an underground blowout, especially if the flow is water. A DOB plug will not maintain strength indefinitely. Cement should be spotted to give a permanent seal once the plug has worked.

The main potential problem with the DOB plug is that it will set up inside the drillstring if it contacts any water. A good diesel spacer ahead and behind are essential to prevent this.

To make the plug slurry, mix three sacks of bentonite per barrel of diesel for an 11.0 ppg slurry. Mica can be added at 15 ppb to increase the final plug strength if desired; use fine mica otherwise the nozzles may get plugged. Volume of slurry will vary between 30 and 150 bbls; more slurry for higher flowing rates and/or more open hole.

Procedure.

1. Line up both rig and cement pumps on the drillstring so that either can be used for displacement if the other fails.
2. Pump 5-10 bbls diesel ahead.
3. Either batch mix or mix/pump the DOB plug on the fly and displace it into the drillstring.
4. Pump 15-20 bbls of diesel behind. Displace with water-based mud with a reasonable rate down the string.
5. Once the slurry reaches the bit, start to pump slowly ($1/_4$-$1/_2$ bbl/min) into the annulus while displacing the slurry and diesel behind out of the string; over-displace by 5 bbls.
6. Wait about 6 hours for temperatures to stabilize. Run a temperature survey through the drillstring (pressure equipment will be needed on the string). The loss zone should appear hotter than normal.
7. Wait another 4-6 hours and run another temperature survey. If the underground blowout has stopped, the temperature in the loss zone will have decreased. If the blowout is still going, set another (larger) plug.

8. If the plug has worked, bullhead cement below it. If the pumping or casing pressure fluctuates significantly, this may indicate that the DOB plug has not held. Under-displace the cement to plug the drillstring to the theoretical top of the plug.

9. Wait on cement; pressure test inside the drillstring. Bleed off and check for backflow.

10. Perforate the drillstring, using the pressure lubricator in case flow starts. Attempt to circulate.

11. If possible to circulate, displace out any formation fluids with a mud density that will stop the losses. Then when the well is static, backoff or blow off the drillstring at or above the perforation depth.

If not possible to circulate, set a wireline plug above the perfs, pressure test, and reperforate higher using steps 10 and 11. Ongoing operations would probably be to P&A the well or to set casing to below the loss zone prior to sidetracking.

$$\left[\begin{array}{c}3.4\end{array}\right]$$

Casing

3.4.1. Conductor Placement

There are three current techniques of setting conductor pipe: jetting in, piledriving or drilling, and cementing to surface. Criteria for drilling, jetting, or driving relate to the desired setting depth and whether the topsoil conditions allow jetting or driving to this depth without damaging the conductor drive shoe. (Refer to the information on individual casing points in Section 1.4.6.)

Offshore conductor; preparation for the job. The job can be done faster and cheaper if the proper preparations are made. Preparations in advance of the rig arrival include the following:

1. Make a decision on what connections will be used and where the conductor will be welded while driving. If welded, then the pipe ends should be prepared by beveling for welding before sending to the rig. Otherwise, a beveling machine needs to be sent out. If connections are to be used then all the information and equipment needed to run the connection has to be organized.

2. For welding, ensure that three or four welding machines plus consumables and welders are available. If required, ultrasonic weld testing equipment and an operator should be sent out. Nondestructive testing (NDT) is normally done only for free-standing conductors but company policies or government regulations may require it.

3. A section of conductor for the splash zone needs proper protection against corrosion. This may involve blasting and painting with special protective paint.

4. The hammer size will depend on the size and weight of pipe to be driven. Check that the rig crane can lift on the complete hammer without having to disassemble and reassemble on the rig site; this will save rig up time.

5. Check that the rotary table diameter is greater than the conductor OD.

6. Conductor may be run on slings attached to padeyes welded on to the conductor. The conductor is landed at the rotary on the padeyes. After making up the next joint, these padeyes are burned off. In this case, preparation will include welding on those padeyes and testing to ensure that they will hold the weight of the conductor string below; ultrasonic and/or load testing. Appropriate length and rating of slings and shackles from the padeyes to the block are needed.

7. If the conductor is to be run using slips, clamp, and elevators rather than padeyes then they need arranging.

8. Generally, a survey is run inside the conductor once driven. This can be done through drillpipe on slickline on the cleanout trip. Either run a Totco survey to confirm that the pipe is at or nearly vertical or a gyrosurvey if it is not expected to be vertical (e.g., a directionally driven conductor) or if the position needs to be accurately known because of close proximity of adjacent conductors. The gyrosurvey kit including winch and surveyor need to be sent out in time.

Jetting in the conductor. The conductor is made up and a jetting sub on HWDP is run about 2 ft inside the shoe that makes up to the conductor running tool. The conductor and jetting assembly is run on drillpipe. During jetting, seawater or spud mud is pumped down the drillstring with returns coming up inside the conductor.

Setting the conductor by jetting can be done quickly since it takes only one trip. To be successful the seabed has to be soft enough to jet. This may be determined by offset records and site surveys. It may be necessary to wait on slack water to ensure that the conductor is vertical before jetting begins.

Drilling and cementing the conductor. In some cases offshore, a hole is drilled with returns to the seabed and the conductor pipe is run and cemented to seabed. The bit will spud directly into the seabed when no supporting structure is present on the seabed (e.g., a template or temporary guide base). It can be tricky to get started; any weight on bit and the drillstring will buckle slightly, causing the bit to walk away across the seabed. Any current may also cause the well to be spudded off vertical. Start with very low parameters but a reasonable flow rate, which will erode the top part of the seabed. Once established then WOB can be increased. Often the hole will be drilled as a smaller diameter pilot, opened up to full size. A large hole requires very high flow rates to keep clean. In addition, if any flow were encountered, a smaller hole will produce less—until it washes out and becomes a large hole!

You may have problems re-entering the hole if no guiding structure is available. An ROV can be used to monitor the position of the bit relative to the hole. Once the conductor is run, it will be cemented via an inner string until cement returns are seen at the seabed, then heavier tail slurry would normally be pumped.

Driving the conductor. Check the conductor driving records for offset wells in the area. This will give you an idea of how much penetration you can expect. It may vary for your new well depending on the pipe size, drive shoe configuration, hammer size/energy setting, and water depth.

Soil-boring surveys can be done. These can be used to analyze the likely depth of freefall and conductor penetration. If using a jack-up rig, these surveys can also be used to estimate spud-can penetration for the specific rig to be used. Often the drilling contractor's insurers will demand that a soil-boring survey be done with a spud-can penetration analysis of the results. Soil borings can be done in advance of rig arrival with a special vessel or it can be done from the rig after arrival, with the rig pinned on bottom before full preloading.

The conductor is made up and run through the rotary. Once the shoe reaches the seabed, check the tide/current conditions. An inclination survey can be run inside the conductor on wireline with a centralizer to check that the conductor is vertical. If necessary, wait on slack water so that the conductor starts off vertical. When ready, continue to run the conductor. Normally it will freely penetrate the top part of the seabed (freefall) before it stands up and driving can begin.

Driven conductors are usually driven to refusal. Refusal is a specified maximum number of hammer blows per foot of penetration at a particular hammer power setting. This maximum is determined by potential damage to the drive pipe and/or hammer, which can occur if driving with insufficient penetration. Consult the hammer manufacturer for recommendations if the company or normal area practices indicate a suitable refusal point for a particular hammer and power setting.

If several wells are drilled closely spaced (such as on a platform), conductors can be driven directionally using a drive shoe that imparts a side force. This technique, if successful, gives better separation of the wells at the shoe. However, once started it is impossible to control and it is quite possible for the conductor to go completely off course. If this happens, the slot may be unusable.

Conductor connections are normally heavy duty quick connect types, such as the Vetco Squnch Joint. The connector has to be suitable for driving operations.

If driving several conductors from a template or platform, stagger the shoe depths a little. If the shoes are all located close together at the same depth, communication between them may occur soon after drilling out the shoe, since this is the shortest leak path. If this is thought likely, spot 30-50 ft of cement inside each conductor before drilling out the first one to prevent such losses from the drilled conductor to undrilled ones. Shallow gas could also follow this path, which is another reason to spot cement.

Sometimes a mudline suspension system will be used. A landing ring is positioned in the conductor string such that at the expected penetration, the landing ring will be 5-10 m below the seabed. It is not usually a problem if the landing ring is a little too deep; however, if it is not deep enough, the suspended well may stick up too much above the seabed. In this case, the conductor can be secured and the conductor shoe drilled out with a smaller diameter bit, usually 8-10 in less than the conductor ID. It is not necessary to drill out too far ahead of the conductor shoe before recommencing driving since the conductor can normally be driven deeper than the drilled hole.

Drilling out may be done even when a mudline suspension system is not used, where penetration/shoe strength is considered to be insufficient for holding hydrostatic pressure during the surface hole section.

Offshore, especially if the flowline is a long way above sea level, a line can be welded in to a hole cut in the conductor close to sea level and a 6 in air-operated valve installed, aligned away from the rig. This will be done after driving is complete. If losses are seen, the valve can be opened so that the hydrostatic head imposed at the conductor shoe is reduced as the returns will flow out of the valve. If the well kicks then the 6 in valve could be closed when the diverter is closed.

After driving the conductor, it should be cleaned out with a large diameter drill bit to the conductor shoe. If this is not done and drilling commences with a smaller diameter bit (say $17\frac{1}{2}$ in in a 30 in conductor), the formation left inside the conductor can cause problems later on. Check that the bit will clear the mudline suspension ring, if run, and take particular care when cleaning out past the landing ring; it can be damaged or torn off by the bit.

3.4.2. Equipment Preparation for Casing

Refer to the checklists in Appendix 4. Use these to assist in ordering the equipment and services, and also for checking the presence and condition of the equipment when it reaches the rig. The following procedures are only guidelines—each job has to be planned individually.

The casing should be laid out, connections cleaned and inspected, drifted, numbered, and measured. Subtract the make-up loss from the total joint length; refer to the *Weatherford Tubing Data Handbook* or a similar publication. For buttress threads, measure from the end of the pin to the base of the triangle for an average make-up loss.

If it will be some time before the casing is run, make up clean dry protectors onto the clean, dry connections. A light coating of oil can be applied to prevent corrosion in damp conditions or where condensation is possible. Do not dope the connections.

Casing accessories such as floats and stage collars can be made up on the pin end of joints by cleaning the threads and applying a thread-locking compound to the pins. They can also be made up using the forklift truck and tongs after running them in by hand with chain tongs. Accessories should not be made up on the box end because, if the casing is dropped, a releasing spear will be used to recover it, and if the accessory is at the top of the fish, it may be impossible to fish it. For casing collars within the shoetrack, backoff (or cut) the collars and threadlock them (or a replacement) on.

If using a solid mandrel type hanger, make up a pin by pin pup joint underneath with threadlock to save time on the rig floor connecting the hanger. Also select and make up the landing joint with a running tool if used and mark on it (from the top of the hanger) the distance from the top of the wellhead spool to the rotary. When the casing is landed this mark should then be level with the rotary table.

Check the diameter of hangers and running tools. Compare to the ID of the bell nipple, stack, and riser to ensure that they will pass through.

Stop collars and centralizers can be applied to the top row of casing on the rack. As the job progresses, the deck crew can apply them as required to the remaining casing. This will save time when running the casing.

Once the casing crews arrive, they should check out their equipment thoroughly. It is worth having them rig up the tongs to the hydraulic power pack and running them, so you can see for yourself that everything works and no leaks are visible. Especially check the casing hand slips; it is very common to find that either the dies are incorrect for the casing to be run or that the slips have more than one type of die, which is very dangerous. Refer to the Varco chapter in the *Composite Catalog* or the Varco catalog for details of how Varco slips should be dressed, or to other manufacturer's details if appropriate.

3.4.3. Job Preparation for Casing

Before pulling out just prior to running casing, condition the mud to the lowest PV and YP and gels. The mud no longer has to clean cuttings out or suspend drilled cuttings and lowering these properties will assist good mud displacement. If you have the tank capacity it would save time if after circulating clean, ready conditioned mud could be displaced into the open hole from a reserve tank rather than circulating to condition mud for one or more circulations.

On the last trip out of the hole, strap the pipe with a steel tape. Confirm that the pipe tally is correct.

Rig up all the equipment, change and pressure test the top pipe rams, if necessary, and when ready to pick up the first joint hold a pre-job safety meeting on the drillfloor. It is better to have the driller organize and lead this meeting (since the driller will be present throughout the job), but the toolpusher/night pusher, drilling supervisor, and drilling engineer should all be present. Following are some ideas for items to cover in this meeting.

Safety precautions (pre-job safety meeting topics). The following can be used as a checklist of areas to cover on the pre-job briefing:

1. A good safe job is required; work efficiently without rushing.
2. All the correct safety equipment is to be used: safety belts at height, etc.; there will be no exceptions.
3. The on-shift driller has overall responsibility for the job. Any problems are to be reported to the driller, and if the problem cannot be solved, the toolpusher/drilling supervisor must be alerted by the driller. The driller should know the maximum pull that can be applied to the casing if it gets stuck.
4. Watch out for pinch points: fingers or hands getting trapped between moving and stationary equipment, such as casing coming up the V-door.
5. Keep the drillfloor reasonably clean to prevent tripping and slipping hazards.
6. No one is to use the V-door stairs when a joint of pipe is being moved between the catwalk and the drillfloor or being picked up by the single joint elevators.
7. When you change shift, make sure you hand over your job to your relief, then stop and watch their work for a few minutes to make sure they are doing the job properly.
8. Any unsafe conditions must be corrected immediately. Watch out for ropes on the V-door and stabbing board getting worn and change in good time. Keep an eye on lifting slings and strops.

9. Joints of casing must have clamp-on protectors in place on the pin end before picking up the V-door. If one drops off, the rig crew must be told right away and the pin must be examined before running.

10. It must be possible to circulate the casing in case of a well control situation, or if the casing has to be washed past a tight spot. The crossover from casing to WECO or other connection must be kept on the drillfloor ready to use.

11. Highlight any special procedures, hazards, etc. with the crews.

3.4.4. Casing Running Procedures

Pick up the shoe joint and lower in the hole. Look into the joint with a flashlight to see if the floats are allowing mud to pass from the well into the casing—they should not. Fill with mud to roughly level with the flowline, then pick up about halfway and lower in again. Look down the joint and check the level again. This checks that the floats allow mud to flow through but hold back pressure. Repeat once the float collar is made up, to check that the float collar allows passage of mud. If the floats do not allow mud to flow through, make up the circulating head and try to circulate through them, then retest as previously described. If they still do not work, the backups will have to be used.

Buttress casing connections are made up to within ±³/₈ in from the base of the triangle stamped on the pin. Make up several joints to the triangle, then use an average from the resultant make-up torques. If in any doubt, check a few more joints visually to the triangle. If you feel a casing collar while being made up, it should warm up but should not get hot. If it does get hot, it may indicate over-torquing or galling and should be backed out and checked.

The shoetrack is normally secured with a pipelock compound on the connections to reduce the likelihood of backing off while drilling out. There is no point in pipelocking only one side of a collar. With the shoe joint in the table, apply the safety clamp then remove the elevators. Using the rig tongs, carefully break out the collar. Clean pin and collar threads thoroughly, applying pipelock to the pin end only. Start to make up the casing collar with a large chain tong until the thread has properly started. Coat the next joint pin end, stab in, and make up with the power tong using a rig tong as backup. Repeat for the remaining shoetrack connections below the float collar. (If possible, breaking out the collars and pipelocking them can be done in advance on the pipe rack, which will save time.) When making up the joints in the rotary table using pipelock, make them up to the triangle and record the make-up torque for later when the shoetrack is drilled out. (To drill out the shoetrack, the off-bottom rotating torque with the drilling assembly at the casing plugs is taken. The torque while drilling the shoetrack is then restricted to 50% of the make-up torque plus the off-bottom rotating torque.)

Extra casing collars should be ordered in case a collar is damaged backing off or galled when made up again.

Apply the safety clamp (or use a spare single joint elevator above the slips) before removing the elevators until there is 25,000 lbs hook load (not including block weight). Fill each joint with mud. Monitor pit gain while running casing to spot losses or gains from the hole. Change to spider slips and elevators before entering open hole.

Once at depth, count the joints remaining (including any unused due to joint damage) and check that this is correct against the running list. Rig up the loaded cement head and wash down the last joint if necessary. Note up and down drags and any fill. Establish circulation and slowly increase, monitoring for losses. Circulate 120% of the casing contents at the highest no-losses rate to ensure that the casing is clear of debris, which would block the float. If using a slip and seal-type hanger, the casing can be reciprocated during circulating and displacing cement to assist mud removal by preventing gelling in washouts or eccentric annuli. This can also be done with a solid hanger, as long as the hole is in good condition with little risk of getting stuck with the casing not landed. Once the solid hanger has been finally landed, close the annular preventer and pressure test above to 500 psi for 10 minutes. If this test is good then the stack can be nippled down after cementing without waiting on surface samples.

Running list example. Prepare the casing running list from the tally. Mark the locations of centralizers, accessories, crossovers, etc. The tally should also give, for each joint, depth in the hole while running and final depth in the well when landed. Refer to the example running list shown in Figure 3-6. If using premium connections, the make-up torque can also be noted on the tally. This tally is easily made up on the rig computer using a spreadsheet, which also makes any amendments quick, accurate, and easy. Pit gain calculations can be made automatically by formulae in the cells.

Casing running list example				* = Spring Centralizer, [||] = Rigid Centralizer		
18.875" OD BTC						
Joint	Length	Total in hole	Run Depth	Centralizers, comments		Pit gain with casing full, bbls
Shoe	2.00	2.00	112.73			0.29
1	11.34	13.34	110.73	* *		1.90
2	11.44	24.78	99.39	* *		3.54
3	10.99	35.77	87.95			5.11
4	10.89	46.66	76.96	*		6.66
5	2.72	49.38	66.07			7.05
6	3.40	52.79	63.34	*		7.54
7	4.25	57.04	59.94	[||]		8.14
8	11.34	68.38	55.69	Enter open hole		9.76
9	5.43	73.81	44.35			10.54
10	6.11	79.92	38.92			11.41
11	10.97	90.89	32.81			12.97
12	11.97	102.86	21.84	[||]		14.68
13	11.45	114.31	9.87			16.32
			-1.58	Overstand		
Joints left after the job						
14	11.32					
15	10.99					
16	11.33					
17	11.23					

Fig. 3-6 Example of Casing Running List Tally

Copies of the tally should be given to the driller, deck crew, crane driver, and anyone else who needs one. The driller should mark off the joints as they are run in the hole, checking the numbers against the tally as they are run. If joints have to be laid down these can be marked and the tally amended with spare joints.

Single-stage casing and cementation: crew preparation checklist. The following preparation work should be done in advance of the casing job:

1. Make up and Baker-lock a PxP crossover under the casing hanger, if a solid (mandrel) hanger is to be used. Make up the running tool or landing joint on the hanger: measure the distance from RT to the top of the wellhead spool and mark this distance on the landing joint from the top of the hanger. Check the landing joint box for damage and check that the thread matches the cement head.
2. Baker-lock the float shoe and float collar onto the bottom of two joints after visually checking the float.
3. Number, measure, and inspect the casing. Produce running list for distribution, once final depth known. See Figure 3-6.
4. Centralizers can be applied to the top row of casing on the rack. If annular clearance and centralizer type permits, put spring centralizers over a stop collar rather than between two; this reduces drags and wellhead wear.
5. Drift all crossovers and pups that will form part of the casing string.
6. Rig up cement return hoses from the wellhead side outlets, if a solid hanger is to be used.
7. Have the emergency hanger (slip and seal) standing by, checked, and ready to use.
8. Check carefully that slips are correctly dressed with the right number of segments and type of dies. The dies should be clean and sharp. Slips dressed incorrectly could lead to injury or a damaged or lost casing string.
9. Check that the safety clamp is correctly dressed, that the springs behind the dies are all intact, and that the dies are properly moving up and down. The number of segments should be correct and dies in good shape.
10. Check that the spider slips and elevators are correctly dressed and in good shape.
11. Check the side door and single joint elevators for correct size, rating, and good latch condition.
12. Ensure that chik sans and T-piece are ready on the drillfloor for the cementing operation.

13. Cementer must check the equipment and preferably make up and dump a test batch of cement while running casing to check that all lines are clear, etc.

Single-stage casing running procedure.

1. Pull the wear bushing. Change TPR to _____ and test bonnets to _____ psi (call D/S when ready). Rig up to run casing, test power tongs. Driller makes up a trip sheet to monitor correct displacement. Driller holds a safety meeting on the drillfloor when ready to pick up the shoe joint (call D/S and T/P when ready for meeting).
2. Pick up the shoe joint and test floats. Land-in slips and set safety clamp. Backoff the collar, clean the threads, apply Bakerlock, and make up the collar with a chaintong. Meanwhile pick up joint _____ on the SJ elevator, leave at the V-door on the rope, and clean the threads.
3. Apply Bakerlock to the pin end, stab into shoe joint and make up with the casing tong to _____ ft/lbs to triangle. Lower into slips, apply safety clamp, and fill joint.
4. Repeat to Bakerlock the next collar on both sides, stabbing in the float collar.
5. Run in, fill the casing, and pick up 10 m. Lower again and look inside with a flashlight to check that the float collar allows mud to flow (level dropped). Set slips and apply clamp.
6. Continue to RIH as per running list, applying centralizers where shown (this can be done on the pipe rack). Use the safety clamp (or spare SJ elevator) above slips until _____ lbs string weight. Change to spider slips and elevators before entering open hole on joint _____. Fill every joint and monitor correct amount of mud returned. If losses or gains suspected, F/C and call D/S immediately. Run in hole at about _____ minutes a joint lowering speed. Cementer to prepare mixwater in good time.
7. Make up the casing hanger assembly. Pick out of slips and take up and down weights. Open side outlets on wellhead. Break circulation, lower casing, and land hanger; check that overstand is correct. Close in on TPR and pressure test above hanger to _____ psi. (If emergency hanger has to be used due to casing not getting to bottom, ignore this paragraph.)

8. Load plugs and make up cement plug container on landing joint. Connect chiksans to SPM and break circulation slowly. When returns are seen, slowly increase pump rate to maximum _____ GPM, monitoring for losses. If there are losses, cut pump rate to maximum no-loss rate. Circulate 120% of casing contents (_____ bbls, _____ strokes).

9. Line up to cement pump. Cementer pump _____ bbls of water ahead, close in at cement head, and pressure test lines to _____ psi. Keep personnel clear of lines under pressure.

10. Proceed with cement job according to separate program.

11. Drop top plug, cementer pump _____ bbls of water behind. Switch to rig pumps, displace at max. no-loss rate, slow down to _____ SPM at _____ strokes, and bump plug at max _____ strokes (theoretical plus 50% shoetrack). Test casing to _____ psi for 15 minutes. Bleed off and check for backflow. If no backflow, rig down running gear, back out landing joint, and prepare to nipple down. Otherwise close in on cement head and wait on cement, meanwhile clear drill-floor and prepare to nipple down.

12. Rig up and run gyrosurvey while hammering up BOP bolts after installing wellhead spool.

Stinger casing and cementation crew preparation checklist. Following is the preparation work to be done in advance of the casing job:

1. Bakerlock the float shoe onto the bottom of a joint after visually checking the float and the stab in sealing surface.

2. Number, measure, and inspect the casing. Produce running list for distribution once final depth is known.

3. Centralizers can be applied to the top row of casing on the rack. If annular clearance and centralizer type permits, put spring central-izers over a stop collar rather than between two; this reduces drags and wellhead wear. (See Fig. 3-6.)

4. Drift all crossovers and pups that will form part of the casing string.

5. Carefully check that slips are correctly dressed with the right num-ber of segments and type of dies and that the dies are not mixed with incorrect ones. The dies should be clean and sharp. Slips dressed incorrectly could lead to injury or a damaged or lost casing string.

6. Check that the safety clamp is correctly dressed, the springs behind

the dies are all intact, and that the dies move up and down as they should. The number of segments should be correct and dies in good shape.

7. Check that the spider slips and elevators are correctly dressed and in good shape.
8. Check the side door and single joint elevators for correct size, rating, and good latch condition.
9. Check the seals on the stab-in sub.
10. Ensure chiksans and T-piece are ready on the drillfloor for the cementing operation.
11. Cementer to check equipment and preferably make up and dump a test batch of cement during running casing, to check that all lines are clear, etc.

Stinger and casing running procedure.

1. Rig up to run casing, test power tongs. Driller to make up a trip sheet to monitor correct displacement. Driller to hold a safety meeting on the drillfloor when ready to pick up the shoe joint (call D/S and T/P when ready for meeting).
2. Pick up the shoe joint, run in, fill the casing, and pick up 10 m. Lower again and look inside with a flashlight to check that the float allows mud to flow (level dropped). Set slips and apply clamp. Backoff the collar, clean the threads, apply Bakerlock, and make up the collar with a chain tong. Meanwhile pick up joint _____ on the SJ elevator, leave at the V-door on the rope, and clean the threads.
3. Apply Bakerlock to the pin end, stab into shoe joint, and make up with the casing tong to _____ ft/lbs/to triangle. Lower into slips, apply safety clamp, and fill joint.
4. Continue to RIH as per running list, applying centralizers where shown (this can be done on the pipe rack). Use the safety clamp (or spare SJ elevator) above slips until _____ lbs string weight. Fill every joint and monitor correct amount of mud returned. If losses or gains suspected, F/C and call D/S immediately. Run in hole at about _____ minutes a joint lowering speed. Cementer to prepare mix water in good time.
5. If a screw on spool is to be used, break the top collar so that it can be unscrewed after cementing. Make up the landing joint with a

backup tong on the collar to the triangle. Land casing on slips for the last collar at correct depth for the casing spool (if using screw on type spool). Apply safety clamp.

6. Place slotted plate on top of casing. Pick up first stand of DP, and make up the stab in sub; check the seals. Run into casing, land elevators on the slotted plate, and take links off. Make up second set of elevators, pick up, and run in next stand. Stab into float, spacing out with pups if necessary. [See note below]

7. Make up kelly cock and circulating head on string. Connect chiksans to SPM and cement manifold with a T-piece. Line up on rig pump and break circulation slowly. When returns are seen, slowly increase pump rate to maximum ____ GPM, monitoring for losses. Observe inside casing, if there are returns then the stab-in sub is leaking. If there are losses, cut pump rate to maximum no-loss rate. Circulate 120% of stinger contents (____ bbls, ____ strokes).

8. Line up to cement pump. Cementer pump ____ bbls of water ahead, close in at kelly cock, and pressure test lines to ____ psi. Keep personnel clear of lines under pressure.

9. Proceed with cement job according to separate program.

10. Bleed off and check for backflow. If floats holding, pull back one stand and circulate pipe contents before POH. If floats not holding, then close in surface lines, maintaining backpressure, and checking every 5-10 minutes. As soon as the backflow is minimal, pick up to unstab and try to circulate down string to clear out gelled cement.

11. Wait until surface cement samples are hard. Back out landing joint; cut conductor or nipple down diverter.

(Note: If a 5 in slip type elevator [YC or similar] is available, this can be set around the DP to save time switching elevators. In addition, a special centralizer may be run on the bottom joint to help stab in. Cement is usually pumped until cement returns are seen at the surface, then the stinger contents are displaced. On surface wellhead systems, cement will have to be flushed out of riser/BOP/conductor with small diameter lines. Sometimes a bullplug can be installed in the conductor below where the spool is to be made up so that it can be easily drained of cement.)

Stuck pipe procedures. If the casing should become stuck while running, the two likeliest causes are geometry related (high DLS or a ledge) or differential. If mechanically stuck the only real choice is to pull on the casing. For differential sticking, a Pipelax or similar pill can be displaced. Refer to the stuck pipe topics in Section 3.3.1 for geometry and differential sticking problems.

In high-angle or horizontal wells, the casing may push cavings or cuttings ahead and build up a wall of debris that can then get you packed off. Use a Tam packer or similar model, which will allow you to commence circulating with the minimum of delay. Consider washing every joint down from about 65° inclination, taking care not to initiate lost circulation with high AVs/ECDs.

If you do get packed off while running casing, ensure you do have closing pressure on the spider elevators, then work the casing to maximum pull and slack off to 25,000 lbs plus block weight (to maintain some weight on the elevators). If possible leave some pressure on the casing—this will give you a "pump out" force that will put more force on the stuck point. If the debris starts to move, you will see the pressure drop and eventually return. If you are lucky you may reinitiate full circulation and clean up the annulus.

If you still remain stuck even though circulation was reestablished, the chances of a good cement job in place are very low. You will still have cuttings on the low side around the pipe (you may also be differentially stuck) and the cement will channel along the high side.

Refer to the notes on stuck pipe relating to cuttings beds in Section 3.2.1, "Solids Control."

Cementing

Reference should be made to the well planning aspects of cementing in Section 2.7. Much of the information is applicable to the wellsite operation.

There are two aspects of the cement job that are both vital to a good job and that the drilling supervisor has direct influence over. These are getting full mud displacement by the spacers and having homogenous slurry of the correct weight.

Mud displacement can be improved using the following techniques:
- Proper pre-job conditioning to lowest PV, YP, and gels
- Proper spacer design with sufficient contact time (10 minutes) and turbulent flow
- Casing or liner reciprocation or rotation

Homogenous slurry is more likely with the following techniques in order of preference:

- Batch mixing

- Use of a computer controlled recirculating mixer
- Using a "manual" recirculating mixer no faster than the cement can be accurately mixed

3.5.1. Mud Conditioning for Maximum Displacement

Drilling mud is designed to gel when circulation is stopped. This property prevents solids such as barite and cuttings from dropping out. Unfortunately, when mud gels in a washout or in the narrow part of an eccentric annulus, it is very hard to start it flowing again since the mud preferentially flows through the easiest path.

Good mud displacement is the single most important factor in having a competent cement job. The first step is to condition the mud prior to pulling out for cementing, to obtain the lowest possible PV, YP, and gels (without weighting materials dropping out). Normally the program will include a step for circulating and conditioning

Having properly centralized casing is the second step and minimizing the time that the mud is static is the third. By the time the casing is in the hole, the first two factors should have been addressed.

3.5.2. Slurry Mixing Options

The best method of mixing cement slurry is by batch. If suitable tanks are available then mix the cement in them, displacing only when the slurry properties are exactly according to the program. The thickening time of the slurry is tested at bottom hole temperature and half an hour on surface should not seriously affect thickening time downhole. If there is not enough tank volume for both lead and tail slurries, consider mixing the tail slurry as a batch (this is the most important and usually the least volume) before mixing and pumping the lead in the recirculating mixer. Switch to the batch tank and pump the tail.

Take several samples of each slurry to monitor the setting time. Use a pressure balance to check the actual slurry weight. The lead slurry should ideally be at least 1 ppg above the mud weight and the tail slurry also at least 1 ppg above the lead slurry weight. This helps displacement of the mud ahead and discourages mixing of the different fluids. The spacers should be in between mud and lead slurry density.

3.5.3. Preparation for Cementing

Well before the cement job, calculate the volumes of slurries required and the materials needed to mix these. Have the cementer do the same, then compare the calculations and sort out any discrepancies. Check that all the required materials are on site or can be ordered in time.

The cement recipe needs to be tested in town with samples of mix water and cement from the rig. Ensure that these are sent to town in sufficient quantity. When the bottom hole temperature is available from the logs, check this against the assumed BHST for the slurry design, and if there is more than 3-4° difference then mention this to the drilling office on the next call.

About twelve hours before you anticipate the cement job will start, have the cementer test run the pump and mixing equipment. Do everything possible to minimize delays on the job once cement is being mixed. When this has been done (perhaps while the casing is being run), have a meeting with the cementer, mud technician, drilling engineer, possibly the toolpusher, and/or any other supervisors involved in the cement job. Talk through the whole program and encourage suggestions. Then write out a procedure and give a copy to all the supervisors. This avoids confusion during the job and leaves you as free as possible to supervise the cementing and sort out problems as they occur. It is important to delegate tasks to the supervisors available to you.

Make a final check that the cement plugs, etc. are on the drillfloor ready to use, stored somewhere safe from accidental damage. Take mud density readings with both normal and pressure balance; any difference between the two will be due to mud aeration. You can run the degasser if necessary to reduce this, since mud aeration will reduce pump volumetric efficiency.

If possible, check the volumetric efficiency of the rig pumps by pumping from the suction tank to another tank and comparing actual strokes with actual volume pumped (use 50 bbls or more). On a long casing string, a difference of 0.5% on volumetric efficiency will make a difference of more than the shoetrack volume over the full casing capacity. To get the best chance of bumping the plug without risking over displacement, the true volumetric efficiency of the mud pumps must be known. If this cannot be done then look at the actual efficiency obtained on the last casing job (if the plug were bumped) and use that figure.

The plug dropping head will ideally contain both bottom and top plugs and will be able to drop each without having to open up the cement head. Once the 120% casing contents has been circulated, keep on circulating while the cementer starts mixing cement slurry and testing the cement line up to the cement head. Switch to the spacers, pump the spacers, then drop the bottom plug and start displacing the slurry. Do not pump the cement faster than you can accurately mix it; it is quite common for slurry in a recirculating mixer to vary by up to 2 ppg on the designed weight if care is not taken. This either produces slurry with far too much free water or more viscous slurry, both of which compromise the job.

The spacer volume and displacement speed should ideally provide a contact time of at least 10 minutes with the formation to allow the spacer to work properly.

3.5.4. Cement Displacement

During the cement job a pressure chart should be recorded and annotated so that the progress of the job can be inspected afterwards in the event of problems. A Dowell PACR or similar printout is acceptable.

Two opposing requirements are high displacement rate to attain turbulent flow versus a lower rate to avoid imposing high annular pressures on the open hole. The best displacement of mud with the spacer and cement is when the fluids are in turbulent flow. In some cases, the speed at which this occurs gives unacceptably high annular pressures due to the circulating pressure drop in the annulus. This may lead to losses that will in turn compromise the cement job.

With deeper casings and liners, a simulation program should be used to calculate flow regimes and pressures throughout the job. This service is available from all the major cementing companies.

Displacement of mud during the cement job is improved if casing can be rotated or reciprocated. It is not certain exactly why this is so; two contributing factors may be:

1. On the down stroke, annular velocities are increased.
2. As the casing is moved, it will tend to move laterally in the well-bore. Thus, if the casing is not perfectly centralized, the "narrow" and "larger" flow-by areas will move around, improving displacement in areas that may otherwise be dead. This may happen whether the casing is rotated or reciprocated.

Reciprocation has been found to be more efficient from field experience. It may be because in reciprocation, both of the above mechanisms are present whereas in rotation only the second one is. With a solid mandrel hanger, only attempt this if the hole is completely trouble free. Land the casing shortly before bumping the plug. If using a slip and seal hanger, position the casing on depth and stop movement shortly before bumping the plug.

Liners cannot usually be reciprocated, since the hanger is set before cementing. Some liner hangers are designed to permit rotation during displacement and while this is not as effective as reciprocation, it is better than no motion at all. If a liner is to be rotated, check the torque on the last trip out at liner hanger depth, and limit the torque while cementing to this value plus the make-up torque of the liner.

Cementing casing with plugs. Normally, two slurries are used; an extended lead slurry and a neat or heavy tail. It is vital to get competent, high-strength cement at the shoe for drilling ahead or at the intervals for perforating.

Batch mixing is the preferred method for producing high-quality, homogenous slurry. For normal casing jobs if a batch tank is available on the rig, the tail slurry can be batch mixed first. Since the setting time is applicable at downhole rather than surface temperature, the pumpable time from starting to displace the cement should not be seriously shortened. This allows you to pump the lead slurry in a recirculating mixer, switch lines, and immediately start to displace the tail slurry without having to pump as slowly as you can accurately mix with the recirculating mixer.

After pumping the slurry, drop the top plug, displace the lines from the cement pump to the rig floor with water, then use the rig pumps to pump mud behind. (On smaller jobs the cement pump may be used to displace the slurry.) Do not over-displace by more than half the shoe-track volume because if the top plug leaks, mud will exit the casing shoe and the shoe will not be well cemented. Monitor for losses and reduce the pump rate if significant losses occur. The cement slurry will U-tube due to its greater density, therefore, surface active volume will first show a gain as cement is pumped, then a loss will show until the mud behind catches up with the cement. In a critical displacement, the cement unit can be used to accurately displace the slurry.

Never allow the bottom plug diaphragm to be slit. This used to be done, presumably in case the diaphragm did not rupture when the plug bumped; but apart from being totally unnecessary, it will allow the cement to mix with the mud ahead.

The pumps should be slowed down shortly before theoretical bump after bumping the plug pressure up the casing to test it as per the program. Release the pressure and check for backflow; if no backflow then preparations can be made for the next operation. Where a solid hanger has been used, the running tool can be laid out and the stack nippled down; otherwise, wait on cement before removing BOPs.

If the plug is not bumped, then the casing must be pressure tested prior to drilling out the cement.

Stinger cementing. Where the internal capacity of the casing is very large (such as in large surface casing), a float shoe or collar with a 4 in bore above is normally used. This allows a drillpipe stinger with a stab-in sub to be run after placing the casing on depth and cement to be pumped down the stinger. Lead slurry is pumped until cement returns are seen at surface and then the tail can be pumped. This has clear advantages: in a large surface hole that is not usually calipered and that will be cemented to surface, cement is mixed and pumped without having to know accurately in advance what the hole capacity is.

Offshore a ROV or wireline deployed camera may be used to monitor for returns. It can be quite hard to see when the cement actually starts returning; often a bright dye is used at the end of the spacer or in the first part of the slurry. The dye itself can be hard to see especially in low visibility water; try adding a few handfuls of mica LCM as well or instead; this is easy to see because it glitters when the camera lights hit it.

3.5.5. Post-Job Evaluation

Various logs can be run to evaluate the cement job quality. Where no problems have been seen and/or the cement job is not critical, a simple cement bond log/variable density log is run to evaluate casing to cement and cement to formation bonding. In more critical cases a cement evaluation tool or ultrasonic indicator tool give more information and can show channeling, microannuli, and other problems—but at a much higher cost.

The job itself should be accurately recorded using chart recorders, computers, and text reports. See the next topic for ensuring that all pertinent information is recorded.

3.5.6. Field Cementing Quality Control Procedures

These procedures are designed to meet two objectives. The first objective is to do everything possible to ensure that cement jobs are carried out as intended with no problems. The second objective is to ensure that if problems do arise, the information necessary to analyze what went wrong is available.

The cement recipe sent to the rig should state quantities of cement, mix water, and additives in words as well as figures. This simple check avoids a bad fax from being misread to mix the wrong quantities. The recipe should also state the total mix fluid when all additives have been mixed, as a final check.

Ensure that the rig has available proper sample containers with secure screw-top lids and labels. Pilot tests on the rig site should be done with whatever equipment is available to you.

Pre-job actions.

1. Send in samples of mud, potable, and drill water to the cementing laboratory every two weeks and additional samples as requested.
2. When the recipe fax is received from town, carry out the following:
   a) Check that all quantities to be used are clear in the fax, clarify with the drilling office if any are not clear
   b) Calculate estimated material requirements and ensure that they are available on the rig site, normally with 100% excess available
   c) Check the total mix fluid requirement from your calculations against the recipe (this is given automatically in the "well-calc.xls" spreadsheet [you can copy this from the web site library page at http://www.drillers.com])

3. Where tank space permits, make up the mix water in advance (but no more than 12 hours early). Run a pilot test using this mix water and the cement (blended first if necessary) to check that the programmed pumpable time is close to that seen on your pilot test. Use an oven to heat up the samples in a covered container, but do not let the sample boil since this will reduce water content and seriously affect the thickening time.

4. Ensure samples are taken of the neat mix water, mix fluid (with additives added), spacer(s), (blended) cement, other additives, and mud in the hole. Seal, label, put in a safe place, and keep until the next cement job in case of later queries.

5. Take density readings of the mud with both atmospheric and pressurized balances on the final circulation prior to cementing. Any difference due to aeration will affect displacement efficiency. Consider running the degasser while circulating prior to the job.

6. Condition the mud in advance to lowest practical PV, YP, and gels. At the end of a hole section prior to running casing, having the mud in good shape will improve cement displacement, reduce surge pressures, and improve the chances of successfully cementing in an enlarged hole.

7. Write out your program for the job; check your calculations against the cementers. Pass your program on to the cementer for comment. No changes are to be made to the program issued from the drilling office without consultation with the duty drilling manager.

During the job.

1. Follow your preplanned program during the job as closely as possible; avoid making changes "on the job" except where necessary to meet unforeseen events.

2. Use the batch mixer for quantities of slurry above the capacity of the recirculating mixer, and within the capacity of the batch tank for critical slurries (small cement plugs, tail slurries).

3. When using the recirculating mixer tank to batch mix very small slurry quantities (coiled tubing plugs, etc.), control the speed of adding cement by restricting the cement line. This should ensure that the correct density is obtained and avoid large slugs of cement powder entering the mixer and only partially hydrating.

4. Check out the recirculating mixer for function in advance. In particular, ensure that the agitator and recirculating centrifuges are all working to give maximum mixing energy in the tank.
5. Take samples from the sample point near the bottom of the tank. If one is not fitted and there is nowhere else to take a sample from near the bottom, discuss with the cement contractor about fitting one. Put at least one sample in the oven and check thickening time at temperature.

After the job.

1. Fill out the cement job report as soon as possible to record all details. In particular, make any recommendations you and/or the cementer feel would be advantageous for changes to future programs, procedures, or the drilling manual with full justifications. Also, keep a copy of your calculations and your program/procedures in the file for future reference.
2. Attach to the cement job report copies of charts, etc., made during the job.

Keep the samples safe until the next cement job, or send to a local laboratory, if requested.

$$\begin{bmatrix} \mathbf{3.6} \end{bmatrix}$$

Drillbits

Section 2, "Well Programming," examined offset bit runs: how to analyze them to make the best bit choice for the next well and the effect of BHA choice on bit choice. This section now deals with the practical aspects of the bit run.

Even though we have done as much as we can at the well planning stage, the final bit choice will normally be made on the wellsite. The start of the run may be at a different place than was anticipated in the drilling program. The performance and grading of the previous bit out may cause a revision in thinking.

3.6.1. Alternative Bit Choices

If depth in is not as planned. If a bit is pulled before the planned end of its run, then by reading the recommendations and examining the hole section summary it may be clear what kind of bit will be best to run next. (See Section 1.1.2, "Hole section summaries.") Usually this will be similar to the one pulled or similar to the next bit planned or an intermediate choice. The process that was used to make the bit program can be repeated fairly quickly if the rig has all the following information available: recommendations in the drilling program, hole section summaries, offset bit data, field operational notes, and the end of section reports from offset wells used during well planning. These can be added to the drilling program as appendices and/or to the drilling manual. The rig site PC can have a copy of the BITREX database, or printouts/offset bit records from it can be appended to the program. (See Section 2.4.2, "Comparing bit records using the BITREX database.")

Sometimes the next bit in may have to drill to a particular depth (coring point, for instance), which is considerably less distance than would be expected from a full bit run. It may be possible to run a cheaper (or used, re-useable) bit instead. The rig needs a list of bit prices and the drilling supervisor should consider bit cost when making the selection. For example, it may be that the bit being pulled early has already drilled through an abrasive zone where premium gauge protection was used. The next bit in may not require this expensive feature and so a cheaper alternative may be possible.

If the preferred bit is not available. Occasionally the recommended bit will not be on site. In this case, the drilling program already defines the features required of the bit for a particular interval and an alternative can be chosen from what is available by applying those criteria. It is not recommended to merely substitute another bit of the same IADC code because this code is a great simplification of the bit features.

Where the onsite drilling supervisor runs a bit that is not in accordance with the program recommendations, the choice must be justified in the end of section report. This requirement will encourage careful consideration before making a decision. The basis for this decision is also important when evaluating the performance and planning the next well.

Refining bit choice and parameters based on previous bit run. A good bit choice, run correctly and pulled at the end of its economic life, should show worn cutting structure and/or bearings. Severe dull bit features (excessive gauge loss, broken cutters, cones locked, etc.) are warnings that something went wrong, especially if the performance fell below expectations.

Try to ascertain what conditions may have caused the specific dull conditions and evaluate what changes could be made to bit choice, running procedures, drilling parameters, BHA, mud, etc., to reduce the impact of these conditions. For example, a common mistake is to assume that broken teeth equates to a bit that is too soft; there are other more likely causes in most cases. Downhole shock or vibration, hard nodules, or junk could all play a part. Running too hard a bit for the formation is likely to compromise your overall bit performance. (Refer to Table 3-5 in Section 3.6.6, "Post-Drilling Bit Analysis," for information on dull bit features.)

3.6.2. Drilling Parameters

Weight on bit. When drilling, weight is applied to the cutters so that rock is penetrated. Up to certain limits the more weight applied the faster the bit will drill. If too much weight is applied, the cutters may become completely buried (known as bit flounder) and weight will be taken by the cones or bit body. This will reduce ROP and rapidly wear the cones. Increasing weight will also accelerate wear on bearings and cutters.

Deviation is also affected by WOB. A rotary locked or build assembly will have an increasing build tendency with greater weights; where a rotary pendulum is in an established drop then increasing weight will tend to increase drop, up to a point where further increasing the weight may produce unpredictable results. In a vertical borehole with a flexible pendulum or build BHA, increasing weight will deflect the wellpath from vertical.

In a motor-bent sub combination, increased weight will increase side force at the bit, and therefore accelerate the rate of direction change in the direction of toolface azimuth, up to the point where the motor stalls.

When planning to change hole direction, the BHA selected may dictate the approximate WOB to be used, which may affect bit choice.

Refer to offset records including the field operational notes and hole section summaries to see what WOB works best in a particular formation. Regular drill-off tests should be carried out.

Optimum WOB can be run when using locked assemblies. See "BHA considerations related to bits" in Section 2.4.7.

Rotary speed, RPM. Increasing RPM will increase ROP up to a point where the cutters are moving too fast to penetrate the formation before they move on. Excess RPM will cause premature bearing failure or may cause PDC or diamond cutters to overheat.

Deviation is also affected by RPM. Higher rotary speeds tend to stabilize the directional tendencies of rotary BHAs. A rotary BHA has a natural tendency to turn to the right, this tendency is weaker at higher rotary speed.

Rotary speeds that cause string vibrations (critical rotary speeds) must be avoided. The driller should recognize this condition and modify RPM accordingly. Two types of vibration can be related to drillstring rotary speed and the calculated approximate speed of occurrence.

Longitudinal Drillstring Vibration

$$\text{Longitudinal Vcrit} = \frac{78,640}{Lp}$$

where Lp = length of DP string, meters.
Critical vibrations also at 4x and 9x this value.

Transverse Drillstring Vibration

$$\text{Transverse Vcrit} = \frac{47,000 \ \sqrt{(D^2 - d^2)}}{L^2}$$

where
    D = pipe OD
    d = pipe ID
    L = joint length
    (All measurements are in inches)

Minimizing bit whirl. Bit whirl occurs where the friction at the gauge of the bit makes the center of rotation locate itself at the edge of the bit (where the formation is in contact), instead of the geometric center. Since the forces on the cutters are now in different directions than the designed direction, cutter breakage can result. Bit design seems to be the dominant factor.

Good stabilization probably decreases bit whirl and many bits are already advertised to be an "anti-whirl" design. Whirl is often initiated when the bit just starts drilling, such as after making a connection.

Research has indicated that using the following procedure after making a connection will minimize the chances of bit whirl starting:

> While still off bottom, bring the mud pumps and the rotary table up to speed. Slowly slack off until the bit starts to take weight. Increase the WOB in small increments (say about 20% of planned total WOB) and allow the rotary table to stabilize in between increments for 10-30 seconds (longer for deeper hole).

Hydraulics. There are two current theories for optimum hydraulics. One gives the total nozzle area to maximize hydraulic horsepower. The other calculates for maximum hydraulic impact force.

To maximize bit hydraulic horsepower, the pressure drop across the bit should be 65% of the total pressure loss in the system. To maximize hydraulic impact force on the bottom, the pressure drop across the bit should be 48%.

Of the two methods, maximizing HHP gives greater pump pressure and lesser flow. Erosion is more likely to be a problem, though erosion on the bit is acceptable as long as it does not lead to premature bit failure. Maximizing bit HHP is not guaranteed to lead to improved ROP through better bottom hole cleaning (especially in harder formations).

There are some advantages to optimizing for impact force. These include:

1. Larger nozzles will reduce nozzle plugging and will be better for pumping LCM
2. Lower pressures cause less pump wear, reducing downtime due to pump failure
3. Higher AVs lead to better hole cleaning

If maximizing HHP does not improve ROP, then either maximize impact force or choose nozzles for the flow rate required at a pressure below maximum at section TD. Calculating exact nozzles and flow rates for optimum hole cleaning under the actual conditions at the time cannot be done due to the large number of variables, some of which are unknown.

Where LCM is anticipated, use nozzles of at least $^{16}/_{32}$ in size, even if the optimization calculations suggest smaller nozzles.

Bit parameter optimization and drill-off tests. Perform regular drill-off tests to optimize ROP and bit life. If applying more WOB or RPM does not increase ROP by a roughly linear amount, then bit wear will be accelerated faster than the increase in ROP. When a drill-off test is performed, you are not necessarily looking for the greatest possible ROP, but a good ROP that gives a reasonable bit life. This will lead to the lowest cost per foot.

Unless the drilling parameters are constrained by factors such as deviation performance, the drillers should be told to optimize parameters for best ROP within a range. Telling a driller to "drill with 35,000 WOB and 90 RPM" will not optimize performance.

Read the manufacturer's recommended WOB and RPM. Look at offset bit runs; see what parameters were used on the best runs and the bit grading. Check the design parameters of the BHA. If a vibration analysis was done on the BHA, note any peak stress WOB/RPM combinations that should be avoided.

Decide on a maximum WOB and associated RPM, and a maximum RPM and associated WOB. Tell the driller to work within those parameters, but to conduct regular drill-off tests to determine the best combination.

Experience will sometimes show that a manufacturer's recommendations for maximum parameters are too conservative. The following bit record illustrates this.

An 8¹/₂ in Smith F47OD drilled 271 m in 82 rotating hours. Average WOB 55,000 lbs, RPM 80. This bit graded 3/3/CT/M/E-E-F/3/NO/TQ. Even after this severe run, the bit grading looks good.

To do a drill-off test, set a fixed weight and vary the RPM up to the maximum. Note the lowest RPM that ROP peaked. (You may see that above a certain RPM, ROP does not increase.) Then set that RPM and vary the weight. At the lowest WOB that gave the best ROP, hold the WOB and again try to vary the RPM. This will work towards the optimum parameters.

Field experience in variable lithology should allow the drillers to spot a change in lithology. The drill-off test should then be repeated.

3.6.3. Mud Motors, Steerable Systems, and Turbines

Other considerations apply when planning to run a downhole motor. They include:

1. Will the maximum bit RPM be exceeded if the string is rotated while drilling with a motor?
2. Will the flow through the motor clean the hole at the envisaged flow rates?
3. Are there any limitations on the bit pressure drop imposed by the motor?
4. Is the proposed bit suitable to use on the type of motor to be used?
5. Can LCM be pumped through the motor and if not, should a circulating sub be run above? Might LCM be needed while drilling with the motor?
6. What size liners are needed in the pump for the necessary flow rates and pressures?
7. Are there any problems with the mud properties? Chemical compatibility with seals, sand content, etc. Check with the motor supplier.
8. What is the plan for before and after the motor run? For example, if running in for straight hole turbodrilling with a PDC bit, the previous assembly should be a fairly stiff one to avoid reaming in with the turbine. Any junk in the hole would require a junk run first. Otherwise, if a steerable system run is to terminate at casing point, then a wiper trip with a rotary assembly should be made to ream to bottom and reduce the chance of mechanically stuck casing by reducing doglegs and ledges.

3.6.4. Monitoring Bit Progress While Drilling

Cost per foot calculations should be done while drilling. Once the cost per foot starts to increase, the bit will be nearing the end of its economic life. However, several other factors should be considered when making a decision to pull the bit.

Pull the bit earlier if there are indications of bearing failure (high and/or fluctuating torque on bottom compared to steady, reasonable torque just off bottom).

Leave the bit in longer if offset information indicates that the slow-down is mainly due to decreasing formation drillability. Sometimes a bit is pulled under these circumstances and the next bit in does not drill any faster. Clearly in this case it is better to extend the bit run if there are no concerns as to bit condition. The hole section summary showing offset bit runs at the same place may indicate this.

There are different theories that aim to make a bit pull decision easy, such as by hours on bit or number of revolutions. However, these will lead to below optimum drilling performance and should only be used when bit bearing condition cannot be monitored. Drilling with a downhole motor or in small, deep, or deviated holes where the off-bottom torque is high would qualify for pulling on hours. It is possible to consistently pull bits at the end of their economic lives, maximizing the overall performance without seriously risking leaving cones in the hole. This requires close and skilled supervision of the run.

3.6.5. When to Pull the Bit

"Cost per foot" calculations can help to decide when to pull the bit. If this is done consistently, the chance of having to fish for cones is small and the overall cost per foot will be minimized.

Set up a spreadsheet. By entering how many minutes per foot or meter drilled (which the mud loggers record) against depth, distance drilled, and overall cost per foot can be easily calculated. You can also download the Excel 5.0 workbook "wellcalc.xls" from the web site at *http://www.drillers.com*, which contains a cost per foot sheet.

The point where the cost per foot is consistently increasing is the point that the bit should be pulled. If the indications are to pull the bit, do not waste time drilling the kelly down. This may be modified by other factors.

- Pull the bit early if there are any signs of bearing failure.
- Leave the bit in the hole if you expect a more drillable formation ahead. Monitor carefully for bearing failure.

3.6.6. Post-Drilling Bit Analysis

Proper analysis of the bit run is important to improve future performance. A large problem is that different people will grade a particular bit differently. The IADC 8 point grading scheme is vastly better than the old TBG grading, however, if gradings are not done with care, it will mislead future drillers.

Record grading details and comments on the bit report. Update recommendations in the field operational notes. Make more extensive comments and recommendations in the end of section report for inclusion in the final well report.

The IADC 8 point grading should always be used to grade bits. The first four digits refer to the cutting structure. The last four digits refer to other characteristics.

IADC 8 point grading scheme.

Digits one and two—tooth wear (cutting structure, tooth/cutter wear). The first digit relates to the teeth which do not touch the borehole wall. The second digit relates to the remaining teeth. Grade the tooth with the most wear in eighths, where 0 is no wear and 8 is totally worn down to the cone.

The IADC recommends that tooth breakage be taken into account when grading wear, but this should be done with care. Tooth wear and tooth breakage are two different characteristics that have different causes and should not be confused. If, for example, 50% of the teeth were broken (perhaps due to bit bouncing on bottom) and the other 50% were worn to two-eighths, then under the IADC the grading would be a 6 or 7. Someone looking at this grading later would assume that the 6 or 7 meant that the teeth were almost worn away and may change the bit selection based on that. Knowing that the tooth wear was moderate but breakage was present, it may be concluded that the cutting structure was broadly suitable for the application, but that drilling parameters and/or the BHA should be changed (such as by running a shock sub).

Digits three and four (cutting structure, characteristics, and location). If more than one major cutting structure wear characteristic is present, enter here the most severe (that which has the greatest effect on bit performance). If tooth wear and breakage are both present, do not enter WT here (the fact that the teeth are worn will be obvious from the first two digits); BT would be more important. (See Table 3-3 for bit codes.) The fourth digit shows the location of this characteristic.

Table 3-3 Bit Codes for IADC 8 Point Grading Scheme

Dull Bit Characteristic Codes		Dull Bit Characteristic Codes	
BC	Broken cones	LN	Lost nozzle
BF	Bond failure (PDC bits)	LT	Lost teeth or cutters
BT	Broken teeth or cutters	OC	Off center wear
BU	Balled up bit	PB	Pinched bit
CC	Cracked cone (note cone number[s])	PN	Plugged nozzle or fluid passage
CD	Cone dragged (note cone number[s])	RG	Rounded gauge
CI	Cone interference	RO	Ringed out
CR	Cored bit	SD	Shirttail damage
CT	Chipped teeth or cutters	SS	Self-sharpening wear
ER	Erosion	TR	Tracking
FC	Flat crested wear	WO	Washed out
HC	Heat checking	WT	Worn teeth or cutters
JD	Junk damage	NO	No dull characteristics
LC	Lost cone (note cone number[s])	RO	Ringed out

Roller Cone Bits Location Codes		Fixed Cutter Bits Location Codes	
N	Nose row	C	Cone
M	Middle row	N	Nose
G	Gauge row	T	Taper
A	All rows/all areas	S	Shoulder
#1	Cone 1	G	Gauge
#2	Cone 2	A	All areas
#3	Cone 3		

Digit five (bearings/seals condition). Enter an X for fixed cutter bits since this digit is not applicable to PDCs, diamond bits, or other fixed cutter types.

For nonsealed bearing bits, attempt to judge how many eighths of bearing life are gone. Enter as a number from 0 to 8, with 0 reflecting no wear and 8 reflecting no life left (cone skidded, locked, or lost).

For sealed bearing bits, try to turn the cones. If there is any play or roughness, or if the cones have skidded (bearing locked), then grade F for a failed seal. If the cone is OK, then grade E for an effective seal. If unsure, grade Q for questionable. If unable to grade for any reason, use an N.

The amount of bearing wear actually present can only be accurately gauged by cutting the bit open and measuring the bearing surfaces. Grade each cone individually, if possible, starting with cone #1.

Digit six (gauge wear). Grade the reduction in bit diameter. Enter the number of $1/16$ in. For $1/8$ in undergauge, enter a 2. For $1 1/2$ in undergauge, enter 24.

Digit seven (other major dull characteristic). The third digit allows you to enter a code referring to the most important dull characteristic of the cutting structure. The seventh digit allows you to enter the next most important dull characteristic, whether of the cutting structure or anywhere else on the bit.

For instance, you may have a Tungsten Carbide insert bit with many broken teeth, some heat checking of the cutters, and junk damage. The third digit would be BT (as the most significant cutting structure dull characteristic) and the seventh digit would be the most significant of HC or JD, whichever would have had the greatest impact in reducing the bit life.

The codes used in the seventh digit are the dull bit characteristic codes shown in Table 3-3 under digit three (cutting structure characteristics).

Digit eight (reason pulled). The reason that the bit was pulled is very important for future evaluation. If the bit was pulled at the end of its economic life (usually due to rate of penetration or torque indications of bearing failure), and bit performance was maximized throughout its run, then this will give an excellent guide for future bit choice at the same place in offset wells. It also serves as a good guideline for how many hours this bit could be run given similar size/WOB/RPM. See Table 3-4 for reason pulled codes.

If the bit were pulled for other reasons then the information, although still valuable, has to be considered in light of why it was pulled and how much more performance might have been expected from the bit.

Refer to the topics in Section 3.6.1 for more information on using dull bit evaluations for future bit selection. Table 3-5 provides information on dull bit features.

Table 3-4 Reason Pulled Codes

BHA	Change bottom hole assembly	DP	Drill plug
DMF	Downhole motor failure	FM	Formation change
DTF	Downhole tool failure	HP	Hole problems
DSF	Drillstring failure	HR	Hours on bit
DST	Drill stem test	PP	Pump pressure (suspected washout)
LOG	POH to run logs	PR	Penetration rate (based on CPF)
LIH	Left in hole	TD	Total depth or casing point depth
RIG	Rig repair	TQ	Torque (excessive or fluctuating torque)
CM	Condition mud	TW	Twistoff
CP	Core point	WC	Weather conditions

Table 3-5 Dull Bit Features

Dull Feature	Possible Cause	Possible Remedy
Broken teeth	Junk Excessive string vibration or shock loading Excessive WOB	Fish junk or drill with a mill tooth bit Avoid critical RPM, run shock sub Reduce WOB
Broken outer teeth	Running into ledges Excessive RPM	Run in carefully Reduce RPM
Bit balling	Bad hydraulics Mud properties	Review hydraulics Review mud properties
Cone dragged	Bearing failure Junk jamming cones	Reduce parameters or time in hole; review bit choice Avoid junk in hole

Table 3-5 Dull Bit Features (cont'd)

Dull Feature	Possible Cause	Possible Remedy
Cone interference	Drilling on junk Mechanical damage hitting a ledge Pinched bit from running into undergauge hole	Avoid junk in hole RIH carefully past ledges Ream in undergauge hole
Chipped teeth or cutters	Junk in hole Hard nodules within formation Shock loading	Avoid junk in hole Use high flow rates and reduced WOB and RPM Avoid critical vibrations, run a shock sub
Erosion	Excessive bit hydraulic horsepower per square inch (HSI)	Use larger nozzles if erosion likely to lead to premature failure (lost cutters)
Heat checking	Balling Low flow rate	See "Bit balling" above Increase flow rate (may need larger nozzles)
Lost cone	Too long in hole Inadequate supervision Junk in hole Bit faulty	Use CPF calcs Listen to the hole Avoid junk Send back bit and any junk fished to manufacturer
Pinched bit	Run into undergauge hole	Ream in undergauge hole
Round gauge	Abrasive formation	Use enhanced gauge protection bit
Ring out (PDC or diamond)	Junk Wrong bit choice Excessive WOB	Avoid junk in hole, run a junk sub on previous bit run when PDC or diamond bit anticipated Review bit choice Reduce WOB

$$\begin{bmatrix} 3.7 \end{bmatrix}$$

Directional Drilling

This subject was covered in Section 2 from a well planning viewpoint. There were many topics covered in Section 2 that are also relevant to the wellsite operation (such as selecting the appropriate length of monel drill collars for magnetic surveys). Refer to Section 2.3 for information on directional drilling, or check the Table of Contents or Index for related subjects.

The material presented here only concentrates on the directional aspects of bottom hole assemblies. For questions of preventing downhole failures relating to tool joint damage, casing wear, and drillstring fatigue, refer to Section 1.5.2, "Dogleg Severity Limits—Combined Buildup and Turn Rate."

3.7.1. Rotary Bottom Hole Assemblies—General Points

Most rotary drilling is done in a tangent section. In this situation, designing a suitable BHA is straightforward. Where deviation work is in hand there will usually be a directional driller on site who will recommend what to run in the bottom hole assembly.

Use a nearbit stabilizer, bored for a float, and run it with a float in place. This gives extra protection in the event of a kick, when tripping, or if there is a leak on the standpipe or pump when the well is shut in. It is safer to routinely run one float when drilling with a diverter or on top hole with no secondary well control equipment.

Float valves sometimes have a small hole to allow the string to fill up while RIH and to allow Pdp to be read when the well is shut in. The ported float provides protection against strong backflow or flow through the drillstring. Although it does not stop flow completely, the small port only allows a relatively small flow through it and therefore still performs a useful function in reducing backflow to a manageable level.

For the maximum anticipated WOB, calculate the number of drill collars and HWDP needed to provide this, in mud, at the relevant hole angle. The HWDP can be run in compression providing a shorter, lighter DC section to reduce hook loads, drags, and sticking. The drillpipe can also be run in slight compression so it is not necessary to pick up excess weight on the BHA. For calculating the axial force exerted by a steel pipe in an inclined wellbore, refer to "Tension due to weight in a deviated wellbore" in Section 1.4.13, "Calculating, Axial Loads."

Design of rotary BHAs for drilling straight. Position full gauge stabilizers above bit, 10-15 m above the NB stab and another at 10-15 m above that. (Bit - NB Stab [Full Gauge] - DC - Stab [Full Gauge] - DC - Stab [Full Gauge] - rest of BHA.) Position undergauge stabilizers one per stand above the top full gauge stabilizer, to hold the drill collars off the wall to reduce sticking. More or less can be run if field experience shows this to be necessary. Run a Totco ring above the NB stabilizer; if a magnetic single shot is run then a monel drill collar is needed above the NB stab. Calculate the drill collars and HWDP for weight as outlined in this section.

This BHA will minimize the likelihood of sticking and allow greater overpull than a heavier BHA. Since it is well locked up, maximum WOB can be run on the bit without causing directional problems and bit life should be maximized by reducing wear and whirl. Drags will be kept lower as dogleg severities are kept at a minimum, especially if each kelly is reamed while circulating the cuttings above the BHA before each connection, as recommended.

For vertical wells, locked BHAs should also be used to prevent bit walk and crooked hole. Consider running the Anderdrift sub in the BHA with a Totco ring on top. This tool will give inclination surveys up to 5° when a connection is made by transmitting mud pulses to surface mechanically (the driller can see the pulses on the gauges). This will allow close monitoring of your verticality with virtually no loss of time and eliminates the risk of running slickline tools inside the drillstring to get a Totco survey.

Design of rotary BHAs for building or dropping angle. For a build assembly, run the second stabilizer undergauge or remove it altogether, depending on the severity of build required. (Bit - NB Stab [Full Gauge] - DC - Stab [1/4 in-1/2 in undergauge] - DC - Stab [Full Gauge] - rest of BHA.)

For a drop assembly, run the NB undergauge, remove it, or remove both the NB and next stabilizer depending on the severity of drop required. (Bit - NB Stab [1/4 in-1/2 in in undergauge] or bit sub - DC - Stab [optional] - DC - Stab - rest of BHA.)

Note that the more flexible (severe) a build or drop assembly is, the more likely it will walk sideways. This may cause an unwanted change in azimuth. Soft formation mill tooth bits also tend to walk more on a flexible assembly than most other bit types.

The closer a stabilizer is to the bit, the more it will influence the forces at the bit. A near bit stabilizer undergauge by 1/16 in will have much more effect on the deviation performance than if the next stab at 10 m goes undergauge by the same amount. Ensure that the bottom three stabilizers are gauged on every trip out of the hole and changed out if necessary.

Choose stabilizers compatible with the formations. The stabilizer blades should be long enough to avoid them digging into the wall; a short blade opposite soft formation would be a bad choice. If longer bladed stabs are not available and this causes concern, run two stabs in tandem to spread the load.

3.7.2. Preventing Keyseating

A keyseat could be cut on the inside of a dogleg section by the drillpipe rotating on the wall while being pulled against it. A keyseat wiper run on top of the DCs will help to backream through the keyseat if problems occur while pulling out of the hole. There is a better alternative, however.

If a keyseat is possible, run a string reamer in the drillpipe, if available. Position it at the top of the dogleg section with the bit on bottom. Size it greater than the DC OD and smaller than the bit; 1/2 in undergauge is usually about right. As drilling progresses, the reamer dresses out the inside of the dogleg section, wiping out any keyseat. This prevents having to backream, which can be quite time consuming without a top-drive system. Run the reamer every two or three trips as necessary. No extra time is taken since it works while drilling.

3.7.3. Directional Jetting—Practical Considerations

For an explanation of this technique, see the "Kickoff by Jetting" in section 2.3.3. The directional driller on site should be experienced in jetting to kick off, since jetting can create very severe doglegs if not carefully controlled.

Run a mill tooth bit - near bit stab (bored for and complete with float) - float sub (if no float in the NBS) - UBHO sub - monels - rest of BHA. Align the UBHO with the large nozzle for orientation. Set up the stabilizers for building angle (see "Design of rotary BHAs for building or dropping angle" in Section 3.7.1). Run in, take a single shot survey, mark the pipe, and align with the required azimuth after making up the kelly. Lock the table, bring the pumps up to speed and jet down the distance from the bit to the top NB stab blades. Pick up, ream through, and drill ahead for the rest of the single. Take another survey after adding a single and repeat the process if necessary. If the formation will not wash with just weight on bit and pump rate, the bit can be spudded.

Spudding the bit involves picking up the string a meter or two and dropping it until the bit reaches bottom (mark the string), then applying the main brake hard. The inertia of the string allows the bit to hammer into the formation, helping to exceed the mechanical formation strength. This precludes the use of any bit type except a steel mill tooth bit.

Jet only a small amount at a time, otherwise extremely high dogleg severities can result. In the right application, jetting saves the cost of using motors for kicking off and saves tripping time to bring out the motor and run a build assembly.

There is an exception to the rule of not jetting more than a short distance on each kelly. Large hole (over 16 in) is difficult to jet, but is possible. Jetting may have to be done for a full kelly down followed by a rotary drilled kelly down.

Smith Tool produces a special bit for jetting. This is like a tricone bit with one cone missing and a large circulation slot where the third cone would normally be. This bit has a right-hand walk tendency.

Marine wholesale distributors tools are not used when jetting due to the large flow rates necessary. Single shot surveys are quite adequate (and much cheaper) than MWDs and string orientation is accurate from the table because there will be little difference between bit and kelly direction due to the shallow depth.

Since drilling is fairly fast and the annulus will get loaded up with cuttings, a float in the string is necessary. Otherwise, time will have to be spent on circulating before every survey to prevent strong backflow.

3.7.4. Single Shot Surveys—General Points

The hole should be clean before running the survey and with no serious concerns as to hole condition. It is possible to take surveys in fairly sticky hole with a lot of care, using a timer unit and stopping the bit off bottom about 10 seconds before the survey, starting to move the string down, initially, 60 seconds after stopping.

A kelly cock should always be put on the top of the string and if wireline is to be run through it, use a metal protector with the bottom cut out to prevent the line cutting the thread.

Sandline should not be used for running or pulling surveys if there is an alternative. It is hard to control and the wire will twist up into a tangled mess when the tension is released as the survey barrel lands, unless the drum is stopped immediately. If it is used, do it with care and keep everyone clear of the rotary. Rigs should be equipped with a slick wireline winch.

The azimuth read from the magnetic compass should be corrected for magnetic variation. If the variation is east, add the variation to the magnetic azimuth. If the variation is west, subtract it from the magnetic azimuth to give the corrected azimuth. The variation should be noted on the drilling program and is also marked on maps of the area (which should be recent since variation changes over time).

3.7.5. Magnetic Single Shot Survey Tool

Refer to "Monel drill collars and stabilizers—selection and use" in Section 2.3.2 for information on calculating the number of non-magnetic drill collars needed for taking magnetic surveys.

Alternative camera units. There are three different camera units that use either a timer set at surface, a motion sensor, or a Monel detector that fires one minute after it detects the presence of a Monel drill collar. If the survey will be run on wireline, the motion sensor or Monel detector is preferable if the hole is not sticky, since it will fire 1 minute after reaching bottom. If the survey barrel is to be dropped, either to be fished on wireline or left in the string during a trip out, then use the timer unit allowing at least two minutes per thousand meters plus a minute for surface preparation prior to dropping (for example, at 4000 m allow 9 minutes). By using the correct unit, you will know exactly when the unit will start to expose the film, without wasting time waiting for the timer to go off long after reaching bottom. This allows the string to be worked until a few seconds before the camera will go off— an advantage if sticking is a problem.

Preparing the tool to take a survey. If the kit has not been run recently, test whichever unit is to be used. Either set the timer, start it going and time the bulbs lighting, or stand the unit in a corner with the Monel test ring in place or with the motion sensor unit switched on. All three bulbs should light for approximately 60 seconds when the camera unit fires. If not, check the bulbs and batteries.

Once the camera unit to be used has been tested OK, screw onto the bottom the angle unit relevant to the expected inclination. For low-angle/vertical wells use the 0-10 unit, for higher angles use the one with the maximum reading that is next above the expected angle. Look at the camera loading gate and ensure that it is closed. Pick up the film magazine; hold it in position against the camera loading gate. Pull the loading lever right out and gently push it back in. If there is film in the magazine, it should pick one up and push it against the closed gate.

This verifies that there is film ready to load into the camera. Open the camera gate, push the film magazine loading lever in, and close the camera gate. Check that the indication is "loaded" on the camera gate closing ring; then, if all is OK, remove the film magazine. The camera is now ready to use.

Attach the bottom of the camera to the rubber shock absorber inside the survey outer barrel. Switch on the timer or motion sensor, make up the outer barrel with 18 in pipe wrenches. If running on wireline make up to the line, otherwise drop down the drillstring. Keep working the string until either the survey on wireline reaches bottom or until a minute before the timer goes off (less if the hole is sticky). Position the bit about 2 m off bottom and wait until either: 1 minute after the timer was due to go off, or 1 minute 40 seconds after the monel detector reached bottom, or 1 minute 40 seconds after the string was stopped off bottom with the motion sensor. Note that the exposure time is 1 minute.

In sticky hole conditions, it is good practice to stop moving the string on a down stroke about 5-10 ft off bottom. When the survey has been taken, move down initially and ensure that the string is free before picking up. If the string has started to get stuck, moving down will tend to free it, whereas picking up may embed the string further.

Developing the survey. When the survey barrel is recovered, undo the outer barrel and recover the camera complete. Remove the drillfloor. Prepare the developing tank by making sure it is clean; reassemble it and pour developing fluid into the top until it is exactly level with the top—no more, because it will drain through into the developing chamber.

Hold the camera unit horizontally (no need to remove the angle unit) and place the developing tank entry against the camera loading gate so the pins are located in the holes and a light tight connection is made, with the developing tank underneath. Open the camera gate; the film can be heard as it drops into the chamber. Pull up the top part of the developing tank and the film will drop right through to the developing chamber.

Put the camera down, hold the developing tank upright, and shake gently for one minute. Open the top, remove the film, and wash in running water. The film can be examined with the magnifying examiner.

3.7.6. Totco Single Shot Survey Tool

The Totco tool is much simpler than the Magnetic Single Shot (MSS). No monel collars are needed. A paper disk is pierced by two pins showing inclination but not azimuth. The tool is loaded by opening the angle unit, placing one of the marked paper disks inside (compatible with the angle unit) and making the unit up again. The timer is set and it can be run on wireline or dropped prior to pulling out of the hole.

After recovering the paper disk, it can be examined under the magnifying examiner. The two pinholes should both show the same angle, 180° apart. There are two to act as a quality check of the survey; and as long as they both show the same angle, the survey should be reliable.

The Totco is simpler and more reliable than the MSS. However, the Totco does not give azimuth information.

3.7.7. Gyro Multishot Surveys

A Gyro survey will sometimes be run in casing while nippling up the BOPs. It will be run on wireline by a dedicated unit with a winch operator and a survey technician.

The gyro is set up on surface by reference to a fixed point, usually some distance away, that has a known bearing from the rotary table. When the gyro is recovered to surface, it will again be compared to the reference point to see how far the gyro has drifted in that time.

Quality checks of the gyro drift should be taken by stopping in the wellhead for 5 minutes before leaving and when arriving back at surface, then for a 5-minute stop after every 15 minutes surveying. The camera takes pictures on to a strip of film and by carefully recording times and depths on surface, each picture can be related to a depth. By stopping for 5-minute checks after every 15 minutes, the accuracy of the gyro can be verified and the correction applied to the photographic survey results. Surveys pulling out should match surveys running in at the same depths.

Check that the depth counter on the unit is in good condition. Record the holdup depth and check that the tool reads zero again when back at surface with the survey. The accuracy of the depth counter is crucial to a good survey, and yet this often seems to be forgotten.

[3.8]

Writing the Final Well Report

The purpose of writing a Final Well Report is to record as much information as possible that will lead to improved performance on subsequent wells. A well-researched and honest final well report is extremely important in improving future performance.

The report will be written in sections covering different topics. For each topic there should be the following:

1. A description and analysis of events. This should show what happened, why it happened, and whether this was different from what was planned. Reasons for decisions made must be fully described.
2. Conclusions showing the main learning points.
3. Recommendations that refer to the conclusions in #2 above, which need to be considered when planning and drilling the next well.

The report should be written as a technical report to be concise and easy to understand. Do not include unnecessary or irrelevant details. The report should be paginated and include a table of contents. It must also give a full and frank appraisal of events, including where mistakes were made. It is important that personnel are told that genuine mistakes will not be a cause for punishment as long as they admit to them and learned from them. Note: Do not just concentrate on negative points. Where the job was well planned and executed and performance was good, show this was the case.

Most final well reports are useless for future well planning. Bulky printoffs showing operations every 15 minutes are really just "padding." Information that is of little or no interest in the future is readily available from the well files and serves only to make the report look comprehensive when it may be inadequate. A proper final well report takes time and effort to produce but looks good and is useful. Following is a suggested format for a final well report.

3.8.1. Suggested Final Well Report Format

1. Title page with the name of the well and of the report author
2. Table of contents
3. General information
    a) Well and rig summary information, plus the date that the well was spudded and abandoned, suspended, and handed over to production.
    b) Well schematic showing final well status.
    c) Time vs. depth graph, showing planned vs. actual performance.
    d) Final cost breakdown showing planned vs. actual costs by category.
    e) A summary of good points from drilling the well.
    f) A summary of areas of concern or problems encountered while drilling the well.
4. Drilling operations
    a) Description by hole section (text); use the drilling supervisor's daily diary notes as the basis, along with daily drilling reports, etc. Write out a description of events. Note especially information on how best to drill each formation encountered; include suggested bits, parameters, mud properties, drilling practices, etc.
    b) Bit performance analysis; planned vs. actual performance, constraints on bit performance, notes on how the bit and BHA selection and parameters could be improved.

c) Mud performance analysis. Show planned vs. actual performance, problems encountered (especially borehole stability related), how these problems were handled, and suggestions for improvements on the next well.

d) Wellbore stability analysis. Where wellbore instability has occurred (usually seen as cavings), give as much information as possible on what was seen. Was the extent of it measured (e.g., by carbide, caliper logs, cavings quantity estimation, etc.)? Describe cavings fully—appearance and quantities. Can particular instability events (such as a sudden temporary increase in cavings) be related to drilling events (such as a trip)? Consider what changes could be made to reduce instability on the next well.

e) Casing and cementing performance analysis. Discuss problems encountered; also comment where performance was good.

f) Specific problem areas: fishing jobs, other remedial work, kicks, etc.

g) Miscellaneous items not covered in the other areas listed.

h) Conclusions and recommendations. How time could be saved, problems avoided or reduced, and anything to improve performance next time.

5. Rig: personnel, equipment, procedures, and contract.
   a) Factors affecting drilling performance (related to personnel, equipment, drilling program, procedures, or rig contract).
   b) Solids control equipment performance analysis.
   c) Conclusions and recommendations.

6. Logistics
   a) Equipment lists as shown in the drilling program; show any deficiencies in these lists (i.e., update for future reference).
   b) Highlight any problems regarding vessels, helicopters, road transport, etc.
   c) Conclusions and recommendations.

7. Service companies
   a) Problems with equipment, personnel, procedures, or contracts.
   b) Conclusions and recommendations.

8. Geological results
   a) Prognosed vs. actual lithology encountered.
   b) Conclusions and recommendations.

9. General
   a) Communications and computers.
   b) Anything else not covered above.
   c) Conclusions and recommendations.
10. Appendices
   a) Any special reports, e.g., well control problems, prognosed vs. actual lithology, bit record.
   b) Casing and cementing reports.
   c) Formation integrity test reports.
   d) BHA details.
   e) Directional surveys and well plots.
   f) Printoff showing pore pressures, fracture gradients, mud densities, and overburden gradient against depth (e.g., Sperry Sun PPFG printout).
   g) Principal mud properties against depth.
   h) Hole section summaries showing performance on this well compared to offsets, if relevant.
   i) Other reference information. Personnel involved in planning and executing the well, for future reference.
   j) A copy of the drilling program.
   k) A copy of the technical justification.

[Appendix]

Appendix: 1

Calculating Kick Tolerances

Assumptions

For each hole section the volume of gas influx that can be safely handled has to be calculated to safely reach the next casing point. These calculations take the worst-case scenario and make assumptions about the likely formation pore pressure to calculate the influx volume.

The worst case for a kick occurring is at the greatest depth—when the next casing point has been reached. To determine the minimum shoe strength required to reach this, some assumptions are made.

For an exploration or appraisal well in a relatively or completely unknown field, assume that the kicking formation may have a pore pressure gradient of 10% higher than the mud gradient (increase this if circumstances warrant). Therefore, a planned mud gradient of 0.5 psi/ft will assume a pore pressure gradient at casing point of 0.55 psi/ft.

For a development well in a known area, assume that the kicking formation may have a pore pressure equal to the mud gradient. Therefore, it is assumed that any kick is most likely to be a swabbed influx.

Method

Once the fracture gradient is known, calculate the maximum gas influx volume at the next casing point. For this kick situation, the maximum pressure on the shoe usually occurs when the top of the gas reaches the shoe (assuming one gas bubble). This may not be the case if the height of the influx at the shoe is less than the height of the influx around the drill collars, if the gas expansion is not enough to compensate for the changes in annular capacity.

Assume the driller's method. This is the worst case and would apply if the string was plugged and a volumetric kill was required.

Calculation of the exact pressure at the top of a gas kick bubble is not feasible; there are unknown factors that make little practical difference. The composition of the gas (compressibility factors) is unknown and temperature effects can be ignored. The calculations given below could be slightly more accurate with much greater effort, but this is not worthwhile since even then the calculated figure cannot account for all the small factors.

After calculating the kick tolerance, consider whether this is enough. Kick tolerances may be set by company policy or government regulation. Account for reaction time of men and equipment, how fast a kick may occur (related to likely permeabilities), and the level of training and competence of the rig crews. Extra precautions may be taken in critical areas, such as restricting ROP, setting fixed drilling parameters to allow a drilling break to be identified more easily, etc.

Example: Given a casing shoe at 5000 ft with a fracture gradient of 0.75 psi/ft, plan to drill to the next casing point at 8000 ft with a mud gradient of 0.6 psi/ft in a vertical exploration well. Assume a gas gradient of 0.1 psi/ft at the casing shoe; 12^1/$_4$ in hole, 5 in drillpipe, and 300 ft of 8 in drillcollars.

First, calculate the MAASP.

MAASP = 5000 x (0.75 – 0.6) = 750 psi.

Next, calculate the height of a gas influx at the casing shoe where the pressure at the shoe = formation breakdown pressure, and the formation at casing point has a pore pressure gradient that is 10% greater than the planned mud gradient.

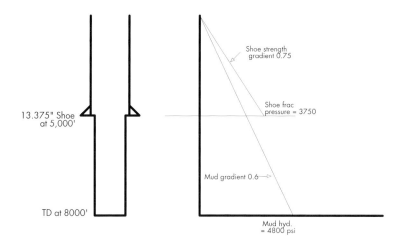

Fig. A-1 Mud and Shoe Strength Gradients

Calculate the maximum pressure at the shoe and the bottom hole pressure.

Pshoe = 5000 x 0.75 = 3750 psi = pressure at the top of the gas bubble.
Pformation = 8000 x 0.6 x 1.1 = 5280 psi.

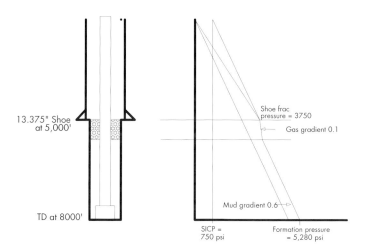

Figure A-2 Solving for Height of Influx with Gas on Bottom

We can now use simultaneous equations to solve for the height of influx, Hi:

a. $Hi + Hm = 3000$ ft
b. $0.1\ Hi + 0.6\ Hm = (5280 - 3750) = 1530$

Multiply equation a by 0.6:

c. $0.6\ Hi + 0.6\ Hm = 1800$

Subtract b from **i**:

d. $0.5\ Hi = 270$; therefore $Hi = 270/0.5 = 540$ ft

Calculate the volume of this height of gas at the casing shoe and the average pressure.
Hydrostatic pressure of the gas = 540 x 0.1 = 54 psi.
Hydrostatic pressure of half of the height of gas = 27 psi.
Annular capacity is 0.1214 bbl/ft; therefore $Vi = 65.6$ bbls.
Pressure in the center of the gas bubble is 3750 + 27 = 3777 psi.
Finally, calculate the volume of this gas at the next casing point.
Pressure in the center of the gas bubble will be 5280 - 27 = 5253 psi.
Using Boyle's Law, $P1V1 = P2V2$ or rearrange using $V2 = P1V1/P2$
Therefore, $V2 = 3795$ x $65.6/5253 = 47.4$ bbls, which is the kick tolerance given the above assumptions.

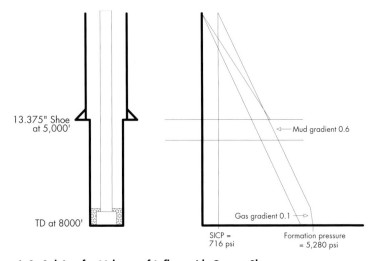

Figure A-3 Solving for Volume of Influx with Gas at Shoe

Several safety factors support this:

- Temperature will drop as the gas expands, reducing the pressure peak at the shoe
- The gas will disperse in the mud, reducing the pressure peak at the shoe
- If surface to bit volume is less than bit to shoe volume, kill mud may exit the bit before the gas reaches the shoe, reducing the peak pressure at the shoe

If the shoe strength is too low for the required kick tolerance, the next casing shoe may have to be higher. If the mud gradient is changed, recalculate the kick tolerance.

The heavier the mud weight used for drilling, the less tolerant the well is (i.e., the shorter distance that can be drilled). A higher mud weight gives less difference to the shoe strength; if mud weight equals shoe strength gradient then $MAASP = 0$. Therefore, it follows that the mud density should not be higher than is necessary for a trip margin or wellbore stability considerations.

The rig should hold sufficient stock of barite to weigh up the complete active system volume to the equivalent mud density achieved by the FIT.

Kick tolerance volume. What determines the acceptable kicking pressure and volume of kick tolerance?

1) The degree of risk. Higher assumed pore pressures should be used for:
   a) Wildcat exploration wells
   b) Areas where short transition zones are expected
2) The speed of reaction. Higher volumes should be used when:
   a) There are larger diameter holes
   b) High permeabilities are expected from kicking formations
   c) Drilling from a floating rig where flow detection may be tricky due to heave
   d) Low standard kick detection equipment is being used
   e) Drill crews are not well trained (some third world countries with local crews)

[Appendix: 2]

Formation Integrity Test
Recommended Procedure

An FIT is performed to determine MAASP and to check the integrity of the seal of cement around the shoe. There are two types: a limit test (where the test is stopped at a predetermined maximum if no leakoff occurs) and a leakoff test (where the test is continued until leakoff is seen). The techniques for both are the same. A limit test is performed to a pressure which confirms the ability of the well to contain a gas kick of a specified influx volume at the next casing depth where formation pressures are known. A leakoff test is used where pore pressures are not well-known or on exploration wells.

A leakoff test yields more information—knowing the pressure to just initiate leakoff gives information relating to the minimum field stresses. A limit test would be appropriate when a very brittle formation is exposed that should not be fractured (i.e., limestone).

An FIT should be carried out after drilling out any casing where a BOP is in use. The drilling program will determine whether it is to be done as a limit or leakoff. An FIT can also be repeated within a hole section to confirm that the well is still able to contain the MAASP as indicated at the casing shoe, if a weaker formation has been penetrated.

The recommended technique allows close control of the operation to prevent formation breakdown. The amount pumped in between each reading will depend on the depth of the casing shoe, since a greater volume will need to be pumped for deeper casings for the same pressure increase.

Test Procedure

Drill out the shoe and approximately 5 m of formation, circulate, and condition mud until mud weight is consistent in the well. Pull the bit back into the shoe. Confirm that the hole is full, then close the TPR around the drillpipe and close in the annulus at the choke manifold. Line up the cement pump on the drillstring.

Fill the cement pump displacement tank and note the volume. Pump slowly until the pressure just starts to rise, stop the pump and note the volume pumped and pressure. Allow pressure to stabilize for a minute or so (longer if necessary). Pump increments of $1/4$ or $1/2$ bbl, stopping to allow pressure stabilization each time and noting volume vs. pressure. Either use a graph or a table to show when the increase in pressure between readings starts to decrease.

Stop the test when either the increase in pressure per unit volume pumped decreases or the stated limit is reached. Check the reading on the choke manifold (i.e., the gauge that will be used during a well kill) and use this as the end pressure. Bleed off back to the cement tank and note the volume returned, except if there is a float in the string in which case the pressure will have to be bled off and returns volume measured at the choke manifold. MAASP at the mud weight in the hole will be the pressure held at the end of the test, either at leakoff or limit. Calculate the equivalent mud density for this pressure.

Open up the well, line up back to the mud pumps. Check that there is enough barite on site to weigh up the entire active system to this density and, if so, resume drilling.

Table A-1 provides an example of a tabular record for a leakoff test.

Table A-1 Leakoff Test Example

Well: Test 1			
Volume Pumped, bbl	Stabilized Pressure, psi	Pressure Increase, psi	Comments
0	0	0	
0.5	50	50	
1.0	150	100	
1.5	250	100	
2.0	350	100	
2.5	420	70	Leakoff
Bled off; 2.1 bbls returned			

$\begin{bmatrix} \text{Appendix: 3} \end{bmatrix}$

Information Sources

No book or manual can ever be complete and no person can remember everything. It is useful to point the way to repositories of useful information or advice.

Internet Resources

Access to the Internet is now essential for drillers. There are many sites that can give you information, connect you to technical expertise, and provide links to other related web sites (which may not have been set up for drilling purposes). Following are the URLs of a few sites which should be of interest.

http://www.drillers.com - I put this first because it is my site! Contains files of interest such as spreadsheets, databases, Word templates, etc., as referenced in this book. There are links to other drilling web sites. Also has information on jobs currently available. No charge for accessing or downloading from the site.

http://mfginfo.com/htm/pet_resources.htm - Not regularly updated but contains an incredible list of petroleum related Internet resources, including links to hundreds of web sites and Internet news groups.

http://www.iadc.org - Home page of the International Association of Drilling Contractors.

http://www.api.org - Home page of the American Petroleum Institute.

http://www.spe.org - Home page of the Society of Petroleum Engineers.

http://www.slb.com - Schlumberger's home page. Contains an index of hundreds of technical papers published by companies within the Schlumberger group.

http://www.ogjonline.com - Home page of PennWell's *Oil & Gas Journal* online magazine, with weekly *OGJ* articles and forums, and links to a wide variety of up-to-date industry information.

Office/Technical Library Resources

API publishes a vast amount of technical reference literature. A catalog is available from API. Copies of the current bulletins, recommended practices, and specifications related to drilling operations should be in every drilling office.

Journal of Petroleum Technology, published monthly by the SPE for all SPE Members.

Technology Review, published quarterly by Schlumberger. Contains in-depth technical articles dealing with cutting-edge technology.

World Oil, published monthly by Gulf Publishing Company.

CD-ROMs

"Practical Well Planning and Drilling Manual on CD," available from PennWell. Demo version available for download from the web site at http://www.drillers.com.

"SPE Image Library," available direct from the Society of Petroleum Engineers. Contains bitmapped images of technical papers (including unsolicited and unpublished papers), which can be viewed onscreen or printed off. Text of the first page is held in a master disk index that can be searched by title, author, keyword, etc.

"Composite Catalog" on CD, available from Gulf Publishing Company. Contains details of hundreds of oilfield products, includes search features. Complements/replaces the printed *Composite Catalog* several-volume sets.

[Appendix: 4]

Drilling Equipment Lists by Operation

The equipment lists shown in Table A-2 is meant as a guide to assist the on-site drilling supervisor and onshore base manager. It should not be assumed that the list is complete, and it is the responsibility of the drilling supervisor to ensure that all required equipment has been obtained.

Note that the table shows the estimated quantity for each operation; in some cases these items will already be on the rig for a previous operation so only the deficient amount need be sent.

Table A-2 Drilling Equipment

		Rig Positioning and Preparation for Operations		
Item	Qty	Description		Comments
1	A/R	Positioning equipment as required for rig and anchor running vessels		
2	1	Rig communications package		
3	1	ROV, if required		

		36 in Hole
Item	Qty	Description
1	2 ea	36 in and 26 in drillbits c/w nozzles
2	1 ea	36 in and 26 in bit breakers
3	2	Bit sub, bored for float c/w 7-5/8 in x 7-5/8 in reg box/box
4	2	Float valve for item #3
5	1	36 in string stabilizer
6	2	Totco ring 7-5/8 in R
7	1	36 in hole opener c/w spare cutters
8	3	9-1/2 in DC c/w 7-5/8 in reg conns
9	9	8 in DC c/w 6-5/8 in reg conns
10	2 ea	8 in and 9-1/2 in DC slips and elevators
11	2	X/Over 7-5/8 in R pin x 6-5/8 in R box
12	2	X/Over 6-5/8 in R pin x NC50 box
13	300 m	5 in HWDP
14	1 lot	5 in DP 19.5 lb/ft, premium grade G
15	3	5 in pup joints, 5Ft, 10 ft, and 15 ft premium grade S
16	2 ea	5 in DP slips and elevators
17	1 lot	Spare dies for items #9 and #14
18	1	Totco kit c/w running and retrieving gear
19	1 lot	DP and DC dope
20	1 set	DC safety clamps
21	250 m	Tubing cement stinger; steel, or GRP if available
22	1	X/Over to run stinger on DP
23	1 lot	Handling equipment for item #19
24	1 lot	Drill water
25	1 lot	Bulk bentonite and barite
26	1 lot	Cement and additives
27	1 lot	Mud chemicals
28	1 lot	LCM fine, medium, and coarse
29	1 lot	Fishing equipment for all downhole tubulars

26 in Hole (Subsea Wellhead)		
Item	Qty	Description
1	1	30 in hydraulic latch c/w H4 profile
2	1 lot	21 in riser c/w pup joints and handling equipment
3	1	Telescopic joint
4	1	Ball joint
5	1	Diverter body c/w inserts and handling equipment
6	2 ea	17-1/2 in and 26 in drillbits c/w nozzles
7	1	17-1/2 in and 26 in bit breakers
8	1	26 in under reamer c/w two redress kits
9	1	Bullnose to fit below item #8
10	1	17-1/2 in sleeve type FG NB stab c/w sleeve breaker
11	3	17-1/2 in sleeve type string stabs c/w 3 FG and 3 UG spiral sleeves
12	2	Bit sub, bored for float c/w 6-5/8 in x 6-5/8 in reg conns
13	2	Float valve for items #10 and #12
14	2	X/Over 7-5/8 in R pin x 6-5/8 in R box
15	2	X/Over 6-5/8 in R pin x NC50 box
16	2	9-1/2 in DC type monel
17	6	9-1/2 in DC c/w 7-5/8 in reg conn
18	6	8 in DC c/w 6-5/8 in reg conn
19	2	8 in drilling jars, Bowen hydromechanical c/w 6-5/8 in reg conn
20	300 m	5 in HWDP
21	1 lot	5 in DP 19.5 lb/ft, premium grade G
22	3	5 in pup joints, 5 ft, 10 ft, and 15 ft premium grade S
23	1	Totco ring 7-5/8 in R
24	1set	Totco kit c/w running and retrieving gear
25	3 sets	8 in and 9-1/2 in DC safety clamp, slips, and elevators
26	2 sets	5 in DP slips and elevators
27	1 lot	Spare dies for items #31 and #32
28	1 lot	DP and DC dope
29	1 lot	ROV c/w launching equipment and spares
30	1 lot	Drill water
31	1 lot	Bulk bentonite and barite
32	1 lot	Cement and additives
33	1 lot	Mud chemicals
34	1 lot	LCM fine, medium, and coarse
35	1 lot	Fishing equipment for all downhole tubulars

		171/2 in Hole (Subsea Wellhead)
Item	Qty	Description
1	2	18-3/4 in seat protector c/w running tool
2	1	BOP test tool c/w spare seals
3	A/R	17-1/2 in drill bits c/w nozzles
4	1	17-1/2 in bit breaker
5	1	17-1/2 in sleeve type FG NB stab c/w sleeve breaker
6	3	1-71/2 in sleeve type string stabs c/w 3 FG and 3 UG spiral sleeves
7	2	Bit sub, bored for float c/w 6-5/8 in x 6-5/8 in reg conns
8	2	Float valve for items #5 and #7
9	2	X/Over 7-5/8 in R pin x 6-5/8 in R box
10	2	X/Over 6-5/8 in R pin x NC50 box
11	2	9-1/2 in DC type monel
12	6	9-1/2 in DC c/w 7-5/8 in reg conn
13	6	8 in DC c/w 6-5/8 in reg conn
14	2	8 in drilling jars, Bowen hydromechanical c/w 6-5/8 in reg conn
15	300 m	5 in HWDP
16	1 lot	5 in DP 19.5 lb/ft, premium grade G
17	3	5 in pup joints, 5 ft, 10 ft, and 15 ft premium grade S
18	1	Totco ring 7-5/8 in R
19	1 set	Totco kit c/w running and retrieving gear
20	2 ea	8 in and 9-1/2 in DC safety clamp, slips, and elevators
21	2 ea	5 in DP slips and elevators
22	1 lot	Spare dies for items #19 and #20
23	1 lot	DP and DC dope
24	1	9-1/2 in shock sub (check conns and X/O to fit above NB stab)
25	1	Circulating sub 6-5/8 in R conns c/w redress kit and ball
26	1	Test sub NC50 pin x 2 in Weco (for testing kelly, etc.)
27	30	DP-casing protectors
28	A/R	Logging tools
29	1 lot	ROV c/w launching equipment and spares
30	1 lot	Drillwater
31	1 lot	Bulk bentonite and barite
32	1 lot	Cement and additives
33	1 lot	Mud chemicals
34	1 lot	LCM fine, medium, and coarse
35	1 lot	Fishing equipment for all downhole tubulars

		12¼ in Hole (Subsea Wellhead)
Item	Qty	Description
1	2	18-3/4 in x 13-3/8 in wear bushing c/w running tool
2	1	BOP test tool c/w spare seals
3	A/R	12-1/4 in drill bits c/w nozzles
4	1	12-1/4 in bit breaker
5	1	12-1/4 in sleeve type FG NB stab c/w sleeve breaker
6	4	12-1/4 in sleeve type string stabs c/w 3 FG and 3 UG spiral sleeves
7	2	Bit sub, bored for float c/w 6-5/8 in x 6-5/8 in reg conns
8	2	Float valve for items #5 and #7
9	2	X/Over 6-5/8 in R pin x NC50 box
10	2	8 in DC type monel
11	12	8 in DC c/w 6-5/8 in reg conn
12	1	8 in pony DC c/w 6-5/8 in reg conn
13	2	8 in drilling jars, Bowen hydromechanical c/w 6-5/8 in reg conn
14	300 m	5 in HWDP
15	1 lot	5 in DP 19.5 lb/ft, premium grade G
16	3	5 in pup joints, 5 ft, 10 ft, and 15 ft premium grade S
17	2	Totco ring 6-5/8 in R
18	1 set	Totco kit c/w running and retrieving gear
19	1 set	Magnetic single shot kit c/w running and retrieving equipment
20	1 set	Magnetic multishot kit c/w running and retrieving equipment
21	1 set	Gyro multishot c/w winch and compensator
22	2 ea	8 in DC safety clamp, slips, and elevators
23	2 ea	5 in DP slips and elevators
24	1 lot	Spare dies for items #19 and #20
25	1 lot	DP and DC dope
26	2	8 in shock sub (check conns and X/O to fit above NB stab)
27	1	Circulating sub 6-5/8 in R conns c/w redress kit and ball
28	1	Test sub NC50 pin x 2 in Weco (for testing kelly, etc.)
29	50	DP-casing protectors
30	A/R	Logging tools
31	1 lot	Drillwater
32	1 lot	Bulk bentonite and barite
33	1 lot	Cement and additives
34	1 lot	Mud chemicals
35	1 lot	LCM fine, medium, and coarse
36	1 lot	Fishing equipment for all downhole tubulars

		81/2 in Hole (Subsea Wellhead)
Item	Qty	Description
1	2	18-3/4 in x 9-5/8 in wear bushing c/w running tool
2	1	BOP test tool c/w spare seals
3	A/R	8-1/2 in drill bits c/w nozzles
4	1	8-1/2 in bit breaker
5	1	8-1/2 in integral blade FG NB stab, BFF
6	1	8-1/2 in integral blade UG NB stab, BFF
7	3	8-1/2 in integral blade string stabs, FG
8	2	8-1/2 in integral blade string stabs, UG
9	2	Bit sub, bored for float c/w 4IF x 4IF conns
10	2	Float valve for items #5 and #9
11	2	8-1/2 in roller reamers c/w 2 redress kits
12	2	X/Over 4IF pin x NC50 box
13	2	6-1/2 in DC type monel
14	12	6-1/2 in DC c/w 4IF conn
15	1	6-1/2 in pony DC 4IF conn
16	2	6-1/2 in drilling jars, Bowen hydromechanical c/w 4IF conn
17	300 m	5 in HWDP
18	1 lot	5 in DP 19.5 lb/ft, premium grade G
19	3	5 in pup joints, 5 ft, 10 ft, and 15 ft premium grade S
20	2	Totco ring 4IF
21	1 set	Totco kit c/w running and retrieving gear
22	1 set	Magnetic single shot kit c/w running and retrieving equipment
23	1 set	Magnetic multishot kit c/w running and retrieving equipment
24	2 ea	6-1/2 in DC safety clamp, slips, and elevators
25	2 ea	5 in DP slips and elevators
26	1 lot	Spare dies for items #22 and #23
27	1 lot	DP and DC dope
28	2	6-1/2 in shock sub (check conns and X/O to fit above NB stab)
29	1	Circulating sub 4IF conns c/w redress kit and ball
30	1	Test sub NC50 pin x 2 in Weco (for testing kelly, etc.)
31	1 ea	4IF kelly cock and inside BOP c/w redress kits
32	A/R	8-1/2 in coring equipment including coreheads, handling equipment, fishing gear, etc.
33	1 lot	Bulk bentonite and barite
34	1 lot	Cement and additives
35	1 lot	Mud chemicals
36	1 lot	LCM fine, medium, and coarse
37	1 lot	Fishing equipment for all downhole tubulars

		30 in Conductor (Subsea Wellhead)
Item	Qty	Description
1	1 jt	30 in x 1 in WT X52 conductor shoe joint
2	10 jts	30 in x 1 in WT X 52 conductor pipe
3	2 jts	30 in crossover joint c/w ST-2 x Rl-4
4	1	30 in wellhead housing and extension c/w st2 conns
5	1	30 in cam actuated wellhead running tool
6	1	Temporary guide base
7	2	Slope indicators
8	1	30 in x 150t side door elevator
9	1 set	30 in bowl and slips
10	1 set	30 in single joint elevators
11	1 set	30 in safety clamp
12	1	Stab in sub NC50 box c/w redress kit
13	1 set	30 in bull tongs
14	1	Hilti gun c/w cartridges and spares or chisel/hammer
15	1	Circulating head c/w NC50 pin x 2 in Weco
16	1 lot	Cement and additives
17	1 lot	Dunnage
18	1	Centralizer for 5 in DP inside casing

		20 in Casing (Subsea Wellhead)	
Item	Qty	Description	Comments
1	2 jts	20 in x 129.3# X56 casing c/w stab in float shoe, FL4S box up	
2	70 jts	20 in x 129.3# X56 casing c/w FL4S connections	
3	1	Stab in sub NC50 box c/w redress kit	
4	2 pcs	18-3/4 in wellhead housing and extension c/w ALT-2 pin	
5	2 pcs	20 in crossover joint ALT-2 x FL4S	
6	2 pcs	18-3/4 in wellhead running tool	
7	1	Hilti gun c/w cartridges and spares	
8	A/R	20 in spring centralizers c/w stop collars and nails	
9	2	20 in rigid centralizers c/w stop collars and nails	
10	1	20 in wellhead running tool	
11	1 set	20 in hand slips	
12	1	20 in x 150t side door elevators	
13	2	20 in 250t spider/elevator	
14	1 set	20 in safety clamp	
15	2	20 in single joint elevators	

		20 in Casing (Subsea Wellhead), continued
16	1 set	20 in manual tongs
17	1	Circulating head NC50 pin x 2 in Weco
18	A/R	Cement and additives
19	1 lot	Dunnage
20	1	Centralizer for 5 in DP inside casing

		133/8 in Casing (Subsea Wellhead)
Item	Qty	Description
1	A/R	13-3/8 in 72 and 68# N80 buttress casing
2	2 sets	13-3/8 in float shoe and float collar c/w BTC thread
3	2 pcs	18-3/4 in x 13-3/8 in casing hanger c/w pup jnt btc pin down
4	2 pcs	Seal assembly for item #3
5	2 pcs	18-3/4 in x 13-3/8 in hanger running tool c/w NC50 conns
6	2 sets	Non-rotating top and bottom cement plugs
7	A/R	13-3/8 in spring centralizers c/w stop collars and nails
8	2	13-3/8 in rigid centralizers c/w stop collars and nails
9	1 ea	13-3/8 in 350t spider/elevator
10	2 pcs	13-3/8 in 150t side door elevators
11	2 pcs	13-3/8 in single joint elevators
12	1	13-3/8 in safety clamp
13	2 pcs	13-3/8 in stabbing guides
14	1 set	13-3/8 in lug jaws
15	2 pcs	13-3/8 in slip inserts
16	1 ea	13-3/8 in top and bottom guides
17	1 set	13-3/8 in klepo protectors
18	2 pcs	Power tongs
19	2 pcs	Power pack
20	2 pcs	13-3/8 in 72 and 68# casing drift
21	1	Casing fillup tool dressed 13-3/8 in
22	1	Plug launching cement head x NC50 pin
23	1	Subsea release head c/w equalizing sub
24	2 sets	Subsea cement plugs c/w drillpipe dart
25	1	13-3/8 in circulating swedge c/w 2 in Weco conn
26	1 lot	Cement and additives
27	1	13-3/8 in casing spear to fit 68 and 72# casing
28	1 lot	Casing dope
29		Bakerlock
30		Dunnage

		95/8 in Casing. (Subsea Wellhead)
Item	Qty	Description
1	A/R	9-5/8 in 47# N80 buttress casing
2	2 sets	9-5/8 in float shoe
3	2 sets	9-5/8 in float collar
4	2 pcs	18-3/4 in x 9-5/8 in casing hanger c/w pup jnt btc pin down
5	2 pcs	Seal assembly for item #4
6	2 pcs	9-5/8 in hanger running tool c/w NC50 conns
7	A/R	9-5/8 in spring centralizers c/w stop collars and nails
8	2	9-5/8 in rigid centralizers c/w stop collars and nails
9	2 sets	Non-rotating top and bottom cement plugs
10	2	Radioactive marker
11	1 ea	350t spider/elevator
12	2 pcs	9-5/8 in 150t side door elevators
13	2 pcs	9-5/8 in single joint elevators
14	2 pcs	9-5/8 in safety clamp
15	2 pcs	9-5/8 in stabbing guides
16	1 set	9 x 10-3/4 in lug jaws
17	2 pcs	9-5/8 in slip inserts
18	1 ea	9-5/8 in top and bottom guides
19	1 set	9-5/8 in klepo protectors
20	2 pcs	Power tongs
21	2 pcs	Power pack
22	2 pcs	9-5/8 in casing drift
23	1 lot	Cement and additives
24	1	9-5/8 in casing spear
25	1 lot	Casing dope
26	A/R	Bakerlock
27	A/R	Dunnage
28	1	Casing fillup tool dressed for 9-5/8 in casing

Appendix: 5

Conductor Setting Depth for Taking Returns to the Flowline

The minimum depth for the conductor shoe to allow returns to the flowline can be calculated:

$$Penetration = \frac{([x + y]\ pReturns)\text{-}(y\ x\ pSea)}{pFormation\text{-}pReturns}$$

where

Penetration = depth of shoe below seabed or surface, feet.
x = height of flowline above sea, feet.
y = water depth, feet.
rSea = sea density, psi/ft.
rReturns = returns density, psi/ft.
rFormation = formation density, psi/ft.

Example

$$Penetration = \frac{(180 \text{x} 0.55)\text{-}(100 \text{x} 0.045)}{0.85\text{-}0.55} = \frac{54}{0.3} = 180'$$

where

Height of flowline = 80 ft MSL.
Water depth = 100 ft.
Formation density gradient = 0.85 psi/ft.
Seawater in the area = 0.45 psi/ft.
Expected returns gradient = 0.55 psi/ft
Calculate the required minimum penetration for the conductor shoe.

[GLOSSARY]

A, Area. Usually given in square inches (eg TFA = Total Flow Area, area of bit nozzle configuration in square inches).

Absorption. The penetration or apparent disappearance of molecules or ions of one or more substances into the interior of a solid or liquid. For example in hydrated bentonite, the planar water that is held between the mica-like layers is the result of absorption.

Acidity. The relative acid strength of liquids as measured by pH. A pH value below 7. See pH.

Adaptive Electromagnetic Propagation Tool. Schlumberger wireline tool that measures phase shift and attenuation of a 1100 MHz wave. Used for Hydrocarbon identification independent of formation water salinity, thin bed detection, Hydrocarbon saturation and mobility and to evaluate invaded zones.

Adhesion. The force which holds together unlike molecules.

Adsorption. A surface phenomenon exhibited by a solid (adsorbent) to hold or concentrate gases, liquids or dissolved substances (adsorptive) upon its surface, due to adhesion. For example, water held to the outside surface of hydrated bentonite is adsorbed water.

Aeration. The technique of injecting air or gas in varying amounts into a drilling fluid for the purpose of reducing hydrostatic head. See also Air Cutting.

Aggregate. A group of two or more individual particles held together by strong forces. Aggregates are stable to normal stirring, shaking or handling as powder or a suspension. They may be broken by drastic treatment such as ball milling a powder or by shearing a suspension.

Air Cutting. The inadvertent mechanical incorporation and dispersion of air into a drilling fluid system. See also Aeration.

Alkalinity. The combining power of a base measured by the maximum number of equivalents of an acid with which it can react to form a salt. In water analysis, it represents the carbonates, bicarbonates, hydroxides and occasionally the borates, silicates and phosphates in the water. It is determined by titration with standard acid to certain datum points.

Analysis, of Mud or Drilling Fluid. Examination and testing of the drilling fluid to determine its physical and chemical properties and condition.

Anchor, Deadline. Means of holding the deadline to the derrick or substructure. Usually this is the primary element of the weight indicator.

Annular Velocity. The velocity of a fluid moving in the annulus. Usually expressed in feet per minute (fpm).

Annulus, or Annular Space. The space between a pipe and a hole. The space between drillstring and hole or casing is an Annulus.

Apparent Viscosity. The viscosity a fluid appears to have on a given instrument at a stated rate of shear. It is a function of the plastic viscosity and the yield point. The apparent viscosity in centipoises, as determined by the direct-indicating viscometer, is equal to the 600 rpm reading. In a newtonian fluid, the apparent viscosity is numerically equal to the plastic viscosity.

Array Induction Imager Tool. Schlumberger wireline tool that gives a resistivity image of the formation. Can give resistivity logs with vertical resolutions of 1 ft in smooth boreholes, also true resistivity, detailed description of invasion resistivity, Hydrocarbon saturation and imaging.

Array Seismic Imager. Schlumberger wireline tool designed to acquire multilevel triaxial seismic data in cased holes. Provides high quality seismic images in relatively close proximity to the wellbore.

Array Sonic. Wireline logging tool.

Array Sonic. Schlumberger wireline tool that digitizes compression-al, shear, and Stoneley waveforms and is designed to minimize the discrepancies between sonic and seismic data. Used for porosity evaluation, shear seismic correlation, thin bed detection, fracture prediction and detection, sand strength analysis, cement bond log-ging and through-casing sonic logging. Can calculate rock mechani-cal properties such as compressive strength and Poisson's Ratio to aid bit selection.

Attapulgite Clay. A colloidal, viscosity-building clay used principally in salt-water muds. Attapulgite, a special fullers earth, is a hydrous magnesium aluminum silicate.

Authority for Expenditure. A document signed by management that authorizes funds to be spent against an approved budget.

Automatic Driller. Mechanism to automatically control the weight on the bit.

Auxiliary Measurements Service. This Schlumberger module pro-vides direct measurements of wellbore fluid resistivity and tempera-ture, also cable tension below the toolstring head. Useful to run in poor hole conditions; should allow differentiation of differential stick-ing and keyseat sticking. Logs temperature gradients, allows detec-tion of lost circulation zones and cement tops.

Back-Up. Refers to the act of holding one section of pipe or a bolt while the other section of pipe or a nut is being turned.

Back-Up Tong. A tong suspended in the derrick, normally on the driller's right, used to hold box end of the tool joint while the pin end is loosened and unscrewed. Also called a make-up tong, as it is moved to pin end to tighten the joint to the recommended torque after a joint is spun in.

Balance, Mud. A beam-type balance used in determining mud densi-ty. It consists primarily of a base, graduated beam and constant-vol-ume cup, lid, rider, knife edge and counterweight. Two types are available; an Atmospheric balance and a Pressurized balance. In the pressurized balance, the top screws down and more mud can be injected by a hand pump through a non-return valve in the lid. A pressurized balance will give a more accurate figure for downhole mud density if any aeration is present.

Barite or Baryte. Natural barium sulphate used for increasing the density of drilling fluids. If required, it is usually upgraded to a specific gravity of 4.20.

Barrel. A volumetric unit of measure used in the petroleum industry consisting of 42 US gallons or 38 Imperial gallons.

Basicity. pH value above 7. Ability to neutralize or accept protons from acids.

Bell Nipple. A short piece of pipe, expanded or belled at the top, to guide tools into the hole. Usually has side connections for the fill-up and mud return lines.

Bentonite. A plastic, colloidal clay, largely made up of the mineral sodium montmorillonite, a hydrated aluminum silicate. For use in drilling fluids, bentonite has a yield in excess of 85 lb/ton. The generic term "bentonite" is neither an exact mineralogical name, nor is the clay of definite mineralogical composition.

Bicarb. Sodium Bicarbonate. Used for treating cement contamination in drilling fluid.

Bit Breaker. A heavy plate which fits in the rotary table and holds the bit while it is being unscrewed from the drill collar.

Block line. Wire rope used to raise and lower the block by means of the drawworks.

Blowout. An uncontrolled escape of drilling fluid, gas, oil or water from the well, caused by the formation pressure being greater than the hydrostatic head of the fluid in the hole.

Blowout Preventer. A device attached immediately above the casing to control pressures and prevent escape of fluids from the annular space between the drill pipe and casing. A BOP can also shut off the hole if no drill pipe is in the hole.

Blowout Preventer. A set of valves designed to shut off the annulus around pipe in the hole or to shut off open hole.

Boilerhouse. To make up a report on a condition as fact without knowledge of its accuracy.

Boll-Weevil (Worm). An inexperienced rig or oil field worker.

Box. The female part of a threaded connection.

Boyles Law. This states that if the temperature of a body of gas is kept constant, pressure is proportional to volume. Usually expressed as $P1V1 = P2V2$.

Break Circulation. To start movement of the drilling fluid after it has been quiescent in the hole.

Breakout. Refers to the act of unscrewing one section of pipe from another section, especially in the case of drill pipe while it is being withdrawn from the well bore. During this operation the breakout tongs are used to start the unscrewing operation. Also refers to pro-motion of a crew member to a position of driller or of a driller to a toolpusher.

Bridge. An obstruction in the drill hole. A bridge is usually formed by caving of the wall of the well bore.

Bridging. The effect of solids forming an obstruction or 'bridge' in the fluid passages. Bridging in reservoir pores may take place with solids over about a third of pore size. Bridging may be deliberate (additives in fluids to reduce fluid loss) or a problem (reservoir permeability reduction due to a precipitate or mud solids bridging flow paths).

Buoyancy. A body immersed in a fluid receives an upthrust equal to the weight of the displaced liquid. The buoyant force is therefore proportional to the volume of the immersed (part of the body) and to the density of the supporting fluid.

By-Pass. Usually refers to a pipe connection around a valve or other control mechanism. A by-pass is installed in such cases to permit passage of fluid through the line while adjustments or repairs are made on the control that is by-passed.

Cake Consistency. According to API RP 13B, such notations as "hard," "soft," "tough," "rubbery," "firm," etc., may be used to convey some idea of cake consistency.

Cake Thickness. The measurement of the thickness of the filter cake deposited by a drilling fluid against a porous medium, most often following the standard API filtration test. Cake thickness is usually reported in 32nd of an inch.

Calcium-Treated Muds. Calcium-treated muds are drilling fluids to which quantities of soluble calcium compounds have been added or allowed to remain from the formation drilled in order to impart special properties.

Catching Samples. Obtaining of samples by the drilling crew from the drilling fluid as it emerges from the wellbore. Cuttings so obtained should be carefully washed until free of foreign matter, dried and accurately labeled with the depth at which they were drilled for evaluation by the geologist/paleontologist.

Cathead. A spool-shaped extension of the drawworks shaft used to lift heavy equipment around the rig and make-up or break-out the drill pipe.

Cation Exchange Capacity. A measure of how reactive to water a clay mineral is.

Catline. A hoisting or pulling line used in conjunction with the cathead to lift heavy equipment or pipe.

Cave-In. A severe form of sloughing (See also Sloughing).

Cavernous Formations. A formation having columnar voids, usually the result of dissolving by formation waters which may or may not be still present.

Cement Bond Tool. Schlumberger wireline tool that provides a measure of casing to cement and cement to formation cement bond.

Cement Evaluation Tool. Schlumberger wireline tool that determines cement distribution and quality from sonic response. Also measures casing thickness and so can detect damage or corrosion.

Cementing. The operation by which cement slurry is forced down through the casing and out at the lower end in such a way that it fills the space between the casing and the sides of the well bore to a predetermined height above the bottom of the well. This is for the purpose of securing the casing in place and excluding water and other fluids from the well bore.

Centralizers, Spring and Rigid. Spring steel guides are attached to casing and which serve to keep it centered in the open hole. Rigid centralizers have no "give" and are only run inside a larger casing.

Centrifuge. A machine using centrifugal force for separating substances of different densities, for removing solids or moisture or for simulating gravitational effects.

Charles Law. This states that if the volume of a body of gas is kept constant, pressure is inversely proportional to temperature. Usually expressed as P1/T1 = P2/T2.

Chemical Barrel. A container in which various chemicals are mixed prior to addition to the drilling fluid.

Chemicals. In drilling-fluid terminology a chemical is any material that produces changes in the viscosity, yield point, gel strength and fluid loss, as well as surface tension.

Choke Line. Extension from the blowout presenter stack used to direct and control well fluids from the annulus when the BOPs are closed.

Choke Manifold. A series of chokes and valves used to restrict and direct the flow of fluid and/or gas.

Christmas Tree. A term applied to the valves and fittings assembled at the top of a well to control the flow of oil or gas.

Chrome Lignite. Mined lignite, usually leonardite, to which chromate has been added and/or reacted. The lignite can also be causticized with either sodium or potassium hydroxide.

Chronological Sample Taker. Schlumberger wireline tool that can recover up to 90 core samples per trip. An explosive charge fires a hollow bullet into the formation, which is attached to the tool with wires. When the tool is moved, the bullet and sample are pulled out of the formation.

Circulation. The movement of drilling fluid from the suction pit through pump, drill pipe, bit, annular space in the hole and back again to the suction pit. The time involved is usually referred to as circulation time.

Circulation Rate. The volume flow rate of the circulating drilling fluid usually expressed in gallons or barrels per minute.

Circulation, Loss Of or Lost Circulation. The result of drilling fluid escaping into the formation by way of crevices or porous media.

Clay. A plastic, soft, variously colored earth, commonly a hydrous silicate of alumina, formed by the decomposition of feldspar and other aluminum silicates. Clay minerals are essentially insoluble in water but disperse under hydration, shearing forces such as grinding, velocity effects, etc., into the extremely small particles varying from submicron to 100-micron sizes.

Coagulation. In drilling-fluid terminology, a synonym for Flocculation.

Cohesion. The attractive force between the same kind of molecules, i.e., the force which holds the molecules of a substance together.

Colloid. A substance that is in a state of division preventing passage through a semi-permeable membrane and in suspension or solution fails to settle out and diffracts a beam of light.

Compensated Density Neutron Tool. Schlumberger/Anadrill LWD tool which combines density and neutron porosity in one tool. Useful for gas detection, porosity evaluation, and lithology determination.

Compensated Dual Resistivity Tool. Schlumberger/Anadrill LWD tool that measures resistivity at two depths of investigation. Also measures spectral gamma ray to give relative proportions of potassium, thorium, and uranium.

Compensated Neutron Log. Schlumberger wireline tool that bombards the formation with fast neutrons. Detectors measure the slowed neutrons that are bounced back by Hydrogen atoms. Used to deduce porosity, lithology, clay analysis, and gas detection.

Conductivity. A measure of the quantity of electricity transferred across unit area per unit time. It is the reciprocal of resistivity. Electrolytes may be added to the drilling fluid to alter its conductivity for logging purposes.

Connate Water. Water that probably was laid down and entrapped with sedimentary deposits, as distinguished from migratory waters that have flowed into deposits after they were laid down.

Consistency. The viscosity of a non-reversible fluid, in poises, for a certain time interval at a given pressure and temperature.

Contamination. The presence in a drilling fluid of any foreign material that may tend to produce detrimental properties of the drilling fluid. In some cases, contamination can damage producing formations.

Continuous Phase. The fluid phase which completely surrounds the dispersed phase that may be colloids, oil, etc.

Controlled Aggregation. A condition in which the clay platelets are maintained stacked by a polyvalent cation such as calcium and are deflocculated by use of a thinner.

Conventional Mud. A drilling fluid containing essentially clay and water.

Coring. The act of procuring a sample of the formation being drilled for geological information purposes. Conventional coring is done by means of a core barrel put on the bottom of the drill pipe where the bit normally operates. As the cutter head of the core barrel penetrates the formations a continuous sample of the formation is taken in the core barrel and later withdrawn with the drill pipe. The wire line core barrel is used in many areas since it permits coring to be done without withdrawing the drill pipe from the well bore between cores. Sidewall cores are formation samples taken from the wall of the well's borehole by a wireline deployed tool.

Correlated Electromagnetic Retrieval Tool. Schlumberger wireline fishing tool which can retrieve ferrous junk in cased or open hole.

Corrosion. The adverse chemical alteration on a metal or the eating away of the metal by air, moisture or chemicals; usually an oxide is formed.

Covalent. A chemical bond between atoms whereby incomplete electron orbits in the bonded atoms are satisfied by the two atoms' 'sharing' one electron. Some covalent compounds do display some ionic character, where the sharing is somewhat unequal (e.g., water) and the compound is then said to be polar. Polar covalent compounds (e.g., alcohol) are generally soluble in water whereas non-polar covalent compounds (e.g., diesel) are not.

Crater. The formation of a large funnel-shaped cavity at the top of a hole resulting from a blowout or occasionally from caving.

Created Fractures. Induced fractures by means of hydraulic or mechanical pressure exerted on the formation.

Crown. The top of the derrick containing the blockline sheaves.

Crystallization point. The temperature of a brine at which salt crystals begin to fall out of solution. There are three temperatures relative to crystallization occurring: first crystal to appear (FCTA), true crystallization temperature (TCT), and last crystal to dissolve (LCTD).

Cuttings. Small pieces of formation that are the result of the chipping and/or crushing action of the bit. See also Samples.

Cycle Time, Drilling-Fluid. The time of a cycle, or down the hole and back. The time required for the pump to move the drilling fluid in the hole. The cycle time in minutes equals the barrels of mud in the hole divided by barrels per minute being pumped.

Cyclone. A device for the separation of various particles from a drilling fluid, most commonly used as a desander. The fluid is pumped tangentially into a cone and the fluid rotation provides enough centrifugal force to separate particles by mass weight. See also Centrifuge.

D exponent. A measure of formation drillability. Trends are more significant than actual figure.

Daltons Law of Partial Pressures. In a mixture of gases, the effect of each gas constituent is the same as if that gas alone existed in the same volume. In the atmosphere assuming 80% nitrogen and 20% oxygen at a pressure of 1 Bar, the partial pressure of nitrogen is 0.8 Bar and the PP of oxygen is 0.2 Bar. For chemical reactions (such as burning) and physical processes (such as dissolution in water), the reaction will take place as if no nitrogen were present and oxygen at 0.2 Bar were present.

Darcy. A unit of permeability. A porous medium has a permeability of 1 darcy when a pressure of 1 bar on a sample 1 cm long and 1 sq cm in cross section will force a liquid of 1 centipoise viscosity through the sample at the rate of 1 cc per sec.

Dead Line or Deadline. Refers to the end of the drilling line which is not reeled on the hoisting drum of the rotary rig. This end of the drilling line is usually anchored to the derrick substructure and does not move as the traveling block is hoisted, hence the term dead line.

Dead Man. A buried anchor to which guy-wires are tied to steady the derrick, boiler stacks, etc.

Deep Propagation Tool. Schlumberger wireline tool that measures the signal level and relative phase of a 25 MHz electromagnetic wave. Used for measuring resistivity of noninvaded virgin formations, invasion profiling, water saturation and lithology analysis.

Deflocculation. Breakup of flocs of gel structures by use of a thinner.

Dehydration. Removal of free or combined water from a compound.

Density. Matter measured as mass per unit volume expressed in pounds per gallon (ppg), pounds per square inch per 1,000 ft. of depth (psi/1,000 ft), and pounds per cubic ft (lb/cu ft). Density is commonly associated with "mud weight". Mud densities should properly be expressed in psi/ft of hydrostatic pressure but is still commonly expressed as pounds per gallon.

Differential Pressure. The difference in pressure between the hydrostatic head of the drilling-fluid column and the formation pressure at any given depth in the hole. It can be positive, zero, or negative with respect to the hydrostatic head.

Differential-Pressure (Wall) Sticking. Sticking which occurs because part of the drill string (often the drill collars) becomes embedded in the filter cake resulting in a non-uniform distribution of pressure around the circumference of the pipe. The conditions essential for sticking require a permeable formation, a pressure overbalance across a nearly impermeable filter cake and a stationary drill string.

Dipole Shear Sonic Imager tool. Schlumberger wireline tool that records shear, compressional and Stoneley waveforms. Used for amplitude variation with offset (AVO) calibration, shear seismic correlation, wellbore stability measurements, sanding analysis, fracture height estimation, and sand mobility.

Dispersant. Any chemical which promotes dispersion of the dispersed phase.

Dispersed Phase. The scattered phase (solid, liquid, or gas) of a dispersion. The particles are finely divided and completely surrounded by the continuous phase.

Dispersion (Of Aggregates). Subdivision of aggregates. Dispersion increases the specific surface of the particle; hence results in an increase in viscosity and gel strength.

Distillation. Process of first vaporizing a liquid and then condensing the vapor into a liquid (the distillant), leaving behind non-volatile substances, the total solids of a drilling fluid. The distillant is the water and/or oil content of a fluid.

Divalent. Having a valency of two. When a salt is referred to as divalent, the metal ion has a valency of two.

Dogleg. The "elbow" caused by a sharp change of direction in the wellbore.

Dogleg Severity. A measure of how quickly the wellbore changes direction, expressed in degrees (of total course change) per 100' or 30m of wellbore length.

Dope. A generic term applied to all types of thread lubricants. For casing, tubing and line pipe use lubricants specified in API Bulletin 5A. For tool joints and drill collars less than 7 inches OD, use a thread lubricant containing not less than 40-60% by weight of finely powdered zinc or not less than 60% of finely powdered lead and with a coefficient of friction not less than 0.08. For drill collars greater than 7 inches OD, a thread lubricant with a lower coefficient of friction should be used (a "slicker" grease) due to the higher make-up torques required by the larger shouldered connections.

Drawworks. The hoist used to control movement of the travelling block by winding in or out the blockline on a drum. Releasing movement usually controlled by a band or disk brake, sometimes assisted by a movement brake (eddy current or hydromatic).

Drill Stem Test (DST). A test run with the drill string to determine whether oil and/or gas in commercial quantities has been encountered in the well bore.

Drill Stem. The entire drilling assembly from the swivel to the bit; composed of the kelly, drill pipe with tool joints, subs, drill collars, stabilizers, shock absorbers, and reamers. Used to rotate the bit and to carry the mud or circulating fluid to the bit.

Drilling Mud or Drilling Fluid. A circulating fluid used in rotary drilling to perform any or all of various functions required in the drilling operation.

Drilling Out. The operation during the drilling procedure when the cement is drilled out of the casing before further hole is made or completion attempted.

Drilling Under Pressure. Carrying on drilling operations while maintaining a seal at the top of the well bore to prevent the well fluids from blowing out.

Dual Induction Resistivity Log. Schlumberger wireline tool that records a Spontaneous Potential log and three resistivity curves at different depths. Used to deduce true formation resistivity Rt, invasion profiles, and quicklook Hydrocarbon detection.

Dual Laterolog Tool. Schlumberger wireline tool that measures deep and shallow formation resistivity. Used to deduce true resistivity Rt, flushed zone resistivity, invasion profiles, quick-look Hydrocarbon detection, and indication of moved Hydrocarbons.

Dutchman. The portion of a stud or screw which remains in place after the head has been twisted off in an effort to remove the entire stud or screw. Also used to refer to a tool joint pin broken off in the drillpipe or drill collar box.

Dynamic. The state of being active or in motion; opposed to static.

Electric Logging. Electric logs are run on a wire line to obtain information concerning the porosity, permeability, fluid content of the formations drilled, and other information. The drilling-fluid characteristics may need to be altered to obtain good logs.

Electrolyte. A substance which dissociates into charged positive and negative ions when in solution or a fused state and which will then conduct an electric current. Acids, bases, and salts are common electrolytes.

Emulsifier or Emulsifying Agent. A substance used to produce an emulsion of two liquids which do not naturally mix. Emulsifiers may be divided according to their behavior into ionic and non-ionic agents. The ionic types may be further divided into anionic, cationic, and amphoteric, depending upon the nature of the ion-active groups.

Emulsion. A substantially permanent heterogeneous liquid mixture of two or more liquids which do not normally dissolve in each other but which are held in suspension or dispersion by mechanical agitation or, more frequently, by adding emulsifiers. Emulsions may be mechanical, chemical, or a combination of the two. They may commonly be oil-in-water or water-in-oil types for drilling fluids.

End Point. Indicates the end of some operation or when a definite change is observed. In titration this change is frequently a change in color of an indicator which has been added to the solution or the disappearance of a colored reactant.

Equivalent Circulating Density. For a circulating fluid, the equivalent circulating density in lb/gal equals the hydrostatic head (psi) plus the total annular pressure drop (psi) divided by the depth (ft) and by 0.052.

Equivalent Weight or Combining Weight. The atomic or formula weight of an element, compound, or ion divided by its valence. Elements entering into combination always do so in quantities proportional to their equivalent weights.

Eutectic Point. When increasing the concentration of salt in a brine, the temperature at which crystals come out of solution at first lowers. At a certain point, addition of more salt will cause this crystallization point temperature to increase. This point, the lowest temperature at which a brine will precipitate salt crystals, is called the eutectic point.

Fast Line. The end of the drilling line which is fixed to the drawworks drum or reel. It is so called because it apparently travels with greater velocity than any other portion of the drilling line.

Fault. Geological term denoting a formation break, upward or downward, in the subsurface strata. Faults can significantly affect the area mud and casing programs.

Feed-Off. The act of unwinding a cable from a drum. Also a device on a drilling rig that keeps the weight on the bit constant and lowers the drilling line automatically. See also Automatic Driller.

Feet per hour (fph). Used to express rate of drilling penetration.

Feet per minute (fpm). Velocity, e.g., of flow in the annulus or pipe movement.

Feet per second (fps). Velocity, e.g., of flow from the bit nozzles.

Fiber Or Fibrous Materials. Any tough stringy material used to prevent loss of circulation or to restore circulation. In field use, fiber generally refers to the larger fibers of plant origin.

Filling the Hole. Pumping drilling fluid continuously or intermittently into the wellbore to maintain the fluid level in the hole near the surface. The purpose is to avoid danger of blowout, water intrusion and/or caving of the wellbore.

Fill-Up or Fillup Line. The line through which fluid is added to the hole, usually into the flow nipple.

Filter Cake. The suspended solids that are deposited on a porous medium during the process of filtration. See also Cake Thickness.

Filtercake Thickness. A measurement of the solids deposited on filter paper in 32nd of an inch during the standard 30-min API filter test. See also Cake Thickness. In certain areas the filtercake thickness is a measurement of the solids deposited on filter paper for a 7h min. duration.

Filter Paper. Porous unsized paper for filtering liquids. API filtration test specifies one thickness on 9-cm filter paper, Whatman No. 50, S & S No. 576, or equivalent.

Filter Press. A device for determining fluid loss of a drilling fluid having specifications in accordance with API RP 13B.

Filtrate. The liquid that is forced through a porous medium during the filtration process. See also Fluid Loss.

Filtration. The process of separating suspended solids from their liquid by forcing the latter through a porous medium. Two types of fluid filtration occur in a well; dynamic filtration while circulating and static filtration when at rest.

Finger Board. A rack located in the derrick to contain the top of the stands of pipe while they are stacked in the derrick.

Fishing. Operations on the rig for the purpose of retrieving from the well bore sections of pipe, collars, bit cones, junk, or other obstructive items which are in the hole.

Flag. To tie a piece of cloth or other marker on a bailing or swabbing line to enable the operator to know the depth at which the swab or bailer is operating in the hole.

Flipped. In an invert water-in-oil emulsion, the emulsion is said to be flipped when the continuous and dispersed phases reverse.

Flocculation. Loose association of particles in lightly bonded groups, non-parallel association of clay platelets. In concentrated suspensions such as drilling fluids, flocculation may be followed by irreversible precipitation of colloids and certain other substances from the fluid.

Fluid Flow. The state of fluid dynamics of a fluid in motion is determined by the type of fluid (e.g., Newtonian, plastic, pseudoplastic, dilatant), the properties of the fluid such as viscosity and density, the geometry of the system, and the velocity. Thus, under a given set of conditions and fluid properties, the fluid flow can usually be described as plug flow, laminar flow, turbulent flow, or transitional.

Fluid Loss. Measure of the relative amount of fluid lost (filtrate) through permeable formations or membranes when the drilling fluid is subjected to a pressure differential.

Fluid. A fluid is a substance readily assuming the shape of the container in which it is placed. The term includes both liquids and gases. It is a substance in which the application of every system of stresses (other than hydrostatic pressure) will produce a continuously increasing deformation without any relation between time rate of deformation at any instant and the magnitude of stresses at that instant. Drilling fluids are usually Newtonian and plastic, seldom pseudoplastic, and rarely dilatant fluids.

Formation Damage. Damage to the productivity of a well resulting from invasion into the formation by mud particles or mud filtrates. Asphalt from crude oil will also damage some formations.

Formation Micro Imager tool. Wireline logging tool that takes several sets of microresistivity readings and produces an image of the borehole wall. Can show fractures, bedding planes, and differentiate between different lithologies.

Formation MicroScanner tool. Schlumberger wireline tool that provides continuous, oriented borehole images. Applications include structural analysis, fracture identification and analysis, sedimentary analysis, high resolution reservoir analysis, and depth matching / orientation of cores.

Fracture gradient. The gradient from surface showing at what pressure a formation will fracture. Usually measured during leak off tests, can be expressed in the same terms as mudweight - PSI/Ft, PPG, etc.

Free Point Indicator Tool. Schlumberger wireline tool run inside drillpipe when stuck. When on depth, anchors are deployed and the drillstring worked in tension and torsion. Tool allows an estimate of stuck depth prior to attempting an explosive backoff.

Functions of Drilling Fluids. The most important function of drilling fluids in rotary drilling is to bring cuttings from the bottom of the hole to the surface. Some other important functions are: control subsurface pressures, cool and lubricate the bit and drill string, deposition of an impermeable wall cake, etc.

Galena. Lead Sulphate (PbS). Technical grades (specific gravity about 7) are used for increasing density of drilling fluids to points impractical or impossible with barite.

Gamma Ray Spectrometry Tool. Schlumberger wireline tool that emits high energy neutrons into the formation, activating the atomic elements. Used for reservoir evaluation through casing and waterflood monitoring.

Gas Cut. Gas entrained by a drilling fluid.

Gas-Oil Ratio. A measure of the volume of gas produced with the oil. It is expressed in cubic feet per barrel.

Gel. A state of a colloidal suspension in which shearing stresses below a certain finite value fail to produce permanent deformation. The minimum shearing stress that will produce permanent deformation is known as the shear or gel strength of the gel. Gels commonly occur when the dispersed colloidal particles have a great affinity for the dispersing medium. Thus gels commonly occur with bentonite in water. See also Gel Strength, Initial and 10-min.

Gel. A term used to designate highly colloidal, high-yielding, viscosity-building commercial clays, such as bentonite and attapulgite clays.

Gel Strength, 10-Min. The measured 10-min gel strength of a fluid is the maximum reading taken from a direct-reading viscometer after the fluid has been quiescent for 10 min. The reading is reported in lb/100 ft^2.

Gel Strength, Initial. The measured initial gel strength of a fluid is the maximum reading taken from a direct-reading viscometer after the fluid has been quiescent for 10 sec. It is reported in lb/100 ft^2.

Gel Strength. The ability or the measure of the ability of a colloid to form gels. Gel strength is a pressure unit usually reported in lb/100 ft2. It is a measure of the same interparticle forces of a fluid as determined ,by the yield point except that gel strength is measured under static conditions, yield point under dynamic conditions. The common gel-strength measurements are initial (10 second) and the 10-min gels.

Gelation. Association of particles to form a continuous structure.

Gelled Up. Oil-field jargon usually referring to any fluid with high gel strength and/or highly viscous properties. Often a state of severe flocculation.

Geochemical Logging Tool. Schlumberger wireline tool that measures concentrations of 12 elements. The processing quantifies mineral analysis to allow formation analysis in complex lithologies, clay volume and typing, porosity, permeability evaluation, sandstone classification, grain density determination, well to well correlation, dielectric log correlation, and Cation Exchange Capacity study.

Geolograph. Patented device which records the rate of penetration during drilling operations. Sometimes referred to as a "tattletale."

Gravity. The ratio of weights of equal volume of two substances, one of which is taken as a standard. Water is taken as a standard of comparison for liquids and solids ("Specific Gravity"). For gas, air is usually taken, although hydrogen is sometimes used.

Gun Perforating. A previously common method of completing a well was to set casing through the oil bearing formation and cement it at depth. The casing is then gun perforated by a device that is lowered in the hole and fires steel projectiles through the casing and into the pay formation.

Gunning The Pits. Mechanical agitation of the drilling fluid in a pit by means of a mud gun, electric mixer, or agitator.

Gyp Or Gypsum. Gypsum is often encountered while drilling. It may occur as thin stringers or massive formations.

Heads, Blowing By. When a well flows intermittently rather than continuously, it is said to be blowing by heads.

Heaving. The partial or complete collapse of the walls of a hole resulting from internal pressures due primarily to swelling from hydration or formation gas pressures. See also Sloughing.

High pH mud. A drilling fluid with a pH range above 10.5. A high-alkalinity mud.

Homogeneous. Of uniform or similar nature throughout; or a substance or fluid that has at all points the same property or composition.

Hooke's Law. This states that for an elastic material, stress () is proportional to strain (). In practical terms if a load is doubled then the stretch in the material due to that load is also doubled, as long as the elastic limit of the material is not exceeded. Thus as there is a proportional relationship, a constant for the material can be calculated. The constant is Youngs Modulus of Elasticity (E).

Hopper, Jet. A device to hold or feed drilling mud additives.

Horsepower (hp). Is a rate of doing work (transferring energy) equivalent to lifting 33,000 pounds at 1 foot per minute (33,000 ftlb/min). This is of course also 550 ftlb/sec.

Hostile Environment Litho Density Tool. Schlumberger wireline tool that measures the density of the formation and also gives short and long spaced photoelectric effect measurements. Can deduce lithology and porosity. A caliper is also given.

Hydration. The act of a substance to take up water by means of absorption and/or adsorption.

Hydraulics. That branch of engineering which treats of liquids in motion, or its action. It is the know-how about the effects of fluid velocities and pressures and the power involved. The pressure of either a standing or a moving column of fluid is directly related to its density (weight per unit volume), but the moving column is also concerned with a pressure relation due to friction of flow or change of velocity as well as density. This friction is related to viscosity (resistance to flow) mainly for smooth (laminar) flow, but it is more concerned with density for turbulent flow which involves most practical situations.

Hydrogen Ion Concentration. A measure of the acidity or alkalinity of a solution, normally expressed as pH. See also pH.

Hydrometer. A floating instrument for determining the specific gravity or density of liquids, solutions, and slurries. A common example is the "Mudwate" hydrometer used to determine the density of mud.

Hydrostatic Head. The pressure exerted by a column of fluid, usually expressed in pounds per square inch. To determine the hydrostatic head at a given depth in psi, multiply the depth in feet by the density in pounds per gallon by 0.052.

Inclinometer. The trade name of an instrument used to determine whether or not the well bore is proceeding in a vertical orientation at any point. In most drilling operations, either regulations of government bodies or contract stipulations, or both, provide a maximum deviation of the well bore from the vertical; commonly this maximum is three degrees. When deviation is in excess of the allowable, it is necessary to modify the drilling procedure to bring it back in line.

Inhibited Mud. A drilling fluid having an aqueous phase with a chemical composition that tends to retard and even prevent (inhibit) appreciable hydration (swelling) or dispersion of formation clays and shales through chemical and/or physical means.

Inhibitor (Corrosion). Any agent which, when added to a system, slows down or prevents a chemical reaction or corrosion. Corrosion inhibitors are used widely in drilling and producing operations to prevent corrosion of metal equipment exposed to hydrogen sulphide, carbon dioxide, oxygen, salt water, etc. Common inhibitors added to drilling fluids are filming amines, chromates, and lime.

Inhibitor (Mud). Substances generally regarded as drilling mud contaminants, such as salt and calcium sulphate, are called inhibitors when purposely added to mud so that the filtrate from the drilling fluid will prevent or retard the hydration of formation clays and shales.

Interfacial Tension. The force required to break the surface between two immiscible liquids. The lower the interfacial tension between the two phases of an emulsion, the greater the ease of emulsification. When the values approach zero, emulsion formation is spontaneous.

Invert Oil-Emulsion Mud or IOEM. An invert emulsion is a water-in-oil emulsion where fresh or salt water is the dispersed phase and diesel, crude, or some other oil is the continuous phase. Water increases the viscosity and oil reduces the viscosity.

Ion. Acids, bases, and salts (electrolytes) which, when dissolved in certain solvents, especially water, are more or less dissociated into electrically charged ions or parts of the molecules due to loss or gain of one or more electrons. Loss of electrons results in positive charges producing a cation. A gain of electrons results in the formation of an anion with negative charges. The valence of an ion is equal to the number of charges borne by it.

Ionic. A chemical bond between atoms whereby an incomplete electron orbit in one atom gains an electron from the other bonded atom. This creates two 'ions' where one is an atom with a positive charge (lost electrons) and the other is an atom with a negative charge (gained electrons). Groups of atoms can also lose or gain electrons to form ionic compounds, such as Magnesium Carbonate - Magnesium (Mg) donates two electrons, Carbonate (CO_3) gains two electrons to form an ionic compound, $MgCO_3$. Some Ionic compounds are partially covalent as one of the atoms will retain some control over its donated electron. Ionic compounds with a low valency (e.g., sodium chloride, NaCl) are usually soluble in water, whereas ionic compounds with a high valency (e.g., Calcium Carbonate, $CaCO_3$) are not.

Jet Perforating. An operation similar to gun perforating except that a shaped charge of high explosives is used to burn a hole through the casing instead of the gun which fires a projectile in gun perforating.

Jetting. The process of kicking off a well in soft formations by using strongly directional mud flow at the bit, oriented in the direction of kickoff. Also known as Badgering.

Jetting. The process of periodically removing a portion of or all of the water, mud and/or solids, from the pits, usually by means of pumping through a jet nozzle arrangement.

Kelly or Kelly Joint. A heavy square, triangular or hexagonal pipe which transmits torque from the rotary table to the drillstring by means of a drive bushing.

Kelly. A heavy square or hexagonal pipe or other configuration that works through a hole in the rotary table and rotates the drill stem.

Keyseat Wiper. A short joint on which are fixed either spiral or straight blades that are approximately Ω" larger in diameter than the largest drill collar in the string and is attached to the top drill collar. The wiper can be rotated or jarred through a keyseat, enlarging it sufficiently to allow the passage of the drill collars.

Keyseat or Key Seat. That section of a hole, usually of abnormal deviation and relatively soft formation, which has been eroded or worn by drill pipe to a size smaller than the tool joints or collars. This keyhole type configuration will not allow these members to pass when pulling out of the hole. A keyseat can also be worn in a casing shoe if the shoe depth coincides with a dogleg.

Kill Line. A line connected to the annulus below a blowout preventer for the purpose of pumping into the annulus while the preventers are closed.

Killing a Well. Bringing a well under control that is blowing out. Also the procedure of circulating water and mud into a completed well before starting well-service operations.

Knowledge Box. A cupboard or desk in which the driller keeps the various records pertaining to drilling operations.

Laminar Flow. Fluid elements flowing along fixed streamlines which are parallel to the walls of the channel of flow. In laminar flow, the fluid moves in plates or sections with a differential velocity across the front which varies from zero at the wall to a maximum toward the center of flow. Laminar flow is the first stage of flow in a Newtonian fluid; it is the second stage in a Bingham plastic fluid. This type of motion is also called parallel, streamline, or viscous flow. See also Plug Flow and Turbulent Flow.

Lead Tong. A pipe tong suspended in the derrick, normally on the left of the driller, used when coming out of the hole to break the connection operated through wireline tied to breakout cathead. In this operation it is on pin end of joint to be broken. During makeup when going in the hole it is used on lower or box end as backup to makeup tong.

Lignosulfonates. Organic drilling fluid additives derived from by-products of sulphite paper manufacturing process from coniferous woods. Some of the common salts, such as the ferrochrome, chrome, calcium, and sodium, are used as universal dispersants while others are used selectively for calcium treated systems. In large quantities, the ferrochrome and chrome salts are used for fluid-loss control and shale inhibition.

Lime. Commercial form of calcium hydroxide.

Lime-Treated Muds. Commonly referred to as "lime-base" muds. These high-pH systems contain most of the conventional freshwater additives to which sacked lime has been added to impart spacial properties. The alkalinities and lime contents vary from low to high.

Liner. Any string of casing whose top is situated at any point below the surface inside of another casing.

Litho Density Log. Schlumberger wireline tool that measures bulk density and photoelectric effect of the formation using a pad mounted gamma ray source and two detectors. Can deduce porosity, lithology, and abnormal pressures.

Live Oil. Crude oil that contains gas and has not been stabilized or weathered. This oil can cause gas cutting when added to mud and is a potential fire hazard.

Logging while drilling. Tools that are run as part of the drilling assembly which measure downhole parameters. Data can be transmitted real time to surface and/or may be recorded downhole for transmission to a workstation once the tool is back on surface.

Lost Circulation Additives or Lost Circulation Material or LCM. Materials added to the mud to control or prevent lost circulation. These materials are added in varying amounts and are classified as fiber, flake, or granular.

Lost Returns or Lost Circulation. To encounter an interruption in

the circulation of drilling fluid due to the fact that the fluid is entering into a porous or fractured formation underground rather than
returning to the surface.

Low-Solids Muds. A designation given to any type of mud where
high performing additives, e.g., CMC, have been partially or wholly
substituted for commercial or natural clays. For comparable viscosity
and densities (weighted with barite), a low-solids mud will have a
lower volume percent solids content. Higher ROP's are often seen
with low-solids muds.

Making a Trip. Consists of hoisting the drill pipe to the surface and
returning it to the bottom of the well bore. This is done for the purpose of changing bits, preparing to take a core, etc.

Marsh Funnel. An instrument used in determining the Marsh funnel
viscosity. The Marsh funnel is a container with a fixed orifice at the
bottom so that when filled with 1,500 cc fresh water, 1 qt (946 ml)
will flow out in 26 ± 0.5 sec. For 1,000 cc out, the efflux time for
water is 27.5± 0.5 sec. See API RP 13B for specifications.

Master Bushing. Adapter used to reduce the size of the rotary table
opening to accommodate bushings, slips, etc.

Mechanical Powered Rig
1. Mechanical Clutch Type Rig - one which connects the internal
   combustion engines to the load by means of friction clutches
   which can be "slipped" a moderate amount to get the load started
   while the engines are operating at a moderate speed.
2. Fluid drive, fluid coupling, and torque converter - consists of a
   pump and turbine and fluid combination for transmitting power
   from engine to load permitting considerably more "slipping" and
   flexibility than a friction clutch and minimize shock loads getting
   back to engine. The older fluid coupling involves little slip and the
   same torque in the load as in the engine. The torque converter
   involves a multistage turbine (sometimes adjustable) which, at low
   speeds of the load, can develop several times as much torque in
   the load as in the engine, obviously useful for accelerating heavy
   loads. (Slightly less efficient.)

Mechanical Sidewall Coring Tool. Schlumberger wireline tool that recovers either 20 or 50 core plug samples in one trip. The cores are cut with a special core bit and barrel driven by an electric motor. Useful to recover high quality core plugs without the damage that explosive core sampling causes. See also CST.

Mesh. A measure of fineness of a woven material, screen or sieve, e.g., a 200 mesh sieve has 200 opening per linear inch. A 200-mesh screen with a wire diameter of 0.0021 in (0.0533 mm) has an opening of .074 mm, or will pass a particle of 74 microns. See also Micron.

Mica. A naturally occurring flake material of varying size used in combatting lost circulation. Chemically, an alkali aluminum silicate.

Micron. A unit of length equal to one millionth part of a meter, or one thousandth part of a millimeter.

Milliliter. A metric system unit for the measure of volume. Literally 1/1000th of a liter. In drilling-mud analysis work, this term is used interchangeably with cubic centimeter (cc). One quart is about equal to 946 ml.

Modular Formation Dynamics Tester. Schlumberger wireline tool that allows pressure testing and sampling of formation fluids by a probe which seals against the wellbore. Sample chambers are available in three sizes and many pressure tests can be made in a single trip. Used for anisotropic permeability estimates, pressure gradients, and PVT testing.

Molecular Weight. The sum of the atomic weights of all the constituent atoms in the molecule of an element of compound.

Monkey Board. A platform on which the derrickman works during the time the crew is making a trip.

Monobore Completion. A well completion design where the completed interval can be accessed at its full bore by wireline or coiled tubing tools. Advantages include smaller hole sizes can be drilled and the entire completion can be accessed by wireline tools or coiled tubing for more efficient production well intervention.

Montmorillonite. A clay mineral commonly used as an additive to drilling muds. Sodium Montmorillonite is the main constituent in bentonite. Calcium Montmorillonite is the main constituent in low-yield clays.

Mousehole. A shallow cased hole close to the rotary table through the derrick floor in which a joint of drill string can be suspended to facilitate connecting the joint to the kelly.

Mud. A water- or oil-base drilling fluid whose properties have been altered by solids, commercial and/or native, dissolved and/or suspended. Used for circulating out cuttings and many other functions while drilling a well. Mud is the term most commonly given to drilling fluids.

Mud Additive. Any material added to a drilling fluid to achieve a particular purpose.

Mud Balance. An instrument consisting of a cup and a graduated arm with a sliding weight and resting on a fulcrum, used to measure weight of the mud.

Mud Gun. A pipe that shoots a jet of drilling mud under high pressure into the mud pit to mix the additives and stir the mud for other reasons.

Mud Logging. A method of determining the presence or absence of oil or gas in the various formations penetrated by the drill bit. The drilling fluid and the cuttings are continuously tested on their return to the surface and the results of these tests are correlated with the depth or origin.

Mud Program. A proposed or followed plan or procedure for the type(s) and properties of drilling fluid(s) used in drilling a well with respect to depth. Some factors that influence the mud program are the casing program and such formation characteristics as type, competence, solubility, temperature, pressure, etc.

Mud Pumps. Pumps at the rig used to circulate drilling fluids.

Mud Still. An instrument used to distill oil, water, and other volatile material in a mud to determine oil, water, and total solids contents in volume-percent.

Mud Weight. In mud terminology, this refers to the density of a drilling fluid. This is normally expressed in either lb/gal, lb/cuft, psi hydrostatic pressure per 1,000 ft of depth.

Mud-Off. In drilling, to seal the hole off from the formation water or oil by using mud. Applies especially to the undesirable blocking off of the flow of oil from the formation into the well bore. Special care is given to the treatment of drilling fluid to avoid this.

Natural Clays. Natural clays, as opposed to commercial clays, are clays that are encountered when drilling various formations. The yield of these clays varies greatly and they may or may not be purposely incorporated into the mud system.

Natural Gamma Ray Spectrometry Log. Schlumberger wireline tool that breaks down the natural gamma ray spectrum to differentiate between minerals such as potassium, thorium, and uranium. Useful for reservoir delineation, well to well correlation, Cation Exchange Capacity studies, igneous rock recognition, clay content definition, estimates of potassium and uranium, and recognition of radioactive materials.

Newtonian Fluid. The basic and simplest fluids from the standpoint of viscosity consideration in which the shear force is directly proportional to the shear rate. These fluids will immediately begin to move when a pressure or force in excess of zero is applied. Examples of Newtonian fluids are water, diesel oil, and glycerine. The yield point as determined by direct indicating viscometer is zero.

Non-Conductive Mud. Any drilling fluid, usually oil-base or invert-emulsion muds, whose continuous phase does not conduct electricity, e.g., oil. The spontaneous potential (SP) and normal resistivity cannot be logged, although such other logs as the induction, acoustic velocity, etc. can be run.

Nuclear Magnetism Log. Schlumberger wireline tool that applies a large magnetic field to the formation, which causes the spin axis of Hydrogen nuclei to align with the field. When the field is removed, the relaxation of the free protons causes a signal in the measurement coil. Used to estimate free porosity and produceability, estimate permeability, irreducible water saturation, residual oil saturation, and to identify heavy oil zones. Cannot be run in combination with other tools.

Nuclear Porosity Lithology Tool. Schlumberger wireline tool that combines gamma density, neutron porosity, and natural gamma ray spectrometry measurements in one tool. Used to deduce accurate bulk density, neutron porosity, thin bed measurements, and density measurements in formations with high natural radioactivity.

Oil Base Dipmeter Tool. Schlumberger wireline tool that uses four microinduction sensors to measure the variations of formation conductivity. Applications include determination of structural dip when non-conductive mud is in use and borehole geometry.

Oil-Base Mud. The term "oil-base" is applied to a special type drilling fluid where oil is the continuous phase and water the dispersed phase. Oil-base mud contains blown asphalt and usually 1 to 5 percent water emulsified into the system with caustic soda or quick lime and an organic acid. Silicate, salt, and phospate may also be present. Oil-base muds are differentiated from invert-emulsion muds by the amounts of water used, method of controlling viscosity and thixotropic properties, wallbuilding materials, and fluid loss.

Packer Fluid. Any fluid placed in the annulus between the tubing and casing above a packer. Along with other functions, the hydrostatic pressure of the packer fluid is utilized to reduce the pressure differentials between the formation and the inside of the casing and across the packer itself.

Partially hydrolyzed polyacrylamide. See also the topic Anionic Polymers; PHPA and PAC

Penetration, Rate Of or ROP. The rate in feet or meters per hour at which the drill proceeds to deepen the well bore.

Permeability. Normal permeability is a measure of ability of a rock to transmit a one-phase fluid under conditions of laminar flow. Unit of permeability is the darcy.

pH. An abbreviation for potential hydrogen ion. The pH numbers range from 0 to 14, 7 being neutral, and are indices of the acidity (below 7) or alkalinity (above 7) of the fluid. The numbers are a function of the hydrogen ion concentration in gram ionic weights per liter. The pH may be expressed as the logarithm (base 10) of the reciprocal (or the negative logarithm) of the hydrogen ion concentration. The pH of a solution offers valuable information as to the immediate acidity or alkalinity, as contrasted to the total acidity or alkalinity (which may be titrated).

Phasor Induction - Spherically Focused Log. Schlumberger wireline tool that records resistivity at three investigation depths. Used for thin bed resolution, interpretation of deeply invaded zones, true resistivity in medium to high contrast formations, and invasion profiles. Also useful for correlation and reservoir modeling.

Pilot Testing. A method of predicting behavior of mud systems by mixing small quantities of mud and mud additives, then testing the results.

Pin. The male part of a threaded connection.

Plastic Fluid. A complex, non-Newtonian fluid in which the shear force is not proportional to the shear rate. A definite pressure is required to start and maintain movement of the fluid. Plug flow is the initial type of flow and only occurs in plastic fluids. Most drilling muds are plastic fluids. The yield point as determined by direct-indicating viscometer is in excess of zero.

Plastic Viscosity. The plastic viscosity is a measure of the internal resistance of fluid flow attributable to the amount, type and size of solids present in a given fluid. It is expressed as the number of dynes per sq cm of tangential shearing force in excess of the Bingham yield value that will induce a unit rate of shear. This value expressed in centipoises, is proportional to the slope of the consistency curve determined in the region of laminar flow for materials obeying Bingham's Law of Plastic Flow. When using the direct-indicating viscometer, the plastic viscosity is found by subtracting the 300-rpm reading from the 600-rpm reading.

Plug Flow. The movement of a material as a unit without shearing within the mass. Plug flow is the first type of flow exhibited by a plastic fluid after overcoming the initial force required to produce flow.

Poisson's Ratio. Poisson's Ratio describes the effect produced on a material in a direction which is perpendicular to an applied stress. If you imagine in two dimensions a square; applying compression to the top of the square will make the material extrude sideways. If this "induced stress" is divided by the original stress, then Poisson's Ratio is the result. Ration for steel is around 0.3 and for in-situ rock varies between 0.25 and 0.5. Poisson's Ratio in rock can be used in deducing formation fracture gradients and may have some relevance to bit selection.

Polar; Polarity. In nature, molecules consist of groups of atoms bonded together. Overall the electrical charge is zero because the numbers of protons and electrons balance. However some covalent molecules may express polarity, where the electrical charges, though balancing overall, are higher in one part of the molecule. In water, one molecule of Oxygen bonds to two molecules of Hydrogen. The molecule is 'bent' like a V, with the Oxygen at the bottom and the Hydrogen atoms at the top end of each arm. As the Oxygen atom has 8 Protons and the Hydrogen only one each, the shared electrons tend to orbit closer to the Oxygen atom. Thus the Oxygen end has a slightly higher negative charge and the Hydrogen atoms have slightly positive charges, balancing overall. Polarity is important in physics and chemistry; water would be a gas at normal temperatures if the polarity of the molecules didn't tend to hold the molecules together. Polarity also effects attractions with other, different polar molecules, including clays (which have different electrostatic charges on the faces and edges of the crystals) and polymers. Water also dissolves some ionic compounds by attraction due to its polarity.

Polished bore receptacle. A close tolerance bore at the top of a liner; may be used to stab in with a seal assembly, e.g., as part of a mono-bore completion.

Porosity. The amount of void space in a formation rock usually expressed as percent voids per bulk volume. Absolute porosity refers to the total amount of pore space in a rock, regardless of whether or not that space is accessible to fluid penetration. Effective porosity refers to the amount of connected pore spaces, i.e., the space available to fluid penetration. See also Permeability.

Precipitate. Material that separates out of solution or slurry as a solid. Precipitation of solids in a drilling fluid may follow flocculation or coagulation, such as the dispersed red bed clays upon addition of a flocculation agent to the fluid.

Pressure Surge. A sudden, usually short duration increase in pressure. When pipe or casing is run into a hole too rapidly, an increase in the hydrostatic pressure results, which may be great enough to create lost circulation.

Pressure-Drop Loss. The pressure lost in a pipeline or annulus due to the velocity of the liquid in the pipeline, the properties of the fluid, the condition of the pipe wall, and the alignment of the pipe. In certain mud-mixing systems, the loss of head can be substantial.

Prime Mover. As applied to oil well drilling, this is the diesel drive, electric motor, or internal combustion engine which is the source of power for the drilling rig.

Protection Casing. A string of casing set to protect a section of the hole and to permit drilling to continue to a greater depth. Sometimes called "protection string" and "intermediate string."

Quiesence. The state of being quiet or at rest (being still). Static.

Rathole.
1. A 30 to 35 ft. cased hole drilled at one corner of the derrick floor used to store the swivel and kelly while making a trip.
2. Sometimes refers to a hole of reduced size in the bottom of the regular well bore. In some cases the driller "ratholes ahead" to facilitate the taking of a drill stem test when it appears such a test will be desirable.

Rate of Shear or Shear Rate. The rate at which an action, resulting from applied forces, causes or tends to cause two adjacent parts of a body to slide relatively to each other in a direction parallel to their plane of contact. Commonly given in rpm.

Reaming. The operations of smoothing the well bore or enlarging the hole to the desired size, straightening doglegs and assist in directional drilling.

Repeat Formation Tester. Schlumberger wireline tool that can take multiple pressure readings and up to two formation fluid samples on one trip.

Reservoir Saturation Tool. Schlumberger wireline tool that uses a neutron induced GR spectroscopy principle to determine oil saturation in unknown or fresh formation water, borehole oil fraction, and Sigma.

Resistivity At the Bit. Anadril LWD tool.

Resistivity. The electrical resistance offered to the passage of a current, expressed in ohmmeters; the reciprocal of conductivity. Freshwater muds are usually characterized by high resistivity, saltwater muds by a low resistivity.

Reverse Circulate. The method by which the normal flow of a drilling fluid is reversed by circulating down the annulus and up and out the drill string.

Rheology. The science that deals with deformation and flow of fluids.

Rig, Braking Capacity (Performance). The capacity to hold the hook-load and retard the continuous movement of the hook-load within reasonable specified limits compatible with specific requirements.

Rock Pressure. A term used for the initial pressure of gas in a well.

Rotary Drilling. The method of drilling wells that depends on the rotation of a column of drill pipe, to the bottom of which is attached a bit. A fluid is circulated to remove the cuttings.

Saltwater Muds. A drilling fluid containing dissolved salt (brackish to saturated). These fluids may also include native solids, oil and/or such commercial additives as clays, starch, etc.

Samples. Cuttings obtained for geological information from the drilling fluid as it emerges from the hole. They are washed, dried, and labeled as to the depth. Also, a portion of well fluid recovered on a DST.

Seawater Muds. A special class of saltwater muds where seawater is used as the fluid phase.

Shale Shaker. A vibrating screen that removes coarser cuttings from the circulating fluid before it flows into the return mud pit.

Shale. Fine-grained clay rock with slate-like cleavage, sometimes containing an organic oil-yielding substance.

Shear (Shearing Stress). An action, resulting from applied forces, which causes or tends to cause two contiguous parts of a body to slide relative to each other in a direction parallel to their plane of contact.

Shear Rate. This is the relative velocity of the fluid layers, divided by their normal separation distance. See also Viscosity.

Shear Strength. A measure of the shear value of the fluid. The minimum shearing stress that will produce permanent deformation. See Gel Strength.

Shear Stress. This is an expression used in muds to describe the force required to overcome a fluid's resistance to flow, divided by the area that the force acts on. See also Viscosity.

Shoulder Effect. An anomaly on resistivity logs where a low resistance formation and a high resistance formation meet. The change in resistivity "channels" the current causing characteristic horn-shaped signal responses in uncorrected logs. The effects can be corrected by tool design and signal processing.

Side Wall Coring. The taking of geological samples of the formation which constitutes the wall of the well bore. Another term in general use for this operation is "side wall sampling."

Skid. Moving a rig from one location to another, usually on tracks, where little dismantling is required.

Slim-Hole drilling. Slim hole, by current Industry definition, is one where the well is TD'd in 41/2" or smaller hole size. Slim-Hole drilling brings significant cost and environmental benefits but makes well control more difficult due to the high ECDs and small annular capacities.

Slip Velocity. The difference between the annular velocity of the fluid and the rate at which a cutting is removed from the hole.

Sloughing. The partial or complete collapse of the walls of a hole resulting from incompetent, unconsolidated formations, high angle or repose, and wetting along internal bedding planes. See also Heaving and Cave-in.

Slug the Pipe. A procedure before pulling the drill pipe whereby a small quantity of heavy mud is pumped into the top section to cause an unbalanced column. As the pipe is pulled, the heavier column in the drill pipe will fall, thus keeping the inside of the drill pipe dry at the surface when the connection is unscrewed.

Slurry. A plastic mixture of Portland cement and water which is pumped into the well to cement casing or plug back.

Soft-torque rotary table control™. The Soft-torque system was designed to reduce torsional drillstring vibrations by controlling the power input to the electric motors at the rotary. It was proposed by Shell and developed in collaboration with Deutag. The system is marketed by Deutag for fitting onto existing electric rigs. Among the claimed benefits are greater bit life, reduced drillstring vibrations (with lower fatigue failures), and better borehole stability.

Spinner Survey. An operation designed to indicate the point at which fluids are escaping from the well bore into a cavernous or porous formation. Also used to determine point of formation fluid entry.

Spontaneous potential. Measures the potential difference at depth between the probe and the earth's surface.

Spud Mud. The fluid used when drilling starts at the surface, often a thick bentonite slurry.

Spudding In. The start of the drilling operations of a new hole. Usually applies when the first bit penetrates below the conductor shoe.

Squeeze. A procedure where slurries of cement, mud, gunk plug, etc. are forced into the formation by pumping into the hole while maintaining a back pressure, usually by closing the rams and/or running a squeeze tool.

Stabbing Board. A temporary platform in the derrick, 20 to 40 feet above the floor, on which a crewman works while casing is being run to guide a joint while it is being screwed into the joint in the rotary table.

Standpipe. Part of the circulating system. A pipe extending, usually along a derrick leg, to a height suitable for attaching the rotary hose.

Strain. When a material has a force applied to it, the material deforms. Strain is a measure of the expansion divided by the original length L. See also Hooke's Law.

Stratigraphic High Resolution Dipmeter Tool. Schlumberger wireline tool that determines structural and stratigraphic dip, stratigraphic analysis, fracture identification, and borehole geometry.

Stress. When a material has a force applied to it, that force can be expressed as the force exerted over the cross sectional area. For metals it is usually expressed in pounds per square inch (psi). See also Hooke's Law.

Supersaturation. If a solution contains a higher concentration of a solute in a solvent that would normally correspond to its solubility at a given temperature, this constitutes supersaturation. This is an unstable condition, as the excess solute separates when the solution is seeded by introducing a crystal of the solute. The term "supersaturation" is frequently used erroneously for hot salt muds.

Surfactant. A material which tends to concentrate at an interface. Used in drilling fluids to control the degree of emulsification, aggregation, dispersion, interfacial tension, foaming, defoaming, wetting, etc.

Surge Loss. The flux of fluids and solids which occurs in the initial stages of any filtration before pore openings are bridged and a filter cake is formed. Also called "spurt loss."

Swabbing. When pipe is withdrawn from the hole in a viscous mud or if the bit is balled, a suction is created.

Thermal Decay Time log. Schlumberger wireline tool that analyzes the decay of fast neutrons to determine formation and borehole sigma values. Used also for gas detection and discrimination, porosity analysis, reservoir and waterflood monitoring, and through drillpipe formation evaluation.

Thinner. Any of various organic agents (tannins, lignins, lignosulfonates, etc.) and inorganic agents (pyrophosphates, tetraphosphates, etc.) that are added to a drilling fluid to reduce the viscosity and/or thixotropic properties.

Thixotrophy. The ability of fluid to develop gel strength with time. The property of a fluid which causes it to build up a rigid or semi-rigid gel structure if allowed to stand at rest, yet can be returned to a fluid state by mechanical agitation. This change is reversible.

Titration. A method, or the process of using a standard solution for the determination of the amount of some substance in another solution. The known solution is usually added in a definite quantity to the unknown until a reaction is complete.

Tooljoint or Tool Joint. A drill-pipe coupler consisting of a pin and box of various designs and sizes. The internal design of tool joints has an important effect on mud hydrology.

Torque. A measure of the force or effort applied to a shaft causing it to rotate. On a rotary rig this applies especially to the rotation of the drill stem in its action against the bore of the hole. Torque reduction can usually be accomplished by the addition of various drilling fluid additives.

Total Depth or TD. The greatest maximum depth reached by the drill bit.

Tour. A person's turn in an orderly schedule. The word that designates the shift of a drilling crew is pronounced as if it were spelled t-o-w-e-r.

Turbulent Flow. Fluid flow in which the velocity at a given point changes constantly in magnitude and the direction of flow; pursues erratic and continually varying courses. Turbulent flow is the second and final stage of flow in a Newtonian fluid; it is the third and final stage in a Bingham plastic fluid.

Twist-Off. To twist a joint of drill pipe in two by excessive force applied by the rotary table. Many failures which result in parting of the drill pipe in the well bore are erroneously referred to by this term.

Ultrasonic Imager. Schlumberger wireline tool that uses a single rotating ultrasonic sensor to provide a complete image of the casing, cement, and annulus. Can show casing damage/corrosion, cement quality, and channeling.

Under-Ream. To enlarge a drill hole below the casing.

Valency. This expresses the number of electrons that an atom or group of atoms needs to lose, gain or share to attain electrical stability. See also Covalent and Ionic.

Variable bore rams. Pipe rams that can close over a range of pipe sizes. VBRs cannot be used to hang off the drillstring.

Velocity, Critical. That velocity at the transitional point between laminar and turbulent types of fluid flow.

Velocity. Time rate of motion in a given direction and sense. It is a measure of the fluid flow and may be expressed in terms of linear velocity, mass velocity, volumetric velocity, etc. Velocity is one of the factors which contribute to the carrying capacity of a drilling fluid.

Vertical seismic profile. Uses downhole sonic tools that are processed to provide a map of the acoustic response of the earth along the wellbore. Useful for prediction ahead of the bit, dip evaluation, thin bed analysis, seismic evaluation of deep horizons, timing and amplitudes of multiples, and direct wave attenuation with depth. Measurements can be made with a single surface source (zero offset VSP) or with a moving surface source (walkaway VSP), and the waves recorded downhole at predetermined stations.

V-G Meter or Viscosity-Gravity Viscometer. The name commonly used for the direct-indicating viscometer.

Viscometer, Direct-Indicating. Commonly called a "V-G meter." The instrument is a rotational-type device powered by means of an electric motor or handcrank and is used to determine the apparent viscosity, plastic viscosity, yield point, and gel strengths of drilling fluids. The usual speeds are 600 and 300 rpm.

Viscosity. This is a measure of the resistance to flow of a fluid. Viscosity in Centipoises is arrived at by dividing the Shear Stress by the Shear Rate.

Wall Cake. The solid material deposited along the wall of the hole resulting from filtration of the fluid part of the mud into the formation.

Water-Base Mud. Common conventional drilling fluids. Water is the carrying medium for solids and is the continuous phase, whether or not oil is present.

Weight. Refers to pipe weight such as pick up weight, normal weight, slack off weight, weight indicator weight.

Weighting Material. Any of the high specific gravity materials used to increase the density of drilling fluids. This material is most commonly barite but can be galena, calcium carbonate, etc.

Well-Logging. See Electric Logging and Mud Logging.

Wetting Agent. A substance or composition which, when added to a liquid, increases the spreading of the liquid on a surface or the penetration of the liquid into the materials.

Wetting. The adhesion of a liquid to the surface of a solid.

Whipstock. A device inserted in a well bore used for deflecting or for directional drilling.

Wildcat. A well in unproved territory.

Workover Fluid. Any type of fluid used in the workover operation of a well.

Workover. To perform one or more of a variety of remedial operations on a producing oil well with the hope of restoring or increasing production. Examples of workover operations are deepening, plugging back, pulling and resetting the liner, squeeze cementing, shooting, and acidizing.

Yield Point. In drilling fluid terminology, yield point means yield value. Of the two terms, yield point is by far the most commonly used expression.

Yield. A term used to define the quality of a clay by describing the number of barrels of a given centipoise slurry that can be made from a ton of the clay. Based on the yield, clays are classified as bentonite, high-yield, low-yield, etc. Not related to yield value below.

Youngs Modulus of Elasticity (E). Youngs Modulus is calculated for metals by dividing stress by strain. See also Hooke's Law. Youngs Modulus for steel (Imperial units) is approximately 30 x 106.

Zero-Zero Gel. A condition wherein the drilling fluid fails to form measurable gels during a quiescent time interval (usually 10 min).

[INDEX]

[Index]

A

absorption, 460
Abu Roash formation, 13, 15, 185
accumulator testing, surface BOP stack configurations and, 328–332
acidity, 460
acronyms, xx–xxiv
adaptive electromagnetic propagation tool, 460
additives
    brines, 33
    lost circulation, 485–486
    mud, 488
    oil mud, 240–244
        alkalinity control, 241
        bridging agents, 244
        emulsifier, 240–241
        filtration control, 243
        oil-wetting agents, 244
        rheology modifiers, 243–244
        viscosifiers, 241–243
adhesion, 460
adjacent well separation, kicking off and, 173
adsorption, 460
advance detection, of shallow gas, 145
aerated fluid drilling, 248
aeration, 460
aggregate, 460
air cutting, 460
air drilling, 246
Alam el Bueib, 185
Alamein, 185
alkalinity control, 460–461
    oil mud additives, 241
alliancing contracts, xvi
alternative camera units, magnetic single shot survey tool, 426
amine corrosion inhibitor, 250
amplitude variation with bandwidth (AVB), 47
analysis
    drilling fluid, 461
    mud, 461
anchor, 461
anhydrite scale, 231
anionic polymers, 212, 213–214
annular space, 461
annular velocity, 461
annulus, 461
API 13 A specification, 249
API Bulletin 5C2, 37, 61, 71, 73
API C75/90/95 alloy, 89
API H40 alloy, 89
API J55 alloy, 89
API K55 alloy, 89
API L80 13Cr alloy, 89
API L80 alloy, 89
API N80 alloy, 89
API P105/110 alloy, 90

API Specification 5CT, 87
API Specification 6A, 88
API Specification 10, 261–262
API V150 alloy, 90
apparent viscosity, 461
approval signatures, drilling program, 130
area, 460
array induction imager tool, 461
array seismic imager, 461
array sonic (AS), 282, 461, 462
attapulgite, 242
    clay, 462
authorization for expenditure (AFE), 18, 462
    for drilling program, 135
automatic driller, 462
auxiliary measurements service, 462
avoidance planning, 293–310
axial forces
    calculation of, 70
    in casing, from friction, 79–80
    lateral forces and, in inclined wellbore, 71
axial loads
    from bending forces, in deviated wellbore, 77–79
    calculation of, 73–80
    inclination and, vector diagram of, 74
axial stress, pipe bending an, 77

B

back pressure manifold (BPM), 153–155
backing off
    drilling problems, 365–368
    fishing operations after, 371–372
    planning/precautions, 365–366
    procedure, 366–368
    in wrong place, 368
back-up, 462
    tong, 462
badgering, kicking off and, 174–175
Bahariya, 185
balance, mud, 462
ballooning effect, kick detection system, 319
barite plugs, 463
    drilling problems, 376–378
    high pressure low permeability kicking formation, 378
    underground blowout situation, 377–378
barium scale, 231
barrel, 463
basicity, 463
bearings/seals condition, IADC 8 point grading scheme, 416–417
bell nipple, 463
bending forces, axial loads from, in deviated wellbore, 77–79
bentonite, 463
biaxial effects, calculation of, 71–72

directional control, downhole tools affecting,
157–179
directional design, 107–119
casing wear and, 113–114
combined buildup and, 110–114
dogleg severity limits and, 110–114
of horizontal wells, 116–117
multilateral wellbores and, 118
slant rig drilling and, 118–119
targets and, 119
turn rate and, 110–114
wellpath planning and, 108–110, 119
directional drilling, 421–428
directional jetting and, 424–425
gyro multishot surveys, 428
keyseating prevention, 423–424
magnetic single shot survey tool, 426–427
rotary bottom hole assemblies, 421–423
single shot surveys, 425–426
Totco single shot survey tool, 428
directional planning, 157–179
dropping hole angle, 178–179
gyro multishot surveys, 165–166
inclination-only surface readout sub and, 164
logging while drilling, 167–168
magnetic single shot surveys, 165
magnetic surveys, 165–166
measurement while drilling, 167–168
measurement/surveying for, 163–173
potential sources of survey errors, 171–173
survey intervals for, 163
tangent section drilling, 177–178
Totco surveys and, 164–165
universal bottom hole orientating sub and,
163–164
vertical wells and, 164–165
wireline deployed surface readout gyro,
166–167
directional surveys, from wireline logging tools, 169
dispersant, 472
dispersed phase, 472
dispersed water-based muds, 206–210
dispersion, 472
of clays, 204, 205–206
displaced fluids weights, buoyancy calculation with,
66
distillation, 472
divalent, 472
divalent brines, scale from, 30
divalent cations, 205
diverter drilling
shallow gas, 147–148
well control, 147–148
diverter system checklist, BOP equipment, 325–326
dogleg, 472
dogleg severity formula, 111, 472
dogleg severity limits, directional design and,
110–114

dope, 472–473
downhole motor, kicking off with, 175
downhole positive displacement mud motors, downhole tools, 162–163
downhole tools
downhole positive displacement mud motors,
162–163
drill collars, 160–161
affecting directional control, 157–179
drill bits, 158
roller reamers, 158–160
stabilizers, 158–159
sideforce generating tools, 162
drawworks, 473
drill bit(s)
alternate choices, 407–409
bottom hole assemblies and, 200
breaker, 463
codes, for IADC 8 point grading scheme, 416
defining recommended, 194–199
downhole tools, 158
drilling parameters, 409–412
dull, IADC 8 point grading scheme, 417,
418–419
features/selection, 195, 196
mill tooth bits, 192–193
monitoring bit progress, 413–414
mud motors, 413
parameter optimization, drilling parameters,
412
polycrystalline diamond compact, 132, 189,
193–194, 195
post drilling bit analysis, 415–416
pulling the bit and, 414
reason pulled, IADC 8 point grading scheme,
417, 418
rock, 192–193
runs, writing field information notes for, 186
selection
drilling parameters and, 201
log data in, 190–192
references for, 201
structured approach to, 181–183
steerable systems, 413
tungsten carbide insert bits, 193
turbines, 413
types, 192–194
whirl minimizing of, 410–411
drill collars, downhole tools, 160–161
drill stem, 473
drill stem test (DST), 18, 473
drilling
aerated fluid, 248
air, 246
below kick tolerance levels, 319–320
with BOP stack, 151–153
clay formations, 204
criteria

M

[Index]